DESIGN AND RETROFIT OF WASTEWATER TREATMENT PLANTS FOR BIOLOGICAL NUTRIENT REMOVAL

HOW TO ORDER THIS BOOK

BY PHONE: 800-233-9936 or 717-291-5609, 8AM-5PM Eastern Time

BY FAX: 717-295-4538

BY MAIL: Order Department
Technomic Publishing Company, Inc.
851 New Holland Avenue, Box 3535
Lancaster, PA 17604, U.S.A.

BY CREDIT CARD: American Express, VISA, MasterCard

WATER QUALITY MANAGEMENT LIBRARY
VOLUME 5

DESIGN AND RETROFIT OF WASTEWATER TREATMENT PLANTS FOR BIOLOGICAL NUTRIENT REMOVAL

EDITED BY

Clifford W. Randall, Ph.D.
James L. Barnard, Ph.D., Pr.Eng.
H. David Stensel, Ph.D., P.E.

TECHNOMIC PUBLISHING CO., INC.
LANCASTER · BASEL

Water Quality Management Library—Volume 5
a **TECHNOMIC**® publication

Published in the Western Hemisphere by
Technomic Publishing Company, Inc.
851 New Holland Avenue
Box 3535
Lancaster, Pennsylvania 17604 U.S.A.

Distributed in the Rest of the World by
Technomic Publishing AG

Printed in the United States of America

10 9 8 7 6 5 4 3 2 1

Main entry under title:
Water Quality Management Library—Volume 5 / Design and Retrofit of Wastewater Treatment Plants for Biological Nutrient Removal

A Technomic Publishing Company book
Bibliography: p.
Includes index p. 417

Library of Congress Card No. 92-53521
ISBN No. 87762-922-6 (Volume 5)
ISBN No. 87762-944-7 (8-Volume Set)

Table of Contents

Foreword xi
List of Contributors xiii

1. **INTRODUCTION** 1
CLIFFORD W. RANDALL, Ph.D.

Synopsis of Book Contents 1
Benefits of Biological Nutrient Removal 2
Economical Benefits of BNR Processes 4
Process Stability 7
BNR Process Configurations 7
BPR Configurations 10
The Effects of Wastewater Characteristics 16
The Effects of Primary Settling 20
Effluent Limitations 21
References 23

2. **PRINCIPLES OF BIOLOGICAL NUTRIENT REMOVAL** 25
H. DAVID STENSEL, Ph.D., P.E. and JAMES L. BARNARD, Ph.D., Pr.Eng.

Introduction 25
Biological Nitrogen Removal 25
Biological Nitrification 25
Microbiology 25
Oxidation and Synthesis Relationships 26
Biological Kinetics 29
Effect of pH 33
Effect of Dissolved Oxygen 34
Nitrification Inhibition and Toxicity 36

Nitrification Rates in Activated Sludge 40
Nitrification Rates in Single Sludge BNR Systems 45

Biological Denitrification 46
Microbiology 46
Effect of Oxygen 51
Effect of pH 53
Oxidation-Reduction Reactions 53
Biological Kinetics 56

Principles of Biological Phosphate Removal 65
Definitions 65
Chemical Precipitation of Phosphates 66
Biological Removal of Phosphates 66
Biological Model 67
Secondary Release of Phosphates 70
Design Implications 71
Denitrification 73
pH Value of the Mixed Liquor 76
Maintaining Dissolved Oxygen Concentration 77
Temperature 77
Mean Cell Residence Time (MCRT) 77
Toxic Interference 78
Leaching from Waste Sludges 78

References 78

3. **DESIGN OF PREFERMENTATION PROCESSES** **85**
JAMES L. BARNARD, Ph.D, Pr.Eng.

Introduction 85

Mechanism of Prefermentation 87

Design of Systems for Optimizing SCVFA Production 89
Dewatering the Underflow from Gravity Thickeners 89
Return the Thickener Overflow Directly to the Anaerobic Basin 89
Recycle on the PST or Gravity Thickener 89
Completely Mixed Fermenter 90

Batch Systems 90
Some Experimental Information 91

Return of Organic Carbon 92

Monitoring of the Process 92

Some Practical Points about Prefermenters 93

References 95

4. **DESIGN OF ACTIVATED SLUDGE BIOLOGICAL NUTRIENT REMOVAL PLANTS** 97
CLIFFORD W. RANDALL, Ph.D., H. DAVID STENSEL, Ph.D, P.E. and JAMES L. BARNARD, Ph.D., Pr.Eng.

Section I: Two-Stage Biological Phosphorus Removal Systems 97
Phosphorus Limitation versus COD Limitation 99
Anaerobic HRT versus Anaerobic Mass Fraction 100
Factors Affecting Phosphorus Removal 103
Sizing the Anaerobic Zone 113
Aeration Requirements 117
Mixing Requirements 117
Design Steps 119
Design Example 120

Section II: Design of Three-Stage, Four-Stage, and Five-Stage BNR Systems 123
Waste Characteristics 123
Effluent Requirements 125
COD:P Ratio 125
COD:TKN Ratio 126
Selection of Aeration Device 126
Clarifier Selection 127
Selection of Three-, Four-, or Five-Stage Plants 127
Zone Sizing 132
Design Example 134
Mixing/Aeration Design 142
Mixed Liquor Recycle 143
Aeration Devices 143
Dissolved Oxygen (DO) Control Systems 145

Section III: Design of Oxidation Ditch BNR Systems 146
Oxidation Ditch Description 146
Nutrient Removal Performance 150
Nitrogen Removal Design 152

Section IV: Design of Sequencing Batch Reactor BNR Systems 156
Nitrogen Removal Design 159
Nitrification Design 163

Section V: Designer Plant Case Histories 169
Experience Overview 169
Underloading and Overdesign 170
Excessive Aeration of the Influent Wastewater and the RAS 171
Poor Final Clarifier Design 171
Sludge Bulking 172

Denitrification in the Aeration Basin 173
Scum Growth 175
Control of Oxygen Input 176
VFA Generation and Addition 178
Chemical Backup 179
References 180

5. **CLARIFIER DESIGN** **185**
ORRIS E. ALBERTSON, M.S., P.E.

Introduction 185
Theoretical and Practical Considerations 186
Types of Settling Regimes 186
Detention Efficiency 187
Short-Circuiting of Feed 189
Flux Theory and Application 191
Theory 191
Application 192
Clarifier Configurations 199
Circular and Square 199
Rectangular, Square, and Stacked 200
Interchannel 203
Clarifier Configuration and Features 203
Primary Clarifiers 203
Secondary Clarifiers 208
Basis of Design 213
Primary Clarification 216
Secondary Clarification 216
Hydraulic Configuration Design Procedures – Secondary Clarifiers 217
Rectangular 217
Circular 221
Design of Sludge Transport/Collection Equipment 225
Discussion 225
Primary Sludge Collectors 232
Secondary Sludge Collectors 232
Circular Clarifiers 235
Retrofitting the Collectors 244
Process Control Strategies 245
Primary Clarifiers 245
Secondary Clarifiers 246
Bypass Treatment 249

References 251

6. **RETROFIT OF EXISTING PLANTS** **255**
CLIFFORD W. RANDALL, Ph.D., H. DAVID STENSEL, Ph.D, P.E. and JAMES L. BARNARD, Ph.D., Pr.Eng.

Introduction 255
Plant Modification for Biological Phosphorus Removal 255
Retrofit Considerations 256
Case Histories 258
Modifications for Biological Nitrogen Removal 289
Factors Affecting Retrofit Designs 290
Compatibility with Existing Treatment Processes 290
Case Histories 291
Retrofitting Plants for Nitrogen and Phosphorus Removal 304
Retrofitting Surface Aeration Plants 304
Retrofitting Medium Bubble Aeration Plants 307
Miscellaneous Retrofitting 308
References 309

7. **GENERAL ACTIVATED SLUDGE MODEL FOR BIOLOGICAL NITROGEN AND EXCESS PHOSPHORUS REMOVAL** **311**
DIPANKAR SEN, M.S., P.E. and CLIFFORD W. RANDALL, Ph.D., M.E.

Introduction 311
Matrix Format Presentation of the Activated Sludge Model 311
Activated Sludge Model for Organic Carbon Removal, Nitrification, and Denitrification 313
The Nitrogen Source for Cell Synthesis 318
Conversion of Soluble Substrate to SCFA 319
Suggested Alteration to the IAWPRC Model for Nitrification 319
Activated Sludge Model for Nitrification, Denitrification and Biological Excess Phosphorus Removal (NDBEPR) 323
Simplified Enhanced Culture Steady-State Model 323
Development of a Mixed Culture Steady-State Model 325
References 331

8. **CONTROL OF BULKING AND FOAMING ORGANISMS** **335**
ORRIS E. ALBERTSON, M.S., P.E.

Overview 335
The Problem 335
Definitions and Measurements 337
History of the Bulking Sludge Problem 342

Causes of Nuisance Organism Growth 348
Filamentous Bulking 348
Non-Filamentous Bulking 349
Foaming 350
Process Alternatives for Bulking Control 351
Fixed Film 351
Selectors 353
Intermittent Feed/SBR 363
Toxicants 364
Foaming Organism Control 366
References 368

9. **SLUDGE HANDLING AND PROCESSING** 375
JAMES L. BARNARD, Ph.D., Pr.Eng., CLIFFORD W. RANDALL, Ph.D. and DIPANKAR SEN, M.S., P.E.
Dewatering and Disposal 375
Sludge Dewatering 375
Composting 377
Digestion 378
Landfill 378
Incineration 378
Heat Treatment 378
High Alkaline Treatment 379
Impact of Solids Handling Units on BNR Performance 379
Description of WWTPS: Processes and Solids Handling Units 380
Discussion 391
Management of Nutrient Balance 391
Sludge Production 392
References 394

10. **PILOT PLANT TESTING OF BNR SYSTEMS** 397
H. DAVID STENSEL, Ph.D., P.E.
Introduction 397
Wastewater Characteristics 397
Biological Phosphorus Removal Data Collection 400
Evaluation of Nitrification Kinetics 402
Evaluation of Nitrogen Removal 407
Sludge Production and Oxygen Requirements 412
Oxygen Requirements 414

Index 417

Foreword

THE first years of the final decade of the twentieth century have seen dramatic changes in the political structure of the world, especially in Eastern Europe and the countries which constituted the former Union of Soviet Socialist Republics. Equally dramatic have been international upheavals in the field of environmental protection. These range from the uncovering of terrible remnants of industrial pollution in nations of the old Eastern bloc, to the rapidly expanding frontiers of residual sludges management and of nutrients and toxics control here in the West. Such immense environmental challenges can only be met through the marshalling of the talents of the best environmental engineers and scientists around the world and through the design and application of cost-effective solutions.

The International Association on Water Pollution Research and Control, the IAWPRC, is an organization capable of addressing these global challenges. The primary mission of the IAWPRC is to disseminate the results of both fundamental and practical research to the international professional community and to apply this information to the development of innovative and effective solutions to the problems of water pollution control. Technology transfer and reports of research and development are regularly published in the Proceedings of the IAWPRC's international biennial and specialized conferences, as well as in the IAWPRC journals: *Water Research, Water Science and Technology,* and *Water Quality International*.

Within the Association biennial conferences provide a unique forum for international researchers and practitioners to debate and discuss methods to address and resolve increasingly complex environmental problems. In keeping with the missions of the IAWPRC, the United States of America National Committee, USANC, organized eight Specialty Courses offered in conjunction with the 1992 IAWPRC Biennial Conference in Washington, D.C. Designed for the practicing engineer, the Specialty Courses covered critical topics in environmental quality management, water pollution control, wastewater treatment, toxicity reduction, and residuals management.

The eight volumes in the *Water Quality Management Library* represent the texts prepared for the Specialty Courses. Experts from the United States and many other countries contributed their know-how and experience to the preparation of these state-of-the-art texts. Collectively, the volumes in the library are a pertinent and timely compendium of water pollution control and water quality management. They form a unique reference source reflecting international expertise and practice in key aspects of modern water pollution science and technology.

W. WESLEY ECKENFELDER
JOSEPH F. MALINA, JR.
JAMES W. PATTERSON

List of Contributors

ORRIS E. ALBERTSON, M.S., P.E., Enviro Enterprises, Inc., Salt Lake City, Utah

JAMES L. BARNARD, PH.D., PR.ENG., Meiring & Barnard, Inc., Pretoria, South Africa

CLIFFORD W. RANDALL, PH.D., Virginia Polytechnic Institute and State University, Blacksburg, Virginia

DIPANKAR SEN, M.S., P.E., Virginia Polytechnic Institute and State University, Manassas, Virginia

H. DAVID STENSEL, PH.D., P.E., University of Washington, Seattle, Washington

CHAPTER 1

Introduction

SYNOPSIS OF BOOK CONTENTS

THE purpose of this book is to present information that can be used for the design and operation of wastewater treatment plants that utilize biological nutrient removal processes, i.e., processes that utilize biological mechanisms instead of chemical mechanisms, to remove phosphorus and nitrogen from wastewaters. The basic fundamentals, concepts, and theories are presented in Chapters 1 and 2. Chapter 3 discusses and presents design procedures for prefermentation units, i.e., units that utilize acid fermentation processes to produce the volatile fatty acids essential for biological phosphorus removal (BPR) from settled primary solids. These units make it possible to produce low-effluent phosphorus concentrations (less than 1.0 mg/L) from any municipal wastewater.

The design fundamentals and procedures for the various types of biological nutrient removal systems are presented in Chapter 4. The chapter begins with considerations and methods for the design of two-stage BPR activated sludge systems. The second section presents methods for the design of three-stage, four-stage, and five-stage BNR systems, i.e., systems that combine both nitrogen and phosphorus removal processes in the same single sludge activated sludge system. Zone sizing, mixing and aeration requirements, and temperature effects are discussed. The third section presents design procedures for Carrousel-type BNR systems, and the fourth section discusses how to design sequencing batch reactor (SBR) systems. The chapter concludes with actual case histories of plants that were specifically designed to remove phosphorus and/or nitrogen using biological processes.

Chapter 5 is a detailed presentation of the considerations and procedures for the design of secondary clarifiers, with inclusion of special considerations for the settling of BNR sludges.

Methods for the retrofit of conventional activated sludge plants to achieve biological removal of phosphorus and/or nitrogen are presented in Chapter

6. The emphasis of the chapter is on actual case histories, with details of the operating performances achieved and the costs involved. The last section of the chapter presents the various ways to modify both the operation and the biological process configuration of oxidation ditches to accomplish enhanced removal of nitrogen separately, or nitrogen and phosphorus simultaneously.

Chapter 7 is a presentation of modeling considerations for BNR processes with a detailed explanation of the IAWPRC model, expanded to include biological phosphorus removal. The matrix format is used to present the model.

The last three chapters deal with special considerations for BNR systems. Chapter 8 discusses the unique aspects of sludge handling and processing. Case histories are used as illustrations. Methods for the control of filamentous microorganisms are given in Chapter 9, with emphasis on problems most commonly reported for BNR systems. Because wastewater characteristics strongly influence the performance of BNR processes, pilot plant procedures that can be used to determine design parameters and coefficients for specific wastewaters are detailed in Chapter 10.

BENEFITS OF BIOLOGICAL NUTRIENT REMOVAL

There are several reasons, or benefits, for utilizing biological nutrient removal (BNR) processes for the treatment of wastewaters. They may be classified as environmental benefits, economical benefits, and operational benefits. The most important of these is the control of eutrophication in the effluent receiving water, which is an environmental benefit. Historically, treatment requirements were determined by the need to protect the oxygen resources of the receiving water, and this was accomplished primarily through the removal of putrescible solids and dissolved organics from the wastewater before discharge. In more recent years, considerable emphasis has been placed on also reducing the quantities of nutrients discharged, i.e., nitrogen and phosphorus, because they stimulate growth of algae and other photosynthetic aquatic life, which lead to accelerated eutrophication, excessive loss of oxygen resources, and undesirable changes in aquatic populations.

The potential impact of discharged nutrients on the oxygen resources of receiving waters can best be illustrated by looking at the amounts of organic matter that can be generated by the nutrients compared to the amount of organic matter in untreated sewage. The COD of raw sewage in the United States is typically about 400 mg/L, whereas the phosphorus content is 6 to 10 mg/L, depending on whether or not a phosphate detergent ban is in place, and the nitrogen content is 30 to 40 mg/L. If one kilogram of phosphorus was completely assimilated by algae and used to manufacture new biomass from photosynthesis and inorganic elements, biomass of 111 kilograms with a COD of 138 kilograms would be produced, assuming algal composition can be represented by $C_{106}H_{263}O_{110}N_{16}P$. Thus, the discharge of 6 mg/L phosphorus could potentially result in COD production equivalent to 828 mg/L, or more than double the COD of the organic matter in the untreated sewage.

It is recognized that much of the algal biomass will slowly biodegrade, but the organics will build up in the bottom sediments where long-term biodegradation occurs, and it is likely that a high percentage of the oxygen demand will eventually be exerted. The rate of the sediment oxygen demand (SOD) will also accelerate as the algal biomass accumulates. It can be seen from the preceding comparison that the removal of biodegradable organics, only, from wastewaters lessens the potential impacts on the oxygen resources of the receiving water by less than half. Considering that phosphorus is a conservative substance that will accumulate within the system, it is clear that all comprehensive eutrophication control efforts should include phosphorus removal from discharged wastewaters.

It is possible for either nitrogen or phosphorus to be the nutrient that limits algal growth. A kilogram of nitrogen could potentially stimulate the manufacture of 16 kilograms of algal biomass, which would be equivalent to 20 kilograms of COD. Thus, 30 mg/L of discharged nitrogen could result in the production of COD equivalent to 600 mg/L. This is less than the potential production by phosphorus, but still greater than the COD of the organics in untreated sewage.

It is probable that either nitrogen or phosphorus will be the limiting nutrient controlling eutrophication because of the relatively large quantities required for biomass growth compared to other nutrients such as sulfur, potassium, calcium, and magnesium. Conventional wisdom in recent years has been that phosphorus is typically the limiting nutrient in freshwater environments, whereas nitrogen is typically limiting in estuarine and marine waters.

Actually, the relationships are considerably more complex than the above generalization because conditions in most lakes, reservoirs, and estuaries are dominated by bottom sediment conditions and seasonal changes. High concentrations of organic matter in the bottom sediments result in high SODs, which result in rapid depletion of dissolved oxygen in the lower levels of the body of water, particularly when it is well stratified and temperatures are high. These conditions occur every year in North America during the late summer and early fall. The resulting anaerobic conditions result in resolubilization of phosphorus, particularly the iron-bound forms, and, to a lesser extent, ammonia. Subsequent mixing of the water column makes the newly released nutrients available for algal, and other growth.

These dynamics are observed annually in the freshwater Occoquan Reservoir, located near Washington, D.C., and in the Chesapeake Bay estuary. It is probable that they occur in all impoundments and estuaries in urbanized areas.

The seasonal changes in eutrophic systems means that the limiting nutrient, i.e., either nitrogen or phosphorus, tends to change with the season. Typically, phosphorus is limiting in the wet weather periods when large quantities of nitrogen are brought into the water body by non-point sources, which happens every winter and early spring in the eastern United States. Nitrogen depletion begins with warm weather, first through assimilation, and then through denitrification after the DO has been depleted by nitrification and SOD. The resulting anaerobic conditions result in the massive release of phosphorus, which causes extreme nitrogen-limiting conditions. Thus, any introduction of nitrogen into the system during the late growing season, such as from treatment plant effluents, will usually result in a near stoichiometric growth of algal biomass. However, when nitrogen is the limiting nutrient, it may not be possible to reduce eutrophication to an acceptable level because of uncontrollable sources, i.e., atmospheric, groundwater, and surface runoff sources. If this is the case, then it becomes necessary to reduce the more controllable nutrient, phosphorus, to a low enough concentration to make it the limiting nutrient.

Because the limiting nutrient dynamics are generally poorly understood for most bodies of water, the best eutrophication control policy is simultaneous reduction of both nitrogen and phosphorus inputs. Also, because of seasonal changes, it is typically necessary to control both point and non-point sources in a complex watershed to attain the desired water quality. Biological nutrient removal (BNR) processes provide a methodology for the removal of nitrogen and phosphorus from point sources that is both environmentally and economically superior to other options.

Additional environmental benefits of BNR compared to phosphorus precipitation and methanol nitrogen removal are reduced chemical consumption, reduced waste sludge production, and reduced energy consumption by the treatment system. Such reductions lessen overall operational and disposal requirements. They also are responsible for the primary economical benefits of BNR.

ECONOMICAL BENEFITS OF BNR PROCESSES

The utilization of BNR processes for the treatment of wastewaters is potentially more economical than conventional activated sludge treatment or methods that incorporate chemical or physicochemical processes. Conventional wisdom has been that the most economical activated sludge system is one that uses aerobic processes to remove biodegradable COD without nitrification. In fact, this concept is incorrect. Conversion of that same system for biological phosphorus removal (BPR), i.e., the incorporation of an anaerobic zone ahead of the aerobic zone, would result in a substantial reduction of the aeration energy costs of the system, and no additional aeration basin or clarifier capacity would be needed. The energy needs for aeration would be reduced because most of the substrate removal and some stabilization of organics would occur in the anaerobic zone, as discussed by Randall et al. (1991), and because the oxygen transfer driving force would be greater since the mixed liquor entering the aerobic zone would have zero DO. A method for predicting the precise amount of the energy savings for any given plant has not been accurately quantified as yet, but a safe estimate is believed to be 10% or greater.

A biological nitrogen removal process is always more economical to operate than a fully aerobic one accomplishing complete nitrification. Nitrification will typically increase the energy costs of aeration by 50% for the treatment of municipal wastewaters, compared to the aeration requirements for COD removal only. Incorporation of an anoxic zone ahead of the aerobic zone in such a system will always result in a reduction of the total aeration energy costs, i.e., it will result in a partial "recovery" of the nitrification aeration energy costs. This is because many activated sludge bacteria can use oxidized forms of nitrogen as an electron acceptor for COD stabilization in place of DO. Any COD stabilized by denitrification does not have to be stabilized by DO metabolism, and the need to transfer oxygen is reduced. Also, because COD stabilization occurs in the anoxic zone, the aeration volume does not have to be increased for nitrogen removal except at very low temperatures (less than 10°C).

The maximum energy recovery possible by denitrification is 62.5% of that required for nitrification. This percentage will be reduced by failure to obtain complete denitrification, which cannot be achieved when only one anoxic zone is used, but a 50% reduction is very realistic, and this will easily offset any energy requirements for mixing in the anoxic zone(s) and nitrate recycle pumping. The actual energy savings is a function of the operating SRT. The aeration energy costs of a fully aerobic system are compared with mixing energy costs as a function of SRT in Figure 1.1. Complete nitrification is established at an SRT of four days for the reference temperature, after which the increase in energy requirement is a function of the endogenous decay rate. The oxygen requirements increase as endogenous decay increases because oxygen is required for biomass stabilization and for nitrification of ammonia released by the decaying cells.

It can be deduced based on the preceding paragraphs that the most energy efficient activated sludge system is a single sludge system that incorporates both an anaerobic zone and an anoxic zone ahead of the aerobic zone. An aeration energy reduction of 20% has been demonstrated at a full-scale plant

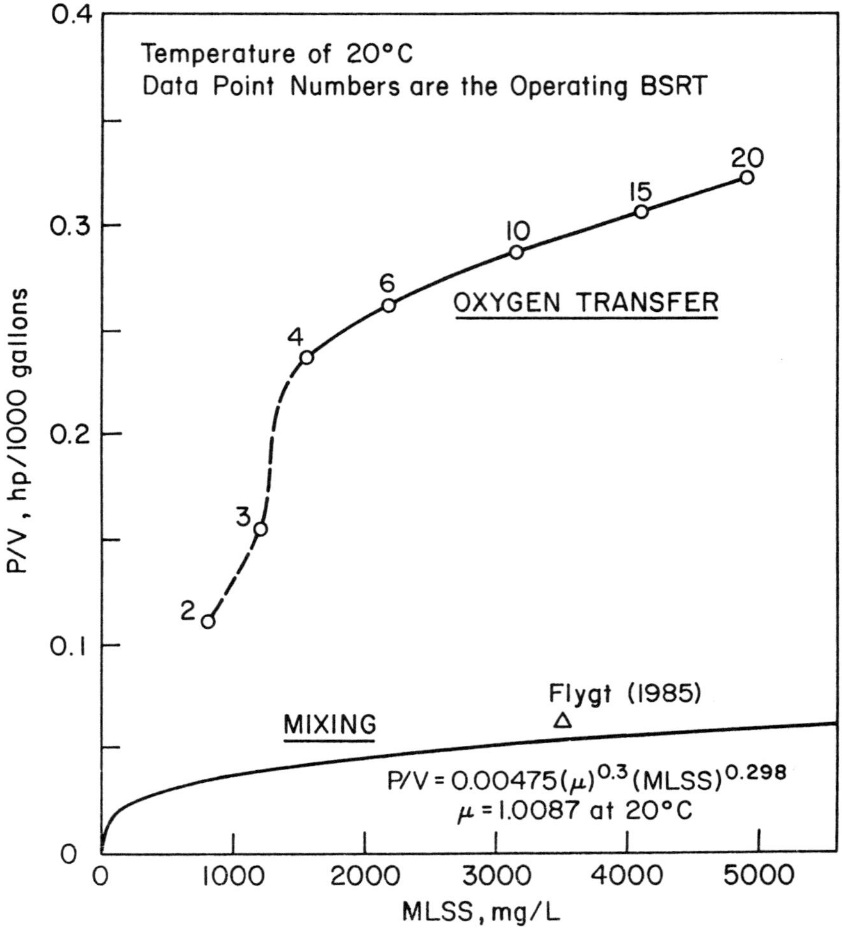

Figure 1.1 Comparison of mixing and oxygen transfer power requirements.

by the incorporation of an anoxic zone for denitrification (Best et al., 1984), and the potential for a 30% reduction with total BNR seems very realistic. Further full-scale comparison is needed to enable accurate incorporation of the savings into design, however.

Biological phosphorus removal provides an obvious economic benefit through the reduction or elimination of chemical addition for phosphorus removal. If the required effluent standard is 1.0 mg/L or more, BPR will eliminate the need for chemical addition to remove phosphorus from most wastewaters. If the required effluent concentration is less than 0.5 mg/L, or if the BOD_5 to TP ratio is significantly less than 20:1, supplemental chemical addition probably will be needed to meet the effluent requirement, unless effluent filtration is practiced. Regardless of how stringent the effluent phosphorus requirements are, however, utilization of BPR will substantially reduce the need for the addition of phosphorus-precipitating chemicals such as iron and aluminum salts. Of course, any reduction in inorganic chemical addition reduces the amount of waste sludge generated by the process, and this provides a further economic benefit by reducing sludge disposal costs.

It has also been shown by both laboratory and full-scale studies that the utilization of anoxic conditions for nitrogen removal reduces waste sludge production (McClintock et al., 1988; Ip et al., 1987).

An additional benefit of denitrification is the recovery of one-half of the alkalinity destroyed during nitrification. This is a significant benefit for poorly buffered wastewaters, such as occur in the piedmont and coastal areas of Virginia and Maryland, as it substantially reduces or eliminates the need to add buffering chemicals to meet effluent pH requirements.

PROCESS STABILITY

It has been shown that anaerobic and/or anoxic zones placed ahead of aerobic zones act as biological selectors that discourage the growth of filamentous microorganisms in activated sludge and generally improve the sludge settling properties. The improvement in sludge settleability is particularly pronounced when a fully aerobic system is switched to BPR, i.e., an anaerobic zone is added. A typical sludge volume index (SVI) for a well-operated BPR system is 50 and the effluent total suspended solids (TSS) concentration is 10 mg/L or less following clarification. Incorporation of nitrogen removal will typically increase the SVI to about 120, but the effluent TSS concentration will not necessarily increase. If the system is properly designed and operated, a BNR process is more stable and produces a better-quality effluent as measured by BOD and TSS than a conventional plug-flow activated sludge process.

It has also been shown through side-by-side comparison that the BPR process can produce an effluent lower in BOD, TSS, and phosphorus than a conventional, fully aerobic process using alum added to the activated sludge for phosphorus removal (Wilson et al., 1990).

BNR PROCESS CONFIGURATIONS

An activated sludge system must include one or more unaerated zones and be a single sludge system in order to accomplish BNR. Single sludge system means that each train contains only one secondary clarifier and settled biomass must recycle through all zones. This is essential for selection

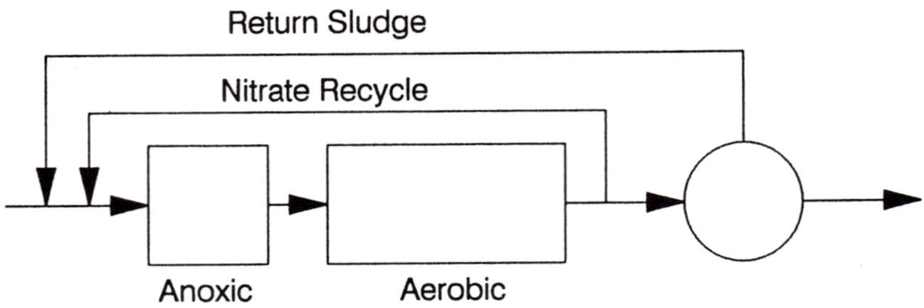

Figure 1.2 Two-stage biological nitrogen removal system.

of the types of bacteria and biochemical processes desired. Phosphorus removal can be maximized by placing the anaerobic zone first so that the phosphorus-removing bacteria have the first opportunity to utilize the organic substrate, thus giving them a competitive edge over those bacteria that cannot utilize or store substrate under anaerobic conditions. Likewise, nitrogen removal can be maximized by placing the anoxic zone first. Phosphorus removal is still possible if the anaerobic zone is placed after an anoxic zone, but only if readily available organics are still present in solution in substantial quantity. Nitrogen removal is less sensitive to the types of organics available, so the anaerobic zone is usually placed first and the anoxic zone follows. However, it is impossible to operate a nutrient removal sequencing batch reactor (SBR) without going through the anoxic phase before the anaerobic phase, if the system is accomplishing nitrification.

The first BNR process configurations were developed in the 1960s. Ludzack and Ettinger (1962) were the first to propose a single sludge nitrification-denitrification process utilizing the biodegradable organics in the influent as the energy source for denitrification. They used two reactors in series, i.e., an anoxic reactor followed by an aerobic reactor, but did not completely separate them and depended on intermixing of the reactor contents to accomplish nitrogen removal. The lack of control of the interchange between the two reactors resulted in variable performance. Barnard,

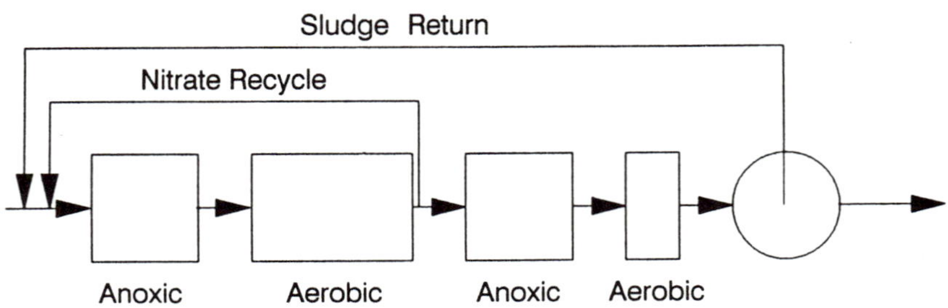

Figure 1.3 Four-stage biological nitrogen removal system (Bardenpho).

while developing the Bardenpho process in 1973, proposed an improvement of the Ludzack-Ettinger process by completely separating the anoxic and aerobic reactors, recycling the settled sludge to the anoxic reactor, and providing an additional recycle from the aerobic to the anoxic reactor for the return of nitrates. The modified Ludzack-Ettinger (MLE) process configuration is shown in Figure 1.2.

The amount of denitrification possible in the MLE process is limited by the total amount of nitrate recycled, i.e., the sum of the quantities in the return activated sludge (RAS) from the clarifier and in the mixed liquor recycle. This also limits the effluent total nitrogen (TN) concentration possible, and the MLE process can seldom produce an effluent concentration of less than 5 mg/L TN when treating municipal wastewater. The actual effluent concentration will be determined by the TN concentration influent to the process and the total recycle flow compared to the influent flow. The economical upper limit for mixed liquor recycle is usually considered to be 4:1, but 2:1 will produce an effluent concentration only 2 to 3 mg/L TN higher for most municipal wastewaters (Brannan and Randall, 1987).

Barnard (1973) invented the Bardenpho process by adding a secondary anoxic reactor after the aerobic reactor to increase total denitrification and decrease the effluent TN concentration possible. The configuration is shown in Figure 1.3 and the process is considered to be capable of producing an effluent TN concentration of 3.0 mg/L or less.

An oxidation ditch can be converted easily to a nitrogen removal system by controlling the amount of air transferred to the mixed liquor. The internal recycle rate in a ditch is very high, on the order of 100 to 200:1, recycle to influent, and this makes it possible to reduce the effluent TN concentration to less than 3 mg/L for almost any wastewater. The required DO concentration for complete nitrification can be very low in a high recycle system because nitrification in such a system is typically not limited by the oxygen

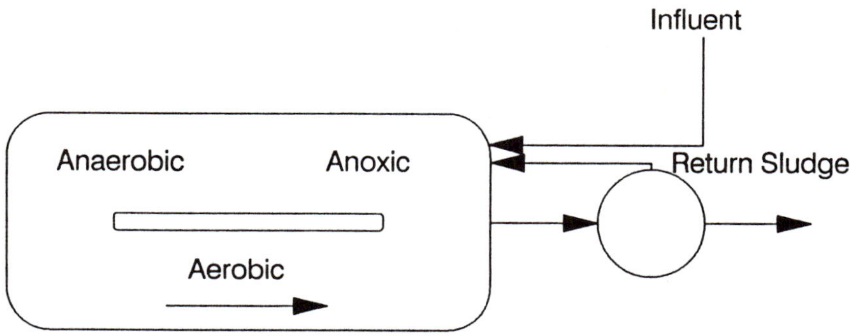

Figure 1.4 Oxidation ditch biological nutrient removal system.

transfer rate. For example, the oxidation ditch system at Bowie, Maryland accomplishes complete nitrification and denitrification with a mixed liquor DO concentration of 0.2 to 0.3 mg/L. Higher levels are detrimental to denitrification, whereas lower levels are detrimental to nitrification. Conceptually, the process can be envisioned as illustrated by Figure 1.4, but performance is actually determined by the mass of oxygen transferred to the mixed liquor rather than by changes in DO concentration.

Sequencing batch reactors (SBRs) can be operated for nitrogen removal by adjusting the time periods of the different phases. A successful sequence for a municipal wastewater is given in Figure 1.5. Very low effluent concentrations can be achieved with proper operation.

BPR CONFIGURATIONS

A process configuration for the utilization of BPR was first developed by Levin and Shapiro (1965). They showed that enhanced phosphorus storage by bacteria was possible and included an anaerobic reactor in the system they developed to accomplish phosphorus release. However, they did not understand the linkage between organic substrate storage and phosphorus release, or fully appreciate the detrimental effects of nitrates in the anaerobic reactor. Consequently, they used a sidestream anaerobic reactor to accomplish phosphorus release from the biomass and then removed the phosphorus from the supernatant using chemical precipitation. The configuration developed, which was termed the PhoStrip process, is shown in Figure 1.6. It is the only sidestream configuration currently in use for biological nutrient removal.

Barnard (1974, 1975) first understood the significance of the anaerobic zone, and he developed the first mainstream systems accomplishing BPR by placing an anaerobic reactor first in front of an aerobic zone, then in front

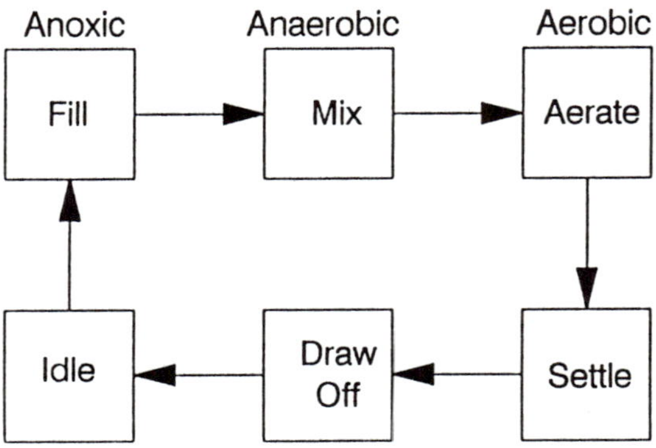

Figure 1.5 Time schematic for sequencing batch reactor biological nutrient removal system.

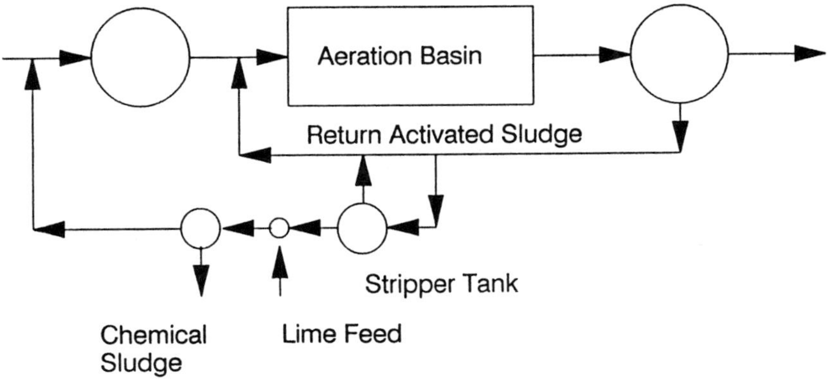

Figure 1.6 Schematic of PhoStrip process.

of the MLE configuration, and then in front of the Bardenpho configuration. The historical development of BPR is briefly summarized in the next paragraph.

It was observed during the late 1960s that several conventional, plug-flow activated sludge systems, such as the Rilling Road plant in San Antonio, Texas, the Tucson, Arizona plant, and the Backwater plant in Baltimore, among others, were accomplishing enhanced phosphorus removal (Vacker et al., 1967; Yall et al., 1970; Milbury, 1971). Barnard reviewed the published information and noted that all of these plants were plug-flow, conventional activated sludge plants operated at high rates to suppress nitrification. He realized the importance of an anaerobic zone to stimulate phosphorus removal and proposed placing an anaerobic stage ahead of the aerobic stage so that the influent organics could be used to create anaerobic conditions instead of relying on endogenous respiration as in the PhoStrip process (Barnard, 1975). The resulting configuration, which he called, "Phoredox," is shown in Figure 1.7. It was first applied on a full-scale basis at Germiston, South Africa in 1974. This same configuration was later

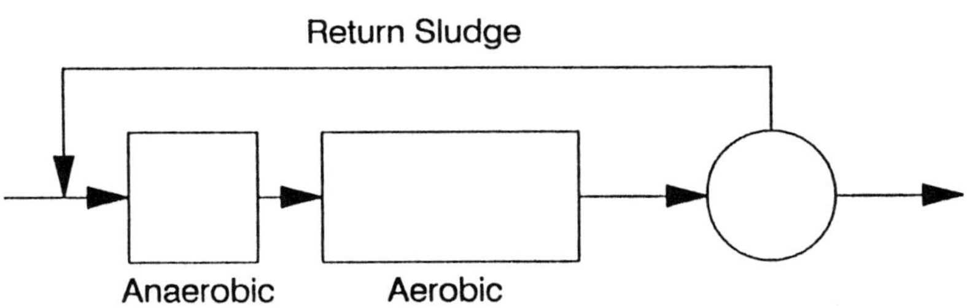

Figure 1.7 Two-stage biological phosphorus removal system.

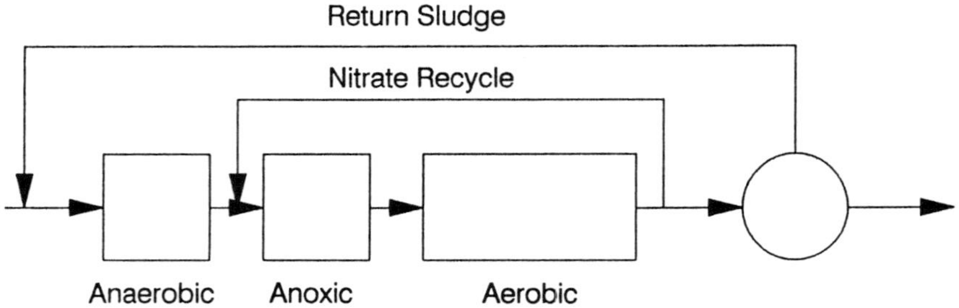

Figure 1.8 Three-stage biological nutrient removal system.

patented by Air Products and Chemicals, Inc. in the United States under the trademark of "Anaerobic/Oxic," or A/O (Spector, 1979).

After pilot-plant and full-scale experimentation, Barnard combined the Phoredox concept with the MLE process at Klerksdorp, South Africa, in 1974. This resulted in the three-stage process shown in Figure 1.8, which also was later patented in the United States by Air Products and Chemicals, Inc. and trademarked as "Anaerobic/Anoxic/Oxic," or A^2/O. The Phoredox concept was next combined with the Bardenpho process for the new Goudkoppies plant in Johannesburg, South Africa, which went on-line in 1978, to produce the five-stage modified Bardenpho configuration, shown in Figure 1.9. Subsequent work by many investigators during the 1980s demonstrated that BPR could be accomplished with a wide variety of system configurations.

Experience has shown that a conventional, plug-flow activated sludge reactor can be easily converted to a BPR process by establishing an anaerobic zone at the influent end of the aeration basin, as illustrated by Figure 1.10. Modifications required are: turning off the air in that section, installation of a baffle that prevents backflow, and installation of mixing equipment. These simple modifications will result in excellent phosphorus removal from most municipal wastewaters and development of an activated sludge with excellent settling and dewatering properties, if true anaerobic conditions are maintained in the anaerobic zone. The system performs best

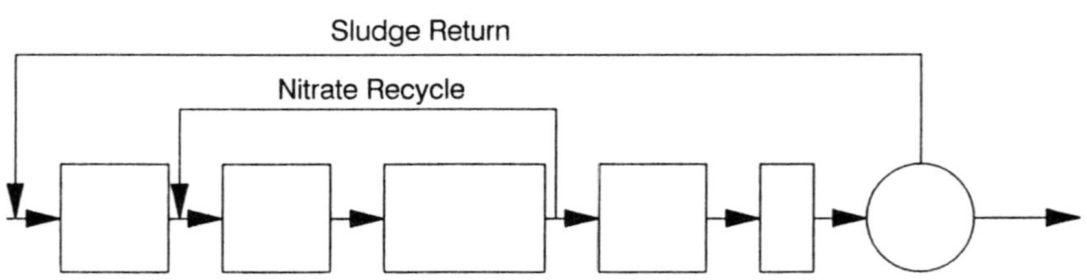

Figure 1.9 Five-stage biological nutrient removal system (modified Bardenpho).

as a high-rate process with no nitrification. The system can be further modified to accomplish nitrogen removal by installing another baffle, additional mixing equipment, and mixed liquor recycle for nitrate return, resulting in a process identical to Figure 1.8.

Experience has also shown that a simple racetrack oxidation ditch can be operated to accomplish both nitrogen and phosphorus removal without chemical addition by simply controlling the amount of oxygen transferred to the mixed liquor (Sen et al., 1990). The resulting process can be visualized as diagrammed in Figure 1.4, although the distinct zones do not exist because of the very high internal recirculation rate. Better phosphorus removal can be obtained by adding an anaerobic reactor upstream of the oxidation ditch, as shown in Figure 1.11. This was also demonstrated by Sen et al. (1990), and they further improved performance by putting two oxidation ditches in series to form the VT2 process shown in Figure 1.12. This system consistently produces effluent TN and TP concentrations of less than 3.0 and 0.7 mg/L, respectively, all months of the year, without chemical addition or effluent filtration.

Configurations with multiple oxidation ditches have also been developed in Denmark for the biological removal of nitrogen alone, or of both nitrogen and phosphorus (Bungaard, 1988). Labeled Bio-Denitro and Bio-Denipho, respectively, the Bio-Denitro configuration and operating sequence is given in Figure 1.13.

Sequencing batch reactors (SBRs) are single reactors that cycle through anoxic, anaerobic, and aerobic conditions to accomplish BNR. Settling is also accomplished in the same reactor, as is supernatant and waste sludge wasting. While all processing steps occur in the same reactor, two or more reactors are essential for wastewater treatment unless a very large holding basin is available. The operating sequence is the same as that previously shown in Figure 1.5. Process operation and performance have been thoroughly described by Irvine and co-workers (1983, 1985, 1987).

Marais and his co-workers determined the detrimental effects of nitrates entering the anaerobic zone on the performance of BPR processes. They

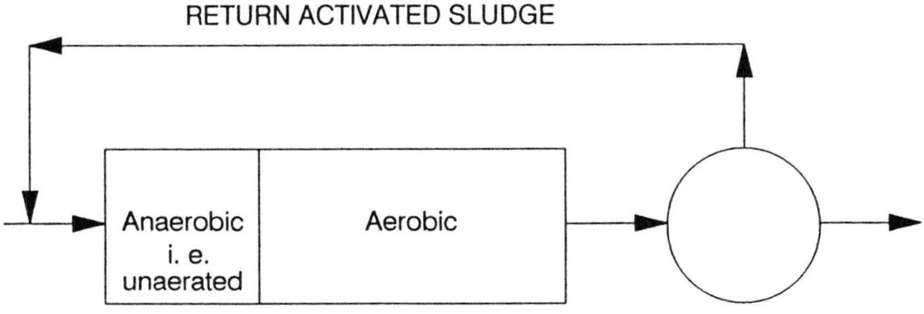

Figure 1.10 Operationally modified activated sludge process.

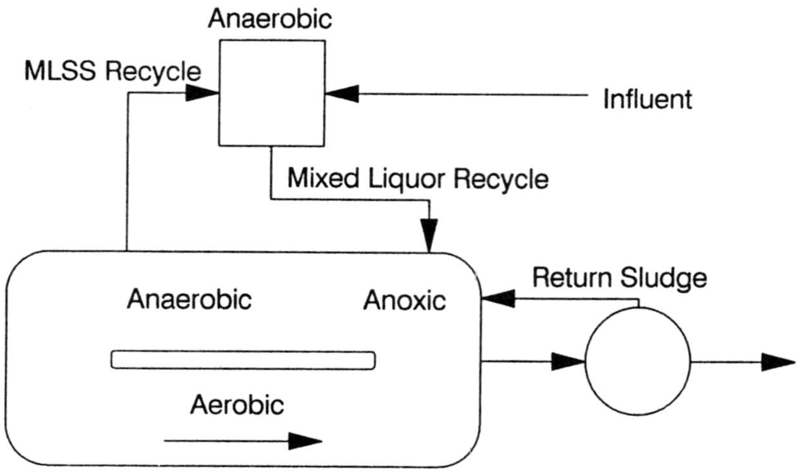

Figure 1.11 VT oxidation ditch biological nutrient removal system.

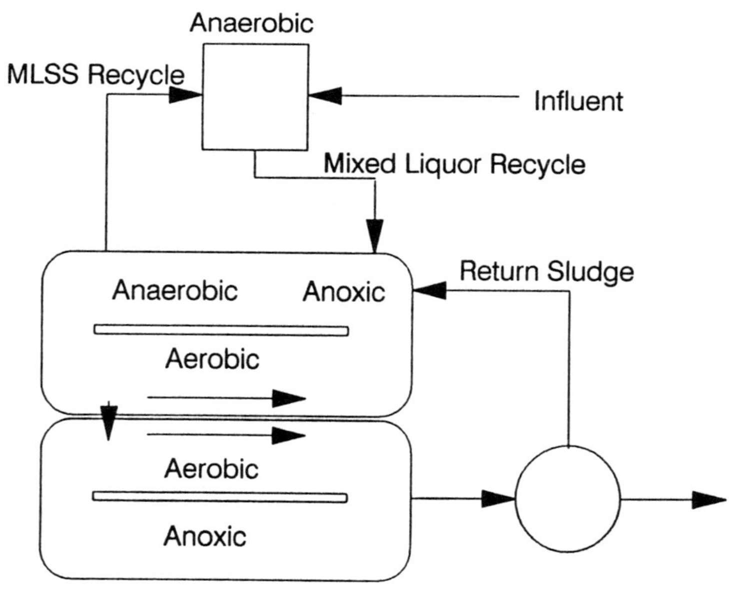

Figure 1.12 VT2 oxidation ditch biological nutrient removal system.

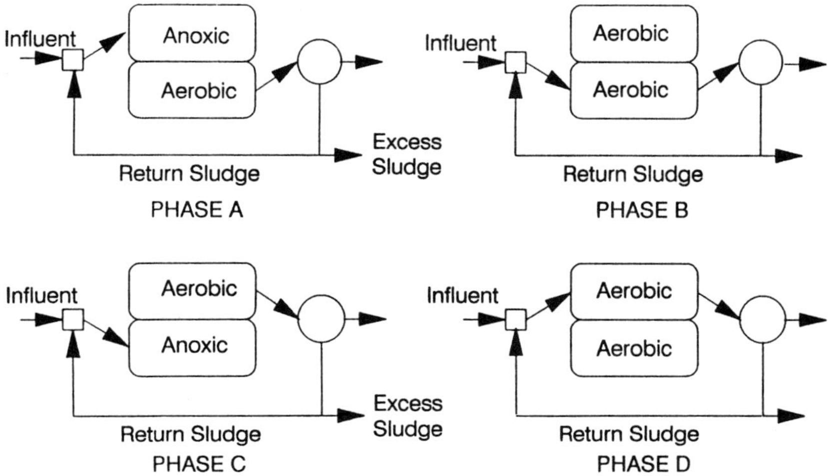

Figure 1.13 Bio-Denitro biological nitrogen removal system.

developed a modification of the three-stage Phoredox process that eliminated the recycle of nitrates in the RAS to the anaerobic zone. This process, which they named the University of Cape Town (UCT) process, is shown schematically in Figure 1.14. They later modified the UCT process to provide even better protection of the anaerobic zone from nitrate recycle, as illustrated by Figure 1.15.

Randall (1984) recommended utilization of the UCT configuration as a high-rate BNR system for economical treatment of municipal wastewaters in the U.S.A. This concept was piloted by CH2M/HILL with consultation from Randall for the Hampton Roads Sanitation District (HRSD), as a new design for replacement of their Lambert's Point wastewater treatment plant, which was providing only primary treatment (Daigger et al., 1987). The

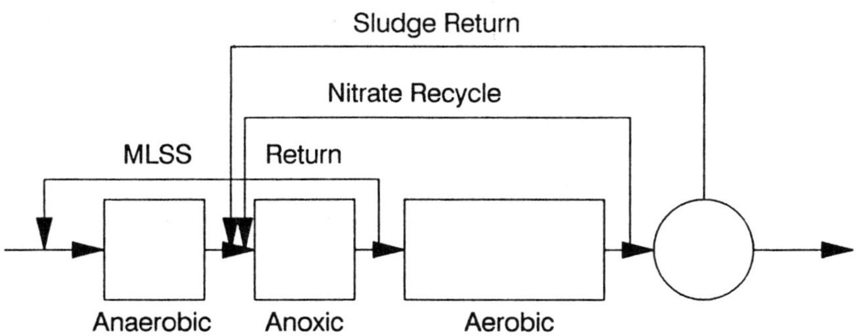

Figure 1.14 UCT-type biological nutrient removal system.

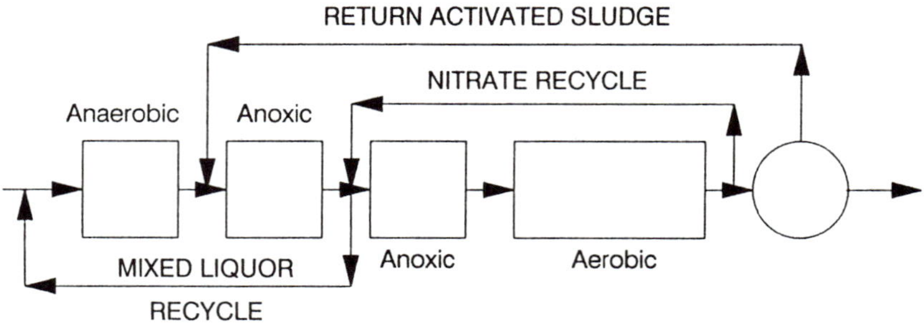

Figure 1.15 Schematic of the modified UCT process.

pilot study was very successful, and the process was patented as a public domain patent by CH2M/HILL and HRSD as the Virginia Initiative Project (VIP) process. The UCT and VIP configurations are identical.

THE EFFECTS OF WASTEWATER CHARACTERISTICS

The performance of a BNR system is strongly affected by the characteristics of the wastewater influent to each zone of the process. This has been discussed in great detail by Marais and his co-workers (Ekama et al., 1983; Siebritz et al., 1983; Ekama and Marais, 1984). The more important aspects may be summarized as follows.

Neither biological nitrogen removal nor biological phosphorus removal can be accomplished without sufficient biodegradable organic substrate, i.e., as measured by COD or BOD. The removal processes are accomplished by heterotrophic bacteria, and they require organic matter for their metabolism. It is possible, then, to state that the removal of one mg/L of nitrogen via denitrification will require a specific quantity of useable COD. The same can be said for the removal of one mg/L of phosphorus by BPR, with the complication that BPR metabolism is strongly affected by the specific organic compound available for assimilation. Therefore, the amount of nitrogen and phosphorus that can be removed by BNR processes in any given system, can be reasonably estimated if the influent organics have been fully characterized. Stated another way, the effluent concentrations of N and P possible by BNR can be predicted for any given wastewater, once the wastewater has been sufficiently characterized. Unfortunately, the characterization required for precise performance modeling is unusual and extensive, and the information is never available unless a specially designed sampling and analysis program has been performed. The sampling-analysis requirements have been discussed in detail by Marais and his co-workers (Ekama and Marais, 1984), and have been more recently updated by Gibson and Dold (1991).

Fortunately, BNR systems can be acceptably designed without the extensive wastewater characterization required for accurate modeling. Although the rates of denitrification are affected by the specific organic compounds available, it has been determined that 8.6 mg COD/L of municipal wastewater is needed to reduce 1 mg/L of nitrate nitrogen to nitrogen gas (Ekama and Marais, 1984). Stated another way, the denitrification of 10 mg/L of nitrate will consume 86 mg/L of biodegradable COD. It can be seen, then, that the denitrification of 15 to 20 mg/L of nitrate nitrogen will consume a large fraction of the organic matter in a typical North American municipal sewage.

The COD consumption during biological phosphorus removal has been estimated by Ekama and Marais (1984) to be 50 to 59 mg/L COD per mg/L of phosphorus removed from municipal wastewater. Experiments at Virginia Tech indicate that 50 mg/L COD consumed per mg/L P removed is a conservative number, and it is recommended for design purposes.

Actually, the efficiency of BPR varies with the specific organic compound available in the anaerobic zone, as demonstrated by Abu-ghararah and Randall (1991). Their results for several volatile fatty acids are shown in Table 1.1, along with the value they obtained for municipal sewage only, using a pilot-plant system. The data show that acetic acid is the most efficient organic for BPR, and is considerably more efficient than propionic acid, another common end product of sewage fermentation. Also note that enhanced BPR cannot be accomplished with formic acid, as predicted by accepted theory. Only valeric acid had a higher COD to P removal ratio than the 50 to 1 ratio recommended for municipal wastewater design. This

TABLE 1.1. Effect of Organic Substrate on Enhanced Biological Phosphorus Removal.

Substrate	mg/L Phosphorus Uptake* / mg/L COD Utilized	mg COD Utilized** / mg P Removed
Formic acid	0	Infinity
Acetic acid	0.37	16.8
Propionic acid	0.10	24.4
Butyric acid	0.12	27.5
Isobutyric acid	0.14	29.1
Valeric acid	0.15	66.1
Isovaleric acid	0.24	18.8
Municipal sewage	0.05	102†

*Total P uptake in aerobic zone.
**COD utilized and P removed in total system.
†Value obtained with highly aerobic sewage.
BSRT = 13 days for all experiments.

indicates that some of the sewage organics are not available for BPR, and this increases the COD utilized:P removal ratio observed for municipal sewage. A surprising result of the experiments was the high efficiency of phosphorus removal accomplished by isovaleric acid. However, it is doubtful that isovaleric acid will be an end product of sewage fermentation.

Fortunately, sewage fermentation results in formation of primarily acetic acid. Sufficient fermentation for BPR frequently occurs in the sewers transporting the sewage to the treatment plant, particularly in the summertime, and additional fermentation can be easily accomplished on-site, if necessary, using specially designed fermenters or through modified operation of thickeners or primary clarifiers. This means that all municipal treatment plants can be retrofitted for BPR, if desired, and that all municipal wastewaters can be treated by BPR. The effluent phosphorus concentration possible with BPR will vary with the organic matter to total phosphorus ratio of the wastewater, however, and supplemental chemical addition may be necessary to meet required effluent limitations. Figure 1.16 shows the relationship between effluent concentration achievable and the BOD_5:TP ratio in the wastewater entering the anaerobic zone. This figure, based on full-scale and pilot-scale experience, indicates that a BOD_5:TP ratio of at least 20:1 is needed to achieve an effluent TP concentration of 1.0 mg/L or less, with secondary clarification as the last treatment step. Of thirty-nine observations, each representing data averaged over time periods of from one month to one year, from nine full-scale plants and two pilot plants, only one had an effluent TP concentration exceeding 1.0 mg/L when the BOD_5:TP ratio of the process influent wastewater was 20:1 or more. Lower effluent TP concentrations could be achieved with the same ratio if polymer-enhanced clarification or effluent filtration was used, as shown by the graph of effluent soluble phosphorus in Figure 1.17.

The efficiency of both nitrogen removal and phosphorus removal can be reduced by conditions that result in the metabolism of useable substrate by other biochemical pathways. For example, design and/or operating conditions can result in the recycle or entrainment of DO and nitrates in the flows to the anaerobic zone, or DO to the anoxic zone. Bacteria that accomplish BNR will use DO and oxidized nitrogen forms as electron acceptors and metabolize the available organics in the anaerobic zone to CO_2 and water before they can be stored for phosphorus removal, to the extent that DO and NO_X are available. The BPR bacteria will use DO and NO_X preferentially, when available, because they obtain more energy for growth through use of these electron acceptors than through anaerobic substrate storage. Similarly, denitrifying bacteria will use DO instead of NO_X in the anoxic zone if DO is available because they obtain more energy for growth using DO. These reactions reduce the organic matter available for nitrogen and phosphorus removal and, consequently, may limit the effluent quality that

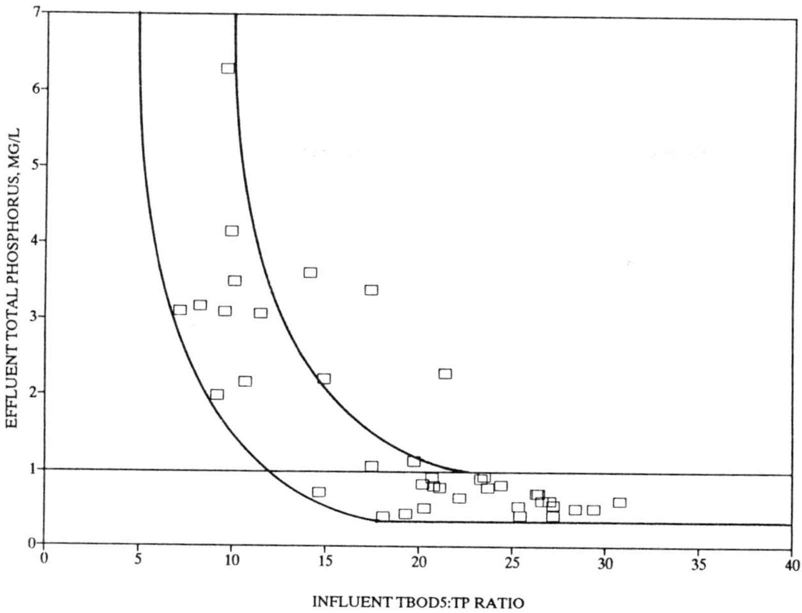

Figure 1.16 Effect of process influent $TBOD_5$:TP ratio on effluent total phosphorus.

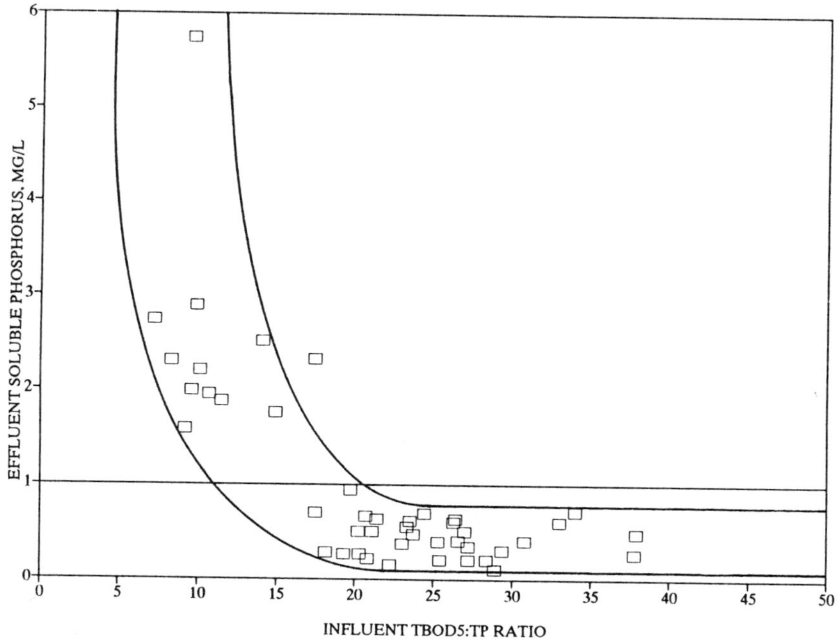

Figure 1.17 Effect of influent $TBOD_5$:TP ratio on effluent soluble phosphorus.

can be achieved. Therefore, good design and operation will minimize the quantities of the unwanted electron acceptors available to the bacteria.

THE EFFECTS OF PRIMARY SETTLING

The utilization of primary settling will substantially reduce the organic matter to phosphorus ratio in the wastewater, and reduce the amount of phosphorus and nitrogen that can be removed by the treatment plant. Recycle streams from sludge processing will have the same effect. For some plants, the inclusion of primary settling and/or the recycle of sludge processing flows will result in effluent phosphorus concentrations greater than those desired. Experience at the York River Plant near Yorktown, Virginia, owned and operated by the Hampton Roads Sanitation District (HRSD), illustrated these effects. The raw wastewater had a BOD_5:TP ratio of 22.6:1. After primary settling, however, the ratio was 12.5:1. The corresponding COD:TP ratios were 41:1 and 22.4:1, respectively.

The ratio decreased so much because, in addition to primary settling, sludge processing flows were recycled back to the headworks. The recycle flows were from the primary sludge gravity thickener, the secondary sludge flotation thickener, and the filtrate from the belt filter press dewatering an anaerobically digested mixture of the two thickened sludges. The combined recycle had a TP concentration of 40.4 mg/L, a TN concentration of 81.6 mg/L, and a BOD_5:TP ratio of 5.8:1. The flow was only 17% of the raw influent, but it contained 75% as much phosphorus and 49% as much nitrogen. The recycle loads would have been higher except for the formation of struvite and other chemical precipitates in the anaerobically digesting sludge. Seventy percent of the TP entering the digesters was wasted from the system in the belt filter press filter cake.

The primary clarifier removed a higher percentage of the organic matter from the combined flow than it removed nutrients. It removed 26% and 34%, respectively, of the BOD_5 and COD, but only 18%, each, of the TP and TKN. The resulting nutrient concentrations to the biological process were TP 11.1 mg/L and TN 29.2 mg/L, whereas the BOD_5 and COD concentrations were 139 and 249 mg/L, respectively. It was also noted that primary settling reduced the organic matter:phosphorus ratios compared to those calculated for the combined flows. The actual results are shown in Table 1.2. Analysis indicated that calcium phosphate precipitates were forming to a limited extent and were being removed with the primary sludge.

The biological process clearly could not produce an effluent TP concentration of 1.0 mg/L or less, given the low BOD_5 to TP ratio, but it removed an average of 9.4 mg/L phosphorus by biological mechanisms, except for high flow or acclimation periods, and produced an average

TABLE 1.2. Changes in Organics-to-Phosphorus Ratios with Recycle and Primary Treatment.

Parameter	Raw Influent	Combined Influent	Change mg/L	Change %	Primary Effluent	Decrease mg/L	Decrease %
$TBOD_5$, mg/L	209	209	0	0	139	70	33
TCOD, mg/L	377	418	41	11	249	169	40
TP, mg/L	9.26	13.76	4.5	49	11.11	2.65	19
TKN, mg/L	28.4	35.7	7.3	26	29.0	6.7	19
$TBOD_5$:TP	22.6	15.2	—	−33	12.5	—	18
TCOD:TP	40.7	30.4	—	−25	22.4	—	26

effluent soluble phosphorus (SP) concentration of 2.4 mg/L. MLVSS phosphorus contents as high as 15% on a dry weight basis were observed. Apparently the process was using about 26.5 mg/L COD to remove 1 mg/L of phosphorus. Furthermore, during the latter stages of the demonstration, it simultaneously removed TN down to a two-month average concentration of 5.7 mg/L, while removing an average of 10.4 mg/L phosphorus and discharging an average SP concentration of 2.3 mg/L. The design factors given earlier would predict a COD requirement of 660 mg/L for these removals, yet the actual COD removal averaged only 234 mg/L. However, the BOD_5 removal averaged 218 mg/L, showing that all of the influent organics were in the readily biodegradable form, probably acetic acid, and much higher than predicted removals were possible. The actual BOD_5:P removal ratio was 21:1.

It appears that substrate-limited BNR systems can remove considerably more nutrient than predicted, but cannot produce an effluent TP concentration of less than 1.0 mg/L at the same time. The results also indicate that stored substrate can be used simultaneously for both nitrogen and phosphorus removal.

EFFLUENT LIMITATIONS

There has been considerable confusion concerning the lower limits of phosphorus and nitrogen concentrations possible with BNR processes. That is to say, how low can the final concentrations be if the microorganisms have an abundance of substrate available compared to the nutrient concentrations, i.e., when the system is nutrient-limited rather than substrate-limited?

Effluent concentrations obtained by a variety of systems treating municipal wastewaters are listed in Table 1.3. The results show that soluble effluent phosphorus concentrations averaging less than 0.35 mg/L have been achieved over long periods of time, and that effluent total phosphorus concentrations of less than 0.50 mg/L are possible with only secondary

TABLE 1.3. Effluent Limitations of BNR Processes.

Treatment System	Operating Phase	Effluent Concentrations, mg/L				
		Soluble P	Total P	Ammonia	Organic SKN	TKN
VIP Pilot	IS	0.28	0.39	0.15	1.13	1.28
	IIS	0.5	0.79	0.05	1.56	1.61
	IVBS	0.4	0.79	0.55	1.26	1.81
	IVE	0.39	0.53	0.10	1.13	1.23
	VB	0.21	0.81	—	—	—
	VD	0.26	0.43	0.29	1.21	1.50
	VE	0.15	0.64	—	—	—
	VIA	0.20	0.42	0.10	1.34	1.44
Bonnybrook	A/O	0.22	0.45	—	—	—
Maryland City	A^2/O	—	—	—	0.89	—
Bowie	VT2	—	0.66	0.15	0.70	0.87
York River	VIP	0.50	0.76	0.50	0.84	1.34
Mauldin Road	A^2/O	0.49	0.52	—	—	—
York River	A^2/O	—	—	0.60	0.50	1.82
Average		0.33	0.60	0.28	1.06	1.43

Final treatment step was secondary clarification for all plants.
Treatment plants range in capacity from 2.85 to 105 ML/d.

clarification for suspended solids removal. Barnard (1988) has stated that effluent soluble phosphorus (SP) concentrations of less than 0.1 mg/L can be reliably obtained by phosphorus-limited BPR processes. When complete nitrification is being achieved, effluent ammonia concentrations as low as 0.05 mg/L are possible, and concentrations less than 0.50 mg/L are typical. However, BNR process effluents will typically contain soluble organic nitrogen concentrations of 1.0 to 1.5 mg/L which cannot be removed without activated carbon treatment, or an equivalent separation technique. Nevertheless, effluent TKN concentrations of less than 1.5 mg/L are possible.

BNR systems using post denitrification units can reduce effluent NO_X concentrations to about 1 mg/L, if desired. When combined with an effluent TKN concentration of less than 1.5 mg/L, effluent total nitrogen concentrations of less than 3.0 mg/L are possible. It should be recognized, however, that effluent TN concentrations of about 6 mg/L can be achieved much more economically than concentrations of less than 3 mg/L. This is because the reactor volumes must be considerably larger to insure complete nitrification at low temperatures. Also, the energy reduction benefits of denitrification are maximized when the nitrate recycle rate is approximately equal to the influent wastewater flow (Brannan and Randall, 1987), and this rate means the effluent nitrate nitrogen concentration will typically be 4 to 5 mg/L.

Effluent nutrient standards for wastewater treatment plants should be set only after full consideration of the seasonal nature of non-point nutrient inputs, the magnitude of atmospheric inputs, and the economics of different levels of wastewater treatment.

REFERENCES

Abu-ghararah, Z. H. and C. W. Randall. 1991. "The Effect of Organic Compounds on Biological Phosphorus Removal," *Wat. Sc. Tech.*, 23:585.

Barnard, J. L. 1973. "Biological Denitrification," *J. Wat. Pol. Con. Fed.*, 72:705.

Barnard, J. L. 1974. "Cut P and N without Chemicals," *Water and Wastes Eng.*, 11(1):33-36 and 11(2):41-44.

Barnard, J. L. 1975. "Biological Nutrient Removal without the Addition of Chemicals," *Wat. Res.*, 9:485-490.

Barnard, J. L. 1988. "Case Studies in Phosphate Removal," *Proceedings International Workshop on Wastewater Treatment Technology, January 11-17, 1988, Copenhagen, Denmark*. The Danish Association of Consulting Engineers.

Best, A. G., G. J. Hatton, A. J. Rachwal and B. Hurley. 1984. "Biological Phosphorus and Nitrogen Removal at an Experimental Full Scale Plant in the U.K.," *Proceedings, IAWPRC Post-Conference Seminar, Enhanced Biological Phosphorus Removal from Wastewater, Vol. 1, Anjou-Recherche, 75389 Paris Cedex 08, France*, pp. 270-289.

Brannan, K. P. and C. W. Randall. November 11, 1987. "The Economics of Nitrogen Removal in a Single-Sludge Activated Sludge System," *Proceedings, Conference on the Impact and Control of Nutrients in Surface Waters*, Charleston, S.C., 30 pages, Environmental Systems Eng. Dept., Clemson University.

Bungaard, E. 1988. "Nitrogen and Phosphorus Removal by the Bio-Denitro and Bio-Denipho Processes – Experience and Development," *Proceedings, International Workshop on Wastewater Treatment Technology, January 11-13, 1988, Copenhagen, Denmark*. Session IV. The Danish Association of Consulting Engineers.

Daigger, G. T., C. W. Randall, G. D. Waltrip, E. D. Romm and L. M. Morales. 1987. "Factors Affecting Biological Phosphorus Removal for a University of Cape Town Type Process Applied in a High Rate Mode," *Proceedings, IAWPRC International Specialized Conference, Biological Phosphate Removal from Wastewater, Rome, Italy, September 28-30*, pp. 177-184 in *Biological Phosphate Removal from Wastewater*, R. Ramadori, Editor. Pergamon Press.

Ekama, G. A. and G. v. R. Marais. 1984. "Biological Nitrogen Removal," in *Theory, Design and Operation of Nutrient Removal Activated Sludge Processes*, Water Research Commission. Published by the Water Research Commission, Pretoria.

Ekama, G. A., I. P. Siebritz and G. v. R. Marais. 1983. "Considerations in the Process Design of Nutrient Removal Activated Sludge Processes," *Water Sc. Tech.*, 15:283.

Gibson, J. and P. L. Dold. 1991. "Wastewater Characterization for Activated Sludge Modelling." Research Paper, Dept. of Civil Engr. & Engr. Mech., McMaster Univ., 1250 Main Street West, Hamilton, Ontario L8S 4L7.

Ip, S. Y., J. S. Bridges and N. F. Mills. 1987. "Effect of Alternating Aerobic and Anaerobic Conditions on the Economics of the Activated Sludge Systems," *Wat. Sc. and Tech.*, 19:911-918.

Irvine, R. L., L. H. Ketchum, Jr., M. L. Arora, E. F. Barth. 1985. "An Organic Loading Study of Full-Scale Sequencing Batch Reactors," *J. Wat. Pol. Con. Fed.* 57(8):847.

Irvine, R. L., L. H. Ketchum, R. Breyfogle and E. F. Barth. 1983. "Municipal Application of Sequencing Batch Treatment," *J. Wat. Pol. Con. Fed.*, 55(5):984.

Irvine, R. L., D. V. S. Murthy, M. L. Arora, J. L. Copeman and J. A. Heihman. 1987. "Analysis of Full-Scale SBR Operation at Grundy Center, Iowa," *J. Wat. Pol. Con. Fed.*, 59(3):132.

Levin, G. V. and J. Shapiro. 1965. "Metabolic Uptake of Phosphorus by Wastewater Organisms," *J. Wat. Pol. Con. Fed.*, 36(6):800-821.

Ludzack, F. L. and M. B. Ettinger. 1962. "Controlling Operation to Minimize Activated Sludge Effluent Nitrogen," *J. Wat. Pol. Con. Fed.*, 34:920.

Manning, J. F. and R. L. Irvine. 1985. "The Biological Removal of Phosphorus in a Sequencing Batch Reactor," *J. Wat. Pol. Con. Fed.*, 57(1):87.

McClintock, S., J. S. Sherrard, J. T. Novak and C. W. Randall. March 1988. "Nitrate versus Oxygen Respiration in the Activated Sludge Process," *J. Wat. Pol. Con. Fed.*, 60(3):342-350.

Milbury, W. F., D. McCauley and C. H. Hawthorne. 1971. "Operation of Conventional Activated Sludge for Maximum Phosphorus Removal," *J. Wat. Pol. Con. Fed.*, 49(9):1890-1891.

Randall, C. W. 1984. "The Development of Energy Efficient Activated Sludge Systems Through Utilization of Nutrient Removal Processes," in *Toxic and Hazardous Wastes, Proceedings of the Sixteenth Mid-Atlantic Industrial Waste Conference*, Edited by M. D. LaGrega and D. A. Long, Technomic Publishing Company, Inc., pp. 52-63.

Randall, C. W., K. L. Brannan, S. A. McClintock and V. M. Pattarkine. 1991. "The Case for Anaerobic Reduction of Oxygen Requirements in Biological Phosphorus Removal Systems." Submitted to *Res. J. Wat. Pol. Con. Fed.*

Sen, D., C. W. Randall and T. J. Grizzard. 1990. "Biological Nitrogen and Phosphorus Removal in Oxidation Ditch and High Nitrate Recycle Systems," CBP/TRS 47/90. Chesapeake Bay Program, U. S. Environmental Protection Agency, 120 pages.

Siebritz, I. P., G. A. Ekama and G. v. R. Marais. 1983. "A Parametric Model for Biological Excess Phosphorus Removal," *Water Sc. Tech.*, 15:127.

Spector, M. L. July 24, 1979. U. S. patent 4,162,153.

Vacker, D., C. H. Connell and W. N. Wells. 1967. "Phosphate Removal through Municipal Wastewater Treatment at San Antonio, Texas," *J. Wat. Pol. Con. Fed.*, 39:750-771.

Wilson, W. A., E. I. Meckelborg and P. Do. 1990. "Full-Scale Biological Phosphorus Removal Trials at Bonnybrook," presented at the *42nd Annual Conference, Western Canada Water and Wastewater Association, Regina, Saskatchewan.* Reid Crowther & Partners, Ltd, 7410 Blackfoot Trail, S. E. Calgary, Alberta T2H 1M5 Canada.

Yall, I., W. H. Boughton, R. C. Knudsen and N. A. Sinclair. 1970. "Biological Uptake of Phosphorus by Activated Sludge," *Appl. Micro.*, 20(1):145-150.

CHAPTER 2

Principles of Biological Nutrient Removal

INTRODUCTION

THE principles of both biological nitrogen removal and biological phosphorus removal are presented in this chapter, with a discussion of the most pertinent literature and experience. Design equations are presented for both nitrification and denitrification, whereas biological phosphorus removal is discussed primarily from a mechanistic and operations perspective.

BIOLOGICAL NITROGEN REMOVAL

Biological nitrogen removal requires a two-step process. In the first step ammonia is oxidized to nitrate (nitrification) and various process configurations are then employed to provide the nitrate as an electron acceptor for biological respiration so that it can be reduced to molecular nitrogen (denitrification). Fundamental considerations will be presented first for the nitrification step, followed by a similar review for the denitrification step in nitrogen removal. Included in this review are the microbiology, basic biochemical reactions, biological kinetics, and the factors that affect the kinetics and performance of biological nitrification and denitrification reactions.

BIOLOGICAL NITRIFICATION

MICROBIOLOGY

Early work by Schloesing and Muntz (1877) investigating the percolation of sewage through a sand column found that the conversion of ammonia to nitrate and nitrite was due to living organisms, since the reaction could be stopped by introducing chloroform vapor into the column. Classical experiments by Winogradsky (1890) led to the isolation of autotrophic nitrifying bacteria, *Nitrosomonas* and *Nitrobacter*, which oxidize ammonia sequentially to nitrite and then to nitrate, respectively. Painter (1970) has listed other autotrophic bacteria genera capable of obtaining energy from the oxidation of ammonia, *Nitrosococcus*, *Nitrospira*, *Nitrosocyctis* and *Nitrosoglea*, and nitrite, *Nitrocystis*, but the nitrification in soil or wastewater treatment processes is attributed primarily to *Nitrosomonas* and *Nitrobacter*.

Painter (1970) summarizes other literature sources that show a large number of heterotrophic bacteria capable of forming nitrite or nitrate. In one study by Eylar and Schmidt (1959) of about 1000 heterotrophic organisms isolated from soil, only fifteen were found to be able to nitrify. A review by Focht and Chang (1975) indicated that heterotrophic nitrification is possible with diverse genera of bacteria, fungi, and actinomycetes. However, it is doubtful that significant quantities of nitrate are generated by heterotrophic organisms, since autotrophic nitrification rates are about ten times greater. Heterotrophic nitrification may be more prominent in atypical environments with either very alkaline or acidic pH conditions.

Bock et al. (1988) reported on the ability of *Nitrobacter* to grow and reduce nitrate while using acetate, formate, pyruvate, or glycerine as organic substrates. They further showed in an aerobic/anoxic biofilm that nitrite could inhibit the growth of the *Nitrobacter* under anoxic conditions. This observation has not been studied in wastewater treatment.

OXIDATION AND SYNTHESIS RELATIONSHIPS

The energy-yielding two-step oxidation of ammonia to nitrate is generally accepted to be as follows:

Nitrosomonas

$$2NH_4^+ + 3O_2 \rightarrow 2NO_2^- + 4H^+ + 2H_2O \qquad (2.1)$$

Nitrobacter

$$2NO_2^- + O_2 \rightarrow 2NO_3^- \qquad (2.2)$$

Total reaction

$$NH_4^+ + 2O_2 \rightarrow NO_3^- + 2H^+ + H_2O \qquad (2.3)$$

The standard free energy release associated with ammonia oxidation has been estimated to be 66 to 84 kcal/mole of ammonia, and for nitrite oxidation it has been estimated to be 17.5 kcal/mole of nitrite (Painter, 1970). If the amount of cell production is proportional to the energy released, a greater amount of biomass should be formed from the oxidation of ammonia to nitrite than from the oxidation of nitrite to nitrate.

Based on the above, the oxygen required for complete oxidation of ammonia is 4.57 g/g N oxidized with 3.43 g/g used for nitrite production and 1.14 g/g used for nitrate production. The amount of oxygen required is less than 4.57 g/g N when synthesis is considered in addition to oxidation due to oxygen obtained from fixation of carbon dioxide and nitrogen into cell mass. Wezernak and Gannon (1967) found that the actual total oxygen consumption was 4.33 g/g N with 3.22 g/g N used for ammonia oxidation and 1.11 g/g N used for nitrite oxidation. The following equation, presented in the *U.S.EPA Nitrogen Control Manual* (1975), accounts for both synthesis and oxidation and shows an oxygen requirement of 4.2 g/g N. The equation was developed using cell yield coefficients of 0.15 g/g NH_4-N oxidized and 0.02 g/g NO_2-N oxidized.

$$NH_4^+ + 1.83O_2 + 1.98HCO_3^- \rightarrow 0.21C_5H_7O_2N + 0.98NO_3^- + 1.041H_2O + 1.88H_2CO_3 \quad (2.4)$$

Table 2.1, prepared by Poduska (1973), shows that cell yields for *Nitrosomonas* are generally higher than those of *Nitrobacter*.

For activated sludge designs employing nitrification, a single yield coefficient that includes the growth of both *Nitrosomonas* and *Nitrobacter* is more convenient and some of the more recently used values are summarized in Table 2.2.

TABLE 2.1. Nitrification Cell Yield Coefficients, g Cells/g Substrate.

Nitrosomonas	Reference	*Nitrobacter*	Reference
0.005	Johnson (1964)	0.04-0.07	Gould (1960)
0.03-0.010	Loveless (1968)	0.02	Downing (1964a)
0.29	Haug (1971)	0.084	Haug (1971)
0.05	Knowles (1965)	0.02	Knowles (1965)
0.05	Downing (1964b)	0.02	Lees (1957)
0.06	Baas-Becking (1927)	0.02	Boon (1962)
0.13	Nelson (1931)		
0.06	Skinner (1961)		

Substrate was NH_3—N for *Nitrosomonas* and NO_2—N for *Nitrobacter*.
Source: Poduska, 1973.

TABLE 2.2. Overall Nitrification Yield Coefficient, g Cells/g NH_4—N Oxidized.

Values	Reference
0.15	U.S.EPA (1975)
0.15	WPCF MOP FD-7 (1983)
0.05	Benefield (1980)
0.22	Beccari (1979)
0.17	Bidstrup (1988)
0.13	Neufield (1980)
0.12	Rozich (1986)

The only higher yield reported which was supported with experimental data is that by Rozich (1986). Rozich observed net yields of nitrifying organisms over steady-state reactor solids retention times (SRT) ranging from 2.5 to 18 days. The observed yield over this range did not show a decrease at longer SRT values to indicate a significant reduction in mass due to endogenous decay. Using data-fitting techniques, he determined an endogenous decay coefficient of 0.0038 d^{-1}. Neufeld et al. (1980) report an endogenous decay coefficient of 0.04, but they did not indicate how the value was derived. An endogenous decay value has not been well established for nitrifying cultures, and many times it is ignored when calculating the net growth of nitrifying organisms. Since nitrifiers derive energy by oxidizing ammonium or nitrite, their decay rate under conditions with these nitrogen compounds lacking could conceivably be different than the classical autooxidation assumed for heterotrophs under starved conditions.

The portion of nitrifiers in the mixed liquor of an activated sludge tank treating municipal wastewater is typically in the 2-5% range due to the low yield and substrate level associated with ammonia oxidation compared to BOD removal. Assuming oxidation of influent ammonia nitrogen and influent BOD concentrations of 20 and 120 mg/L, respectively, and a net nitrifier yield of 0.15 g TSS/g NH_4-N oxidized and a net yield of 0.6 g TSS/g BOD removed for the non-nitrifier solids fraction, the nitrifiers would account for about 4% of the aeration tank mass.

Besides illustrating the stoichiometry between ammonia oxidized, oxygen consumption and biomass production, Equation (2.4) also shows, as does Equation (2.1), that a significant amount of alkalinity is destroyed during ammonia oxidation to nitrite. Approximately 7.14 mg of alkalinity (as $CaCO_3$) are consumed per mg NH_4-N oxidized. This means that a municipal plant with 25 mg/L of ammonia available for oxidation should have an influent alkalinity of about 260 mg/L (as $CaCO_3$) to assure an aeration tank final alkalinity of about 80 mg/L after full nitrification. Some wastewater treatment facilities require the addition of lime or soda ash to maintain desirable pH levels for nitrification.

Poduska (1973) reviewed the results of studies on the effect of micronutrients on the growth of nitrifying bacteria in pure cultures and found that some stimulated growth at certain concentrations. Table 2.3 lists these micronutrients and the stimulatory concentrations reported.

BIOLOGICAL KINETICS

The generally accepted design approach for nitrification systems is to determine the aeration tank volume by selecting an appropriate SRT that will provide the desired effluent ammonia nitrogen concentration at the important operating conditions expected, which include temperature, dissolved oxygen concentration, and pH. The design SRT is related to the expected specific growth rate of the nitrifying bacteria using the Monod kinetic expression and usually includes a safety factor to account for the need to have a greater inventory of nitrifying bacteria to handle peak ammonia loads. Based on the development of a dynamic model and laboratory studies, Poduska (1973) has established a guideline that the safety factor should be equal to the peak to average ammonia loads to maintain the effluent ammonia concentration at or below the selected design value.

$$SRT_d = S.F.(SRT) \tag{2.5}$$

where:

SRT_d = Design solids retention time, d
S.F. = Design safety factor = peak/average NH_3-N load
SRT = Required solids retention time, d

The aeration tank volume for a completely mixed activated sludge system can be calculated as follows using the design SRT:

TABLE 2.3. Stimulatory Concentration of Micronutrients for Nitrifying Bacteria.

Compound	Concentration, mg/L
Calcium	0.5
Copper	0.005-0.03
Iron	7.0
Magnesium	12.5-0.03
Molybdenum	0.001-1.0
Nickel	0.1
Phosphorus	310.0
Zinc	1.0

Source: Poduska, 1973.

$$V = [Y_{nH}(S_o - S) + X_I + Y_{nN}(\mathrm{N}_{ox})]Q(\mathrm{SRT}_d)/X \qquad (2.6)$$

where:

V = Aeration tank volume, m^3
Q = Influent flow, m^3/d
X = Aeration tank mixed liquor concentration, mg/L
S_o = Influent BOD_5, mg/L
S = Effluent BOD_5, mg/L
N_{ox} = Ammonia in influent flow oxidized, mg/L
Y_{nH} = Net yield of heterotrophic organisms at design SRT including endogenous decay, g TSS/g BOD_5 removed
Y_{nN} = Net yield of nitrifying bacteria, g TSS/g N oxidized
X_I = Influent non-biodegradable inert solids concentration, mg/L

Equation (2.6) shows that the design SRT is one of many factors that determine the aeration tank volume. The aeration tank mixed liquor suspended solids concentration is critical, and will depend on the sludge settling characteristics and clarifier design (Metcalf and Eddy, 1991). The specific wastewater characteristics are important, as indicated by the influent BOD_5 and inert solids in the equation.

The required SRT is equal to the reciprocal of the net specific growth of the nitrifying organisms.

$$\mathrm{SRT} = \frac{1}{\mu_n - K_{nd}} \qquad (2.7)$$

where:

μ_n = Nitrifying bacteria specific growth rate, g new cells/g cell-d
K_{nd} = Endogenous decay rate of the nitrifying bacteria, g cells decayed/g cells-d

The experimental data that provides net specific growth rate values has been developed either without measuring endogenous decay rates for nitrification or without finding it to be a significant factor. Thus, since information is not available to substantiate a decay coefficient value, it is ignored and the required SRT is the reciprocal of the specific growth rate determined by Monod kinetics:

$$\mu_n = \frac{(\mu_n, \max)\mathrm{N}}{K_n + \mathrm{N}} \qquad (2.8)$$

where:

μ_n,max = Maximum specific growth rate of nitrifiers, g cell produced/g cell-d
K_n = Half saturation coefficient, mg/L
N = Ammonia nitrogen concentration, mg/L

When lower reactor ammonia nitrogen concentrations are required, a lower specific growth rate and correspondingly higher SRT are required per Equations (2.7) and (2.8).

In active nitrifying systems the nitrite concentration is usually very low compared to the ammonium concentration, due to the higher nitrite oxidation rates by *Nitrobacter* compared to ammonia oxidation rates by *Nitrosomonas*. Thus, the kinetics are based on ammonia-nitrogen utilization rates or the activity of the *Nitrosomonas* bacteria. Note that during the startup of nitrification systems, the nitrite concentration will be higher until the *Nitrobacter* population increases to equilibrium.

The values for μ_n,max are affected by temperature. Various investigators have performed kinetic studies to determine μ_n,max as a function of temperature and the resultant models are summarized in Table 2.4 along with maximum specific growth rate values determined at single temperatures by other studies. A plot of these data in Figure 2.1 illustrates the wide range of maximum specific growth rates reported, indicating that other factors may have affected the maximum specific growth rate measured. These factors could include reactor mixing intensity, activated sludge floc size, pH, and dissolved oxygen level. The higher values from Downing et al. (1964) were obtained using pure cultures developed from Thames River

TABLE 2.4. Effect of Temperature on Nitrification Maximum Specific Growth Rates.

		μ_n,max (d^{-1})		
Source	μ_n,max vs. Temp. °C	10°	15°	20°
Downing (1964a)	$(0.47)e^{0.098(T-15)}$	0.29	0.47	0.77
Downing (1964b)	$(0.18)e^{0.116(T-15)}$	0.10	0.18	0.32
Hultman (1971)	$(0.50)10^{0.033(T-20)}$	0.23	0.34	0.50
Barnard (1975)	$0.33(1.127)^{T-20}$	0.10	0.18	0.37
Painter (1983)	$(0.18)e^{0.0729(T-15)}$	0.12	0.18	0.26
Beccari (1979)				0.27
Bidstrup (1988)				0.65
Hall (1980)				0.46
Lawrence (1976)				0.50

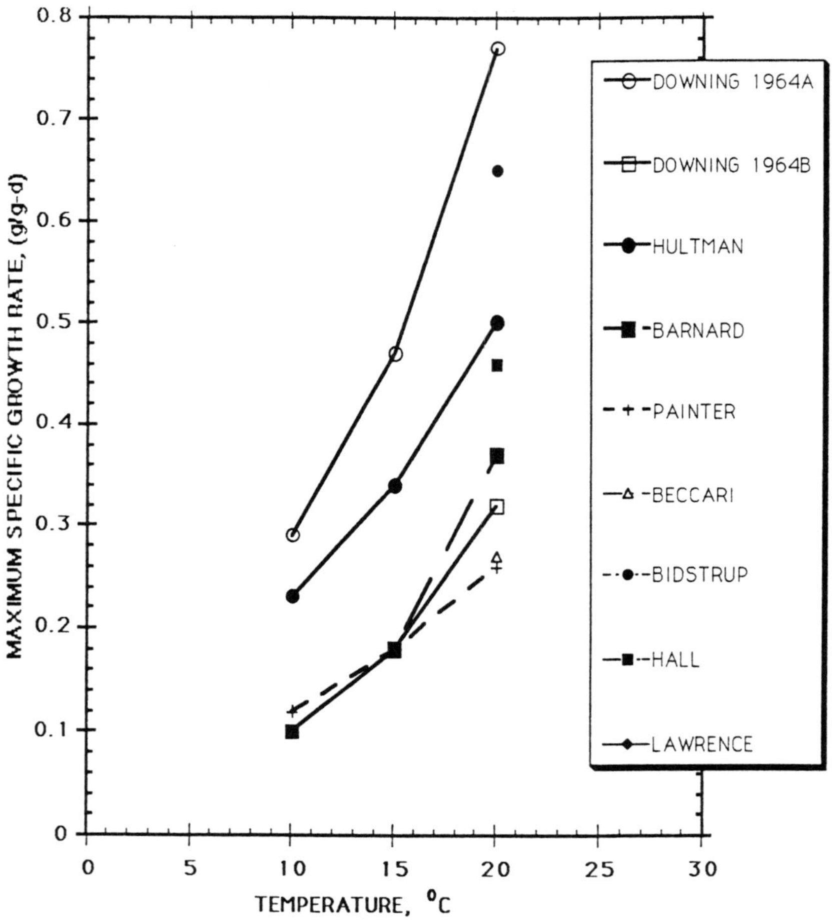

Figure 2.1 Nitrification maximum specific growth rate vs. temperature.

water instead of an activated sludge mixture. This would result in more dispersed growth with little floc formation. This higher relationship has been used in developing a design approach in the *U.S. EPA Nitrogen Control Manual* (1975). It appears that maximum specific growth rates at 20°C in the range of 0.3-0.5 d^{-1} may be more appropriate.

The half-saturation coefficient, K_n, found for nitrification is generally very low compared to values commonly reported for heterotrophic organisms. Knowles et al. (1965) reported the following equation for K_n as a function of temperature.

$$K_n = 10^{0.051T - 1.148} \tag{2.9}$$

where:

T = Temperature, °C

This relationship yields K_n values of 0.23, 0.41, and 0.74 mg/L at 10°, 15°, and 20°C, respectively. Chudoba et al. (1985) reported a K_n value of 0.41 mg/L at 20°C.

EFFECT OF pH

The reported effects of pH have varied for different investigators, and the interpretation of these results is further compounded by the fact that the pH effects were determined in batch tests without any acclimation period. Wild et al. (1975) showed the maximum specific growth rate at pH 7.0 to be about 50% of that at pH 8.0 with a relatively linear relationship in between. Engle and Alexander (1958) and Downing et al. (1964) showed little effect of pH change between 7.2 and 8.0 and a linear decrease in nitrification rate below pH 7.2. Hall (1974) reported complete nitrification between 7.0 and 9.4 pH and no nitrification at pH 6.3. The following equation (U.S.EPA, 1975) was proposed to describe pH effects below pH 7.2:

$$\mu_{n,\text{pH}} = (\mu_n, 7.2)[1 - 0.833(7.2 - \text{pH})] \tag{2.10}$$

A similar relationship has also been proposed (University of Capetown, 1984) to describe pH effects below 7.2:

$$\mu_{n,\text{pH}} = (\mu_n, 7.2)(2.35)^{\text{pH} - 7.2} \tag{2.11}$$

At pH 6.5 the two equations show a nitrification rate at 42 and 55%, respectively, of the rate at pH 7.2. Boon and Laudelout (1962) found the nitrification rate for *Nitrobacter* at pH 6.5, to be 60% of the rate at 7.5. More recently, work by Antoniou et al., (1990) using unacclimated cultures, showed a nitrification rate at pH 6.9 equal to 84% of the rate at pH 7.9 at 20°C. At 15°C they found the rate at pH 6.8 to be 42% of the rate at pH 7.8, indicating a greater pH effect at lower temperatures.

Acclimation may result in less of a pH effect than shown above. Stankewich (1972) and Haug and McCarty (1972) have observed that nitrification rates can be reestablished at the former maximum growth rate after acclimation at lower pH levels. Haug and McCarty found full acclimation as pH was decreased from 7.0 to 6.0 after ten days in a fixed film reactor.

EFFECT OF DISSOLVED OXYGEN

Reports on the effect of dissolved oxygen (DO) concentration on nitrification rates vary widely in the literature from studies that used either enriched cultures in laboratory tests or wastewater plant activated sludge. Downing and Scragg (1958) reported that a DO concentration of at least 0.3 mg/L is needed for nitrification to occur. In laboratory studies Schoberl and Engel (1964) reported that nitrification rates for *Nitrosomonas* were independent above a concentration of 1.0 mg/L and for *Nitrobacter* above 2.0 mg/L DO. Boon and Laudelout (1962) studied growth rates of *Nitrobacter winogradski* at DO concentrations of 1 mg/L and saturation DO at temperatures between 20 and 35°C. They found that the growth rates at the 1.0 mg/L DO concentration were lower than at the saturation DO concentration, and the difference was affected by temperature. The growth rates at 1.0 mg/L were 79, 80, 70, and 58% of that at saturation DO for temperatures of 20, 23.7, 29 and 35°C, respectively.

Using activated sludge mixed liquor, Downing et al. (1964), and Wild et al. (1971) found that nitrification rates were not affected by DO concentrations above 1.0 mg/L. On the other hand, Wuhrman (1963) found that nitrification rates were not affected by DO concentration changes from 4.0 to 7.0 mg/L, but that the rate at 1.0 mg/L was 90% of the rate at the higher DO concentrations. Nagel and Haworth (1969) reported that nitrification rates in activated sludge doubled as the DO concentration increased from 1.0 to 3.0 mg/L.

A generally accepted approach to quantify the effect of DO concentration on nitrification kinetics has been to use double Monod kinetics that show the nitrifier specific growth rate as a function of both ammonia nitrogen and DO concentrations as follows:

$$\mu_n = \frac{(\mu_n, \max)\mathrm{N}}{K_n + \mathrm{N}} \left[\frac{\mathrm{DO}}{K_o + \mathrm{DO}}\right] \tag{2.12}$$

where:

DO = Dissolved oxygen concentration, mg/L
K_o = Half saturation coefficient for dissolved oxygen, mg/L

A task group for the International Association on Water Pollution Research and Control (IAWPRC, 1986) included the same equation to describe nitrification rates in their general and versatile model for the activated sludge process. They recommended a K_o value of 1.0 mg/L. Other half saturation values have been summarized by Stenstrom and Poduska (1980) and are included with other sources in Table 2.5.

The variations in dissolved oxygen effects have been explained by evaluating fundamental mechanisms of oxygen transport and consumption in an activated sludge floc (Stenstrom and Poduska, 1980; Benefield and Randall, 1980; Eckenfelder, 1990) and modeled using such fundamental considerations with experimental data by Stenstrom and Song (1991). The basic concepts of these models are that (1) nitrifying bacteria are distributed throughout an activated sludge floc particle with heterotrophic bacteria; (2) the DO concentration in the floc particle will be lower than the DO concentration of the bulk liquid and will vary with the depth of the floc as a function of DO diffusion and respiration level. Thus, the observed nitrification rate can vary due to BOD loading changes even though the measured DO concentration in the bulk liquid is constant.

When nitrification occurs with significant BOD removal at higher organic loadings, the DO concentration will be depleted more rapidly with floc depth, and the observed nitrification rate would be lower. This has been supported by experimental results by Stenstrom and Song (1991). The decrease in the observed nitrification rate would be greater as the floc size is increased, since a greater fraction of nitrifiers would be located in oxygen-deficient zones in the floc. A larger operating SRT that would

TABLE 2.5. Reported Half-Saturation Values Describing Effect of Dissolved Oxygen on Nitrification Rates.

Organism	K_o, mg/L	Reference
Nitrosomonas	0.3	Loveless and Painter (1968)
	0.25	Peeters et al. (1969)
	0.50	Laudelout et al. (1974)
Nitrobacter	1.84	Peeters et al. (1969)
	0.72	Laudelout et al. (1976)
Activated sludge	0.43	Stankevich (1972)
	2.0	Nagel and Haworth (1969)
	0.45, 0.56	Stenstrom and Song (1991)

provide more nitrifying organisms is needed to compensate for this effect. Stenstrom and Song alternately state that a longer SRT is needed to ensure nitrification at lower operating DO concentrations and higher mass-transport resistances. They claim that the limiting DO concentration for nitrification in activated sludge may be 0.5 mg/L to 2.5 mg/L depending on the plant loading and floc characteristics.

Nitrifying organisms have been reported to be unaffected by high DO concentrations by Okun (1949) and Haug and McCarty (1971). They reported no inhibition at DO concentrations of 33 mg/L and 60 mg/L, respectively. Charley et al. (1980) studied the effect of a high DO concentration at 30°C and found a short-term inhibition during the first four days. By the fifth day the nitrification rate showed no effect from the higher DO concentration.

NITRIFICATION INHIBITION AND TOXICITY

Nitrifying organisms have earned a reputation for being more susceptible to toxic chemicals discharged into municipal and industrial wastewater treatment plants. The toxic effect may occur in many ways. It may be inhibitory to the nitrifiers, so that they continue to grow and oxidize ammonia but at a significantly reduced rate. The toxicity may be sufficient to kill the nitrifiers and thus stop nitrification activity until the toxicity is removed and new growth occurs, or the toxicity effect may be reversible where nitrification stops temporarily, but resumes when the toxicity is removed or diluted. Quantification of toxicity in treatment systems is difficult, due to the number of factors that may also affect nitrification rates. These include temperature, pH, dissolved oxygen concentration, and organic loading, as discussed earlier, sorption of toxics onto biomass, toxic loss by stripping and biodegradation, and the compounding effects when more than one toxic compound is available. Thus, bench-scale studies using activated sludge from operating plants, or more often, laboratory cultures grown with synthetic feed, are performed to assess the effect of selected inorganic and organic compounds at different concentrations. Batch tests may be run with these cultures to assess the effect of wastewater samples or selected compounds on nitrification rates.

Respirometer techniques are also available, and a procedure using an electrolytic respirometer has been described by Stensel et al. (1976) to determine the maximum specific growth rates of nitrifying organisms exposed to selected samples. The test involves preparing two identical sample bottles containing the sample and enough ammonia to allow zero order nitrification rates. A small amount of nitrifying seed is added to one of the bottles. The other bottle receives a known inhibitor to prevent nitrification. The oxygen consumption is followed with time and used to calculate the nitrifying bacteria maximum specific growth rate. The following equations are used to relate the maximum specific growth rate of the nitrifying organisms to the oxygen consumption data. This procedure assumes that the *Nitrobacter* are not inhibited if the *Nitrosomonas* are not, but ammonium nitrogen, nitrate nitrogen, and nitrite nitrogen measurements are made at the end of the test to check this assumption.

At high enough ammonium nitrogen concentrations the nitrifying bacteria will grow at their maximum rate, and this can be defined by the following equation, which is a simplification of the Monod expression at relatively high substrate levels:

$$\frac{dX}{dt} = \mu_m X \tag{2.13}$$

where:

X = Nitrifying bacteria concentration, mg/L
$\frac{dX}{dt}$ = Rate of increase in nitrifying bacteria concentration, mg/L-d
m_m = Maximum net specific growth rate, g/g-d

Integration of Equation (2.13) yields:

$$X = X_o e^{\mu m t} \tag{2.14}$$

where:

t = Time, days
X_o = Initial bacteria concentration at time zero, mg/L

The amount of nitrifying bacteria grown is a function of the ammonia removed, which will require a proportional amount of oxygen consumption. Thus, the change in cell mass can be related to the measured oxygen consumption in the respirometer:

$$\frac{dX}{dt} = Y_n \frac{d\mathrm{N}}{dt} = \frac{Y_n}{a}\frac{d\mathrm{O}}{dt} \tag{2.15}$$

where:

Y_n = Nitrifying bacteria yield coefficient, mg cells/mg NH_4^+-N oxidized
$\frac{d\mathrm{N}}{dt}$ = Rate of ammonium nitrogen oxidation, mg/L-d

a = oxygen consumption coefficient, mg O_2 used/mg $NH_4^+ - N$ oxidized

$\frac{dO}{dt}$ = Rate of oxygen consumption, mg/L-d

The derivative of Equation (2.14) is taken to obtain the rate of change of cell mass under maximum growth conditions and this is substituted into Equation (2.15) to yield:

$$\frac{dO}{dt} = \left[\frac{(\mu_n)X_o a}{Y_n}\right] e^{\mu mt} \tag{2.16}$$

Taking the natural logarithm of both sides results in:

$$\mathrm{Ln}\,\frac{dO}{dt} = \mathrm{Ln}\left[\frac{\mu_m(X_o)a}{Y_n}\right] + \mu_m t \tag{2.17}$$

The respirometer data is used to determine the oxygen consumption rate for nitrification each day, and this is plotted in a semi-log graph vs. time to obtain μ_m. Due to the growth of the nitrifying organisms, the daily oxygen consumption rate shows an exponential increase each day. To provide for exponential growth, the initial population of nitrifying organisms used in the test procedure must be low.

The observed effect of various compounds on nitrification rates reported in literature will be summarized in the remainder of this section. The first area of interest is the inhibiting effects of ammonia and nitrite as researched by Anthonisen et al. (1976). They found that inhibition was due to the un-ionized forms of ammonia and nitrite and the toxicity would thus be a function of pH, which affects the ionic distribution of these compounds. At free ammonia (FA) concentrations in the range of 10 to 150 mg/L, inhibition of *Nitrosomonas* was observed to result in an increase in ammonium concentration in their reactors. Between FA concentrations of 0.1 mg/L to 1.0 mg/L, *Nitrobacter* was inhibited to result in nitrite buildup during ammonium oxidation. *Nitrobacter* was also inhibited at un-ionized or free nitrous acid (FNA) concentrations from 0.2 mg/L to 2.8 mg/L. The availability of FA and FNA are a function of pH and solution concentrations of ammonium and nitrite. The concentrations of FA and FNA can be determined from the following equations presented by Anthonisen et al.:

$$\text{FA, mg/L} = \frac{(\text{TA})(10^{\text{pH}})(17/14)}{K_b/K_w + 10^{\text{pH}}} \tag{2.18}$$

where:

TA = Total ammonia in solution including ammonium and free ammonia as N, mg/L
K_b = Ionization constant of the ammonia equilibrium equation
K_w = Ionization constant of water

Both constants, K_b and K_w, are a function of temperature:

$$\frac{K_b}{K_w} = e^{(6344)/(273 + T)} \tag{2.19}$$

where:

T = Temperature, °C

The FNA concentration is a function of temperature, pH, and total nitrite concentration:

$$\text{FNA}(\text{HNO}_2),\ \text{mg/L} = \frac{(NO)(46/14)}{(K_a)10^{\text{pH}}} \tag{2.20}$$

where:

NO = Total FNA and ionized nitrite as N, mg/L
K_a = Ionization constant of the nitrous acid equilibrium equation

$$K_a = e^{(-2300)/(273 + T)} \tag{2.21}$$

Table 2.6 shows the range of ammonium nitrogen concentrations and nitrite nitrogen concentrations as a function of pH where *Nitrobacter* may be inhibited.

Metals have been found to be toxic to growing *Nitrosomonas* cultures by Skinner and Walker (1961), with complete inhibition for the following metals and concentrations: nickel, 0.25 mg/L; chromium, 0.25 mg/L; and copper, 0.1-0.5 mg/L. Beckman et al. (1972) reported 100% inhibition for nickel and zinc at a 3.0 mg/L concentration. Loveless and Painter (1968) also found complete inhibition with copper at 0.1 mg/L concentration.

There are various reports on organic compound toxicity with Painter (1970) summarizing some earlier observations. He noted that certain chelating agents such as thiourea, allyl-thiourea, 8-hydroxyquinoline, salicyladotime and histidine were toxic to *Nitrosomonas*. Peptone decreased the growth rate of *Nitrosomonas* by 25% at 1.0 mg/L and by 60%

TABLE 2.6. Ammonium Nitrogen and Nitrate Nitrogen Concentration Range for *Nitrobacter* Inhibition as Function of pH (T = 20°C).

pH	NH_4^+—N Range, mg/L	NO_2^-—N Range, mg/L
6.0	210-2100	30-330
6.5	70-700	88-1050
7.0	20-210	260-3320
7.5	7-70	
8.0	2-20	

at 10.0 mg/L. Table 2.7 summarizes data reported by Hockenburg and Grady (1977) from their work and others on nitrification inhibition due to organic compounds.

In addition to Hockenburg and Grady, other compounds identified by Sharma and Ahlert (1977) are listed in Table 2.8.

Blum and Speece (1991) compared IC_{50} concentrations between a laboratory *Nitrosomonas* culture and the microtox test for a number of priority pollutant compounds. There was no consistent pattern between the two toxicity testing results. In some cases they exhibited IC_{50}s at similar concentrations, but in a little over half of the compounds tested, one or the other showed greater toxicity sensitivity. Numerous compounds were tested, but compounds with nitrifier IC_{50} concentrations of 20 mg/L or less are arbitrarily listed in Table 2.9.

NITRIFICATION RATES IN ACTIVATED SLUDGE

The observed nitrification rate in activated sludge will be a function of numerous factors including the mixed liquor temperature, influent nitrogen, BOD and inert solids, the operating SRT, and yield coefficients for the heterotrophic organisms and the nitrifiers. The specific nitrification rate (SNR) is defined as follows and is calculated by using Equation (2.6) to determine (HRT)(X).

$$\text{SNR} = \frac{N_{ox}}{(\text{HRT})(X)} = \frac{N_{ox}}{[Y_{nH}(So - S) + X_I + Y_{nN}(N_{ox})]\ \text{SRT}} \qquad (2.22)$$

where:

SNR = Specific nitrification rate, g/g-d
HRT = Hydraulic retention time, V/Q, d

TABLE 2.7. Organic Compounds That Inhibit Nitrification.

Compound	Concentration (mg/L) at Approximately 75% Inhibition
Acetone	2000
Allyl alcohol	19.5
Allyl chloride	180
Allyl isothiocyanate	1.9
Benzothiazole disulfide	38
Carbon disulfide	35
Chloroform	18
o-Cresol	12.8
Di-allyl ether	100
Dicyandiamide	250
Diguanide	50
2,4-Dinitrophenol	460
Dithio-oxamide	1.1
Ethanol	2400
Guanidine carbonate	16.5
Hydrazine	58
8-Hydroxyquinoline	72.5
Mercaptobenzothiazole	3.0
Methylamine hydrochloride	1550
Methyl isothiocyanate	0.8
Methyl thiuronium sulfate	6.5
Phenol	5.6
Potassium thiocyanate	300
Skatol	7
Sodium dimethyl dithiocarbamate	13.6
Sodium methyl dithiocarbamate	0.9
Tetramethyl thiuram disulfide	30
Thioacetamide	0.53
Thiosemicarbazide	0.18
Thiourea	0.076
Trimethylamine	118

Compound	Concentration (mg/L) at Approximately 50% Inhibition
Dodecylamine	<1
Aniline	<1
n-Methylaniline	<1
Ethylenediamine	15
Napthylethylenediamine diHCl	23
2,2′-Bipyridine	23
p-Nitroaniline	31
p-Aminopropiophenone	43
Benzidine diHCl	45
p-Phenylazoaniline	72
Hexamethylene diamine	85
p-Nitrobenzaldehyde	87

TABLE 2.7. (continued).

Compound	Concentration (mg/L) at Approximately 50% Inhibition
Triethylamine	127
Ninhydrin	>100
Benzocaine	>100
Dimethylgloxime	140
Benzylamine	>100
Tannic acid	>150
Monethanolamine	>200

Source: Hockenburg and Grady, 1977.

and:

$$Y_{nH} = Y/(1 + K_d\,\mathrm{SRT})$$

where:

Y = Heterotrophic synthesis yield coefficient, gTSS/g BOD_5 used
K_d = Endogenous decay coefficient, gTSS lost/gTSS-d

Figure 2.2 illustrates the SNR (mg/gr MLSS-hr) as a function of a

TABLE 2.8. Organic Compounds That Inhibit Nitrification.

Compound	Inhibiting Concentration
Phenol	100 mg/L
Vitamins riboflavin, α-lipoic acid; *b*-pyridoxine HCl	50 mg/L
Thiamine	5 mg/L
Amine acids	
L-Lysine	
L-Threonine	
L-Histidine	
L-Valine	
L-Arginine	
L-Methionine	4 mg/L
2-chloro-6 trichloromethyl-pyridine	10 mg/L
Diethyldithiocarbamate	10^{-5} M
Methylene blue	10^{-4} M
Tanning and tannin derivatives	10^{-6}-10^{-8} M

Source: Sharma and Ahlert, 1977.

TABLE 2.9. *Nitrosomonas* Toxicity Due to Organic Compound for IC_{50} Concentrations of 20 mg/L or Less.

Organic Compound	IC_{50} Concentration, mg/L
4-Aminophinol	0.07
3-Chlorophenol	0.20
2-Aminophenol	0.27
2-Bromophenol	0.35
2,3-Dichlorophenol	0.42
2,3,6-Trichlorophenol	0.42
1,3-Dichloropropene	0.48
5-Chloro-1-pentyne	0.59
2,3-Dichlorophenol	0.61
1,3-Dichloropropene	0.67
Chlorobenzene	0.71
4-Chlorophenol	0.73
2,4-Dichlorophenol	0.79
Trichloroethylene	0.81
4-Bromophenol	0.83
1,1-Dichloroethane	0.91
2,3,5,6-Tetrachlorophenol	1.30
1,1,2,2-Tetrachloroethene	1.40
1,1,2-Trichloroethene	1.90
2,2,2-Trichloroethanol	2.00
4-Nitrophenol	2.60
2-Chlorophenol	2.70
3,5-Dichlorophenol	3.00
2,3,5-Trichlorophenol	3.90
2,4,6-Tribromophenol	7.70
Resurcinol	7.80
2,4,6-Trichlorophenol	7.90
Pentachloroethane	7.90
2,6-Dichlorophenol	8.10
1,1,1,2-Tetrachloroethane	8.70
1,2,4,5-Tetrachlorobenzene	9.80
2-Nitrophenol	11.00
Benzene	13.00
1,5-Dichloropentane	13.00
1,2,3,4-Tetrachlorobenzene	20.00

Source: Blum and Speece, 1991.

steady-state operating SRT, influent ammonia and BOD_5 oxidized, and influent inert solids concentrations as affected by using or not using primary clarification.

The following assumptions were used in Equation (2.22) to develop Figure 2.2.

Parameter	No Primary Trt.	With Primary Trt.
N_{ox}, mg/L	25.0	25.0
X_I, mg/L	80.0	25.0
$S_o - S$, mg/L	200.0	120.0
$N_{ox}/S_o - S$, g/g	0.125	0.208
Y_{nN}, g/g	0.15	0.15
Y, g/g	0.60	0.60
K_d, g/g-d	0.08	0.08

Without primary treatment and with higher operating SRTs, a smaller fraction of the mixed liquor would be nitrifying organisms to result in a much lower observed SNR. Assuming a mixed liquor concentration of 3000 mg/L and SRT of twenty days, the calculated SNR with primary treatment would be 2.7 mg/L-hr compared to 1.2 mg/L-hr for a system without primary treatment.

Limited data are available on full-scale plant operations where critical operating conditions were varied to initiate or limit nitrification. The most classical report in the literature is on experiments at the Chicago Metropolitan Sanitary District (Prakasam et al., 1979) using an aeration

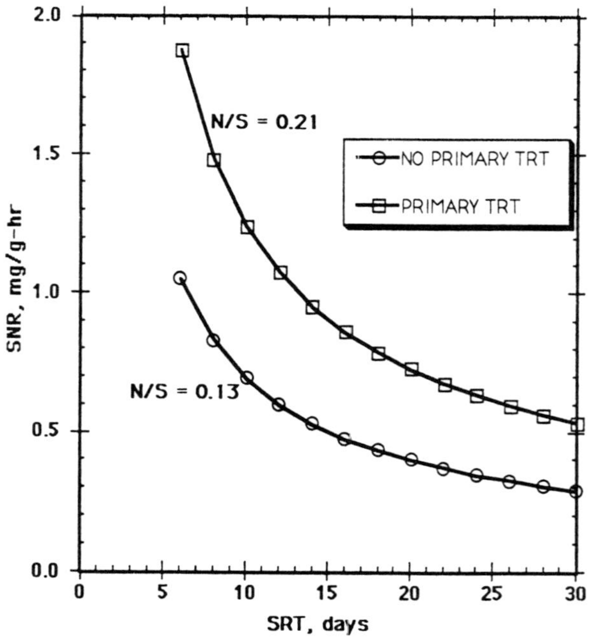

Figure 2.2 Specific nitrification rate (SNR) function of SRT and *N/S*.

tank battery with a nominal capacity of 190,000-227,000 m^3/d (50-60 mgd) following a small-scale pilot-plant study. During winter operation of the small-scale system, over three weeks were required to achieve full nitrification at 9°C using an operating SRT up to nineteen days. At 16-23°C, an SRT of five days was sufficient to maintain nitrification. The full-scale system required an average SRT of ten days at operating temperatures of 11 to 13°C to achieve effluent ammonia nitrogen concentrations below 1.0 mg/L.

Bliss and Barnes (1981) reported unstable nitrification performance at a small plant near Sydney, Australia serving 8000 people with an operating SRT of four to five days, even though the wastewater temperature was 23-24°C. The mixed liquor pH ranged from 6.4 to 7.0. He attributed the lower-than-expected performance to fluctuations in the operating DO concentration and pH.

Young et al. (1979) studied single- and two-stage nitrification at a domestic conventional wastewater treatment facility in Cherokee, Iowa with an average flow of 3360 m^3/d (0.94 mgd). An effluent ammonia nitrogen concentration of less than 1.0 mg/L was consistently met with a single-stage nitrification SRT of ten days for mixed liquor operating temperatures between 15-20°C. At temperatures between 10-13°C, SRT values of sixteen to eighteen days provided the effluent ammonia requirements.

Cold weather nitrification was demonstrated at the Marlborough, MA plant at 5°C with effluent ammonia nitrogen concentrations at 0.5 mg/L or less (Barth and Stensel, 1981). The operating SRT was about seventy-five days and testing was not available to assess the limiting operating SRT.

NITRIFICATION RATES IN SINGLE SLUDGE BNR SYSTEMS

It is generally acknowledged that nitrification rates in second-stage activated sludge reactors are greater than those in single sludge fully aerobic conventional reactors because the organic matter has been removed in the first stage and there is less heterotrophic competition in the second stage. Recent research by Hanaki et al. (1990a, 1990b) clearly demonstrated that the presence of biodegradable organics inhibited ammonia oxidation, apparently because of competition from growing heterotrophic cells. Nitrite oxidation was not inhibited. They also showed that low DO concentrations will substantially increase the growth yield of ammonia oxidizers, which compensates for reduced specific substrate utilization rates at low DO. However, the nitrite oxidizers are strongly inhibited by 0.5 mg/L DO.

Single sludge biological nutrient removal activated sludge systems treating municipal wastewaters typically remove all, or nearly all, of the biodegradable COD in the anaerobic and anoxic zones, resulting in ideal growth conditions for nitrifiers in the aerobic zone. Therefore, higher specific nitrification rates should occur in the aerobic zone of a BNR system than in a fully aerobic conventional system treating the same wastewater at the same total SRT. Research by Randall et al. (1991) has shown that this is, in fact, what happens. The nitrification rates they obtained, based on both total MLVSS and on calculated *Nitrosomonas* biomass are shown in Table 2.10. Furthermore, the ammonium concentration entering the BNR aerobic zone will always be substantially less than that entering the fully aerobic system, resulting in a smaller ammonium loading as shown in Table 2.11. Yet concentrations will be reduced at approximately the same rate because removal is zero order, resulting in substantially smaller aerobic volumes in a single sludge BNR system compared to a fully aerobic conventional system. The volumes are compared in Table 2.12 for systems that accomplished near-complete nitrification, i.e., 1.0 mg/L or less NH_3-N.

BIOLOGICAL DENITRIFICATION

Two modes of nitrate reduction can occur in biological systems: assimilating and dissimilating or denitrification. Assimilating nitrate reduction involves the reduction of nitrate to ammonia for use in cell synthesis. It occurs independently of oxygen tension and when ammonium nitrogen is not available. On the other hand, dissimilating nitrate reduction or denitrification is coupled to the respirating electron chain and involves the reduction of nitrate to nitrite to nitric oxide to nitrous oxide to nitrogen:

$$NO_{3-} \rightarrow NO_{2-} \rightarrow NO \rightarrow N_2O \rightarrow N_2$$

Denitrification is considered to be an anoxic process, occurring in the absence of oxygen, and requires an organic or inorganic electron donor.

MICROBIOLOGY

Interest in denitrification reactions started in the middle 1800s to explain the loss of nitrogen from fertilized soils. In 1876, Meusel (from Payne, 1981) suggested that bacteria were responsible for the destruction of nitrate in soils or water. In 1882, Gayon and Dupetit (from Payne, 1981) introduced the term denitrification to describe gas production during anaerobic activity with nitrate-using bacteria from sewage. A wide range of bacteria has been found to carry out denitrification, but similar microbial capability has not been found in algae or fungi. Bacteria capable of denitrification are both heterotrophic and autotrophic.

Among the heterotrophic organisms Payne (1981) lists the following genera: *Achromobacter, Acinetobacter, Agrobacterium, Alcaligenes,*

TABLE 2.10. Summary of Specific Nitrification Rates and Ammonia Oxidation Rates.

System	MCRT d	Aerobic MCRT d	Temp °C	Total NH_3—N Oxidized mg/L	Aerobic MLVSS mg/L	Specific Nitrification Rate mgN/gMLVSS/h	*Nitrosomonas** VSS mg/L	Ammonia Oxidation Rate mgN/mg* *Nitrosomonas*, d
BNR	15	8.3	20	18.8	2636	1.783	122	0.834
	15	8.9	15	26.5	3457	1.916	191	0.703
	15	8.3	10	20.5	2538	2.019	161	0.687
	5	2.7	20	23.2	1014	5.720	74	1.729
	5	2.8	15	22.4	1207	1.498	75	1.593
	5	2.9	10	8.1	1352	1.498	28	1.494
	2.7	1.5	20	21.6	749	7.210	42	2.695
	1.5	0.9	20	12.3	446	6.895	14	4.382
Conventional	15	15	20	21.2	1348	1.986	101	0.631
	15	15	15	26.6	2177	1.527	143	0.560
	15	15	10	22.5	2092	1.344	133	0.506
	5	5	20	26.5	1284	2.580	72	1.107
	5	5	15	21.9	1050	3.095	75	1.036
	5	5	10	15.0	1373	1.366	46	0.982
	2.7	2.7	20	27.1	658	5.148	47	1.716
	1.5	1.5	20	10.5	424	6.044	22	2.774

*Based on calculated values.

TABLE 2.11. Comparisons of Available Ammonia Loadings and Removals.

% Nitrification	MCRT, d	Temp °C	mg/L Aerobic Influent NH_3—N		mgd Aerobic NH_3—N Loading			mgd Required NH_3—N Oxidation to 1.0 mg/L		
			Conv.	BNR	Conv.	BNR	% Diff.	Conv.	BNR	% Diff.
100	15.0	20	10.2	6.2	3084	2812	8.8	2782	2359	15
100	15.0	15	12.3	8.9	3720	4037	−8.5	3417	3583	−5
100	15.0	10	10.5	7.1	3175	3221	−1.4	2873	2767	4
100	5.0	20	14.6	8.9	4415	4037	8.6	4113	3583	13
93[1]	2.7	20	13.6	6.9	4113	3130	24	3810	(2676)	30
79[1]	5.0	15	12.8	12.5	3871	5670	−46	3568	(5216)	−46
44[2]	1.5	20	15.9	17.6	4808	7983	−66	4506	(7530)	−67
23[3]	5.0	10	16.3	21.0	4929	9526	−93	4627	(9072)	−96

[1]Nitrification in BNR system, 100% nitrification in fully aerobic system (conv.).
[2]Nitrification in BNR system, 79% nitrification in conv. system.
[3]Nitrification in BNR system, 65% nitrification in conv. system.
( ) Nitrification to 1.0 mg/L was not accomplished in experimental system.

TABLE 2.12. Comparison of BNR Hydraulic Retention Times (HRTs) and Conventional HRTs to Accomplish Complete Nitrification—Based on Steady-State Results.

System	MCRT	Temp.	HRT (h)				Additional HRT Needed for BNR
			Anaerobic	Anoxic	Aerobic	Total	
Conventional	15.0	20			3.2	3.2	
BNR	15.0	20	1.0	1.0	1.7	3.7	0.5
Conventional	15.0	15			4.6	4.6	
BNR	15.0	15	1.2	1.2	3.3	5.7	1.1
Conventional	15.0	10			3.2	3.2	
BNR	15.0	10	1.5	1.5	2.2	5.2	2.0
Conventional	5.0	20			3.8	3.8	
BNR	5.0	20	1.0	1.5	2.3	4.8	1.0
Conventional	2.7	20			6.0	6.0	
BNR	2.7	20	1.0	1.5	*5.0	7.5	1.5

*Estimated from plot of available NH_3—N vs. aerobic HRT needed for nitrification for 2.7 d MCRT and 20°C.

( ) %.

Arthrobacter, Bacillus, Chromobacterium, Corynebacterium, Flavobacterium, Hypomicrobium, Moraxella, Neisseria, Paracoccus, Propionibacterium, Pseudomonas, Rhizobium, Rhodopseudomonas, Spirillum and *Vibrio*. In addition, Gayle et al. (1989) list *Halobacterium* and *Methanomonas*. Most of these bacteria can use oxygen as well as nitrate and some can also carry out fermentation in the absence of nitrate or oxygen. Under autotrophic denitrification, bacteria use carbon dioxide or bicarbonate as a carbon source instead of organic carbon. *Paracoccus* denitrifiers and *Thiobacillus denitrificans* use hydrogen and reduced sulfur compounds, respectively, for electron donors during denitrification. Both can grow heterotrophically if an organic carbon source is present (Gayle et al., 1989).

Pseudomonas species are the most common and widely distributed of all the denitrifiers (Payne, 1981). Some of the species do not reduce nitrate, but initiate denitrification with nitrite. Certain species also accumulate nitrous oxide as the terminal product instead of di-nitrogen. *Pseudomonads* have been shown to use a wide array of organic compounds including hydrogen, methanol, carbohydrates, organic acids, alcohols, benzoates, and other aromatic compounds (Payne, 1981).

Hyphomicrobium can be noted for its distinct structure and reproduction mode. They tend to fix themselves to a submerged surface and then extend a tubular prosthesis from the free end. Budding occurs at the end of each prosthesis and cells are released which swim away and eventually lodge elsewhere. *Hyphomicrobium* use simple one- and two-carbon acids and alcohols for carbon sources and do not use more complex organics.

Bacillus contains species that reduce nitrate and nitric oxide but not nitrite or nitrous oxide. One species, *B. azotoformans*, can use any of the three electron acceptors—nitrate, nitrite, and nitrous oxide—which is also possible in many other denitrifiers. These bacteria can use a wide range of organic substrates. Certain *Flavobacterium* cannot reduce nitrate, but reduce nitrite and nitrous oxide to di-nitrogen gas. They also can only use simple carbohydrates for carbon sources.

Moraxella are unique in that they only use aromatic compounds for a carbon source. The reduction of double bonds of the ring has been reported (Williams and Evans, 1975 in Payne, 1981) to occur first, followed by degradation of the resulting cyclic, nonaromatic ring. Ring cleavage is achieved by enzymatic hydrolysis rather than oxidation.

Paracoccus denitrifiers (Payne, 1981) can use a wide range of carbon sources for denitrification ranging from simple substrates such as methanol to more complex substrates. A unique feature of this genera is that the bacteria can grow as autotrophs as well, oxidizing hydrogen for an energy source and using carbon dioxide as the sole carbon source. Early studies with these bacteria showed that they were sensitive to the accumulation of nitrite in the 20.0 to 40.0 mg/L concentration range. The addition of yeast extract enabled the culture to overcome the nitrite toxicity and oxidize hydrogen.

Rhodopseudomonas and *Azospirillum* are bacteria found within the root zones of plants. They have the unique ability to be able to perform nitrogen fixation or denitrification. Both types of bacteria are able to reduce nitrous oxide to di-nitrogen.

Thiobacillus denitrificans can oxidize sulfur, sulfide, or thiosulfate with nitrate as the terminal electron acceptor (Payne, 1981) and sulfate as the by-product. Some cultures reduced only nitrate or nitrite when sulfide was supplied, but di-nitrogen was produced when thiosulfate was the electron donor. Batchelor and Lawrence (1978) have modeled the use of nitrate and sulfur in a denitrifying biofilm.

A unique heterotrophic organism, *Thiosphaera pantotropha*, has been reported by Robertson et al. (1988) to be able to simultaneously nitrify and denitrify under aerobic conditions, using acetate for a carbon source. The denitrification rate was improved when the dissolved oxygen concentration was about 30% of saturation. A hypothetical mechanism presented by the authors was that nitrite production interfered with the ammonia oxidizing electron respiratory chain to activate an alternate respiratory chain that oxidizes NADH by denitrification. The authors further point out that the environmental significance is unknown since the nitrification rate of *T. pantotropha* appears to be about one-tenth that of autotrophic nitrifiers.

Kugelman et al. (1991) claimed that aerobic denitrification may be a significant mechanism to account for aeration tank nitrogen losses they have observed in sequential anaerobic-aerobic activated sludge systems. They showed the addition of nitrite-inhibited oxygen respiration using sludge from anaerobic-aerobic systems and demonstrated that the nitrite was removed under aerobic conditions. On the other hand, they showed no nitrite inhibition using sludge obtained from strictly aerobic systems. The implications of these observations are not fully understood yet. Previous experience with anaerobic-aerobic systems has shown that the incorporation of an anoxic zone between the anaerobic and aerobic zones was necessary to accomplish significant nitrogen removal.

EFFECT OF OXYGEN

A dissolved oxygen concentration of 0.2 mg/L and above has been reported to inhibit denitrification for a *Pseudomonas* culture (Skerman and MacRae, 1957; Terai and Mori, 1975) and by Dawson and Murphy (1972) for activated sludge treating domestic sewage. Nelson and Knowles (1978) reported that denitrification ceased in a highly dispersed growth at a DO concentration of 0.13 mg/L.

The issue of DO concentration effects on denitrification in activated sludge systems is confounded by the fact that the measured bulk liquid DO concentration does not represent the actual DO concentration within the activated sludge floc. The fraction of the floc biomass exposed to zero DO or anoxic conditions depends on the bulk liquid DO concentration, the floc size, and the organic and ammonia loadings which affect oxygen respiration rates. Rittman and Langeland (1985) summarized reports of other investigators on DO concentrations and types of activated sludge systems where denitrification was observed under apparent aerobic conditions. These included 0.3 to 0.8 mg/L in a four-channel oxidation ditch, 0.5 mg/L in activated sludge, and 0.3 to 1.5 mg/L in semi-batch activated sludge.

Besides considering the effect of DO on the initiation of denitrification, the DO concentration may also affect the denitrification rate. Wheatland et al. (1959) found that the denitrification rate at a 0.20 mg/L DO concentration was about one-half of the rate at a zero DO concentration. Focht and Chang (1975) noted a similar effect of DO concentration for denitrification in seawater. Wheatland et al. also noted that as the DO concentration was raised to 2.0 mg/L, the denitrification rate was only 10% of the zero DO concentration rate. Batchelor (1982) has used a multiple Monod kinetic model to include the inhibiting effect of DO concentration on denitrification rates, and the IAWPRC model (1986) also incorporates this method for the anoxic organism growth term. Batchelor used a DO inhibition kinetic coefficient (K_o) of 0.25 mg/L versus 0.10 mg/L in the IAWPRC model.

$$R_{xn} = \mu_m (F_{DN})(X_H) \left[\frac{S}{K_s + S}\right] \left[\frac{K_o}{K_o + O}\right] \left[\frac{NO}{NO_s + NO}\right] \quad (2.23)$$

where:

R_{xn} = Rate of growth of denitrifying organisms, mg/L-d
μ_m = Maximum specific growth rate for heterotrophic organisms, g/g-d
F_{DN} = Fraction of heterotrophic organisms using nitrate for electron acceptor, g/g
X_H = Heterotrophic organism concentration, mg/L
S = Readily degradable substrate concentration, mg/L
K_s = Half saturation coefficient of readily degradable substrate, mg/L
K_o = DO inhibition coefficient, mg/L
O = DO concentration, mg/L
NO = Nitrate nitrogen concentration, mg/L
NO_s = Half saturation coefficient for nitrate nitrogen reduction, mg/L

Many nitrogen removal systems alternate exposure of the activated sludge to anoxic and aerobic conditions. Thus an important design and operating consideration is the time required for the organisms to activate or deactivate their nitrate reductase enzyme at the end of the electron transport respiratory chain. Surprisingly little information is available on this important reaction step change for activated sludge wastewater treatment systems. Though not considered representative of anoxic-aerobic wastewater treatment system observations, Nelson and Knowles (1978) found that *Azospirillum* required a two-hour lag period to shift metabolism to nitrate reduction. Payne and Riley (1969) showed that 40 min. was required for an aerobically grown culture to shift to maximum activity under anoxic conditions. Payne (1981) also notes that not all cells in a culture shift from oxygen respiration to nitrate reduction and some maintain their oxygen respiring capacity under the anoxic environment.

In addition, Payne states that once denitrifying enzymes are produced, they will continue to function for "some time" after aerobic conditions have been established. He further notes that the presence of nitric oxide appears to protect denitrifying cells from the effects of dissolved oxygen. Further research is obviously needed in this area to determine response times and levels of denitrifying cultures to changes in dissolved oxygen conditions. Repeated exposure to anoxic and aerobic conditions may also have an effect on the organism response characteristics.

EFFECT OF pH

Alkalinity is produced in denitrification reactions and the pH is generally elevated, instead of being depressed as in nitrification reactions. In contrast to nitrifying organisms, there has been less concern about pH influences on the denitrification rates, but the results reported do vary, as summarized by Focht and Chang (1975). Dawson and Murphy (1972) found an optimal denitrification rate at a pH of 7.0 with half the rate observed at pH values of 6.0 and 8.0 for the same culture. No acclimation time was provided in this work, as in other studies on pH effects. Other investigators (Nommik, 1956; Wiljer and Delwiche, 1954; Bremner and Shaw, 1958) have shown no effect of pH between a pH of 7.0 and 8.0. A linear decrease in rate was noted from 8.0 to 9.5 and from 7.0 to 4.0. These investigators also noted that neutral to alkaline pH conditions favored the conversion of nitrous oxide to di-nitrogen gas.

OXIDATION-REDUCTION REACTIONS

Major factors affecting the design of denitrification reactors include the amount of substrate necessary to deplete the influent oxidized nitrogen and the denitrification reaction rates. An appreciation of the oxidation-reduction reactions and biological yield coefficients is needed to determine substrate requirements. The following shows oxidation-reduction reactions proposed for various substrates used for denitrification in wastewater treatment. Methanol (McCarty et al., 1969) is shown first, as it has been the most widely studied substrate and most commonly used substrate when an external carbon source has been employed.

Methanol

$$5CH_3OH + 6NO_3^- \rightarrow 3N_2 + 5CO_2 + 7H_2O + 6OH \tag{2.24}$$

Acetic Acid (Frick and Richard, 1985)

$$5CH_3COOH + 8NO_3^- \rightarrow 4Na + 10CO_2 + 6H_2O + 8OH \tag{2.25}$$

Methane (Barrenstein et al., 1986)

$$5CH_4 + 8NO_3^- \rightarrow 4N_2 + 5CO_2 + 6H_2O + 8OH^- \tag{2.26}$$

Sewage (Beer and Wang, 1978)

$$C_{10}H_{19}O_3N + 10NO_3^- \rightarrow 5N_2 + 10CO_2 + 3H_2O + NH_3 + 10OH^- \tag{2.27}$$

In all the above heterotrophic denitrification reactions, one equivalent of alkalinity is produced per equivalent of nitrate-nitrogen reduced. This equates to 3.57 mg (as $CaCO_3$) of alkalinity production per mg of nitrate nitrogen reduced, or one-half the amount destroyed when ammonium is oxidized to nitrate.

Barth et al. (1968) estimated that the ratio of BOD required to nitrate nitrogen reduced is 4.0. McCarty et al. (1969) determined that methanol was a preferred external carbon source based on a low sludge yield and cost. They provided the following equation to relate methanol requirements to nitrate nitrogen removal:

$$M = 2.47\,(No) + 1.53\,Ni + 0.87\,DO_u \tag{2.28}$$

where:

M = Methanol required, mg/L
No = Nitrate nitrogen reduced, mg/L
Ni = Nitrite nitrogen reduced, mg/L
DO_u = Dissolved oxygen reduced, mg/L

Using the above formulation the methanol required as a ratio of the COD used to nitrate nitrogen reduced is 3.71. Stensel et al. (1973) obtained a ratio of 3.45 in laboratory studies and noted that this ratio is related to the net cell yield, which can be affected by the system operating SRT. Figure 2.3 shows how the COD/N ratio for nitrate reduction can vary as a function of the net cell yield coefficient. The net yield can vary as a function of the organic substrate used and the system SRT. As the system SRT is increased, the net cell yield decreases and the amount of COD needed per quantity of nitrate nitrogen is reduced. The COD/N ratio was developed by considering that the COD used can be accounted for by cell synthesis and COD oxidation by nitrate reduction to produce energy for the cell. The derivation of the COD/N ratio is shown by the following equations.

$$COD = COD_{sy} + COD_E \tag{2.29}$$

$$COD_E = 2.86N_E \tag{2.30}$$

$$COD_{sy} = 1.42m \tag{2.31}$$

where:

COD = COD used, g
COD_E = COD of cells produced from COD used, g
COD_E = COD oxidized for energy, g
N_E = Nitrate nitrogen reduced for energy production, g

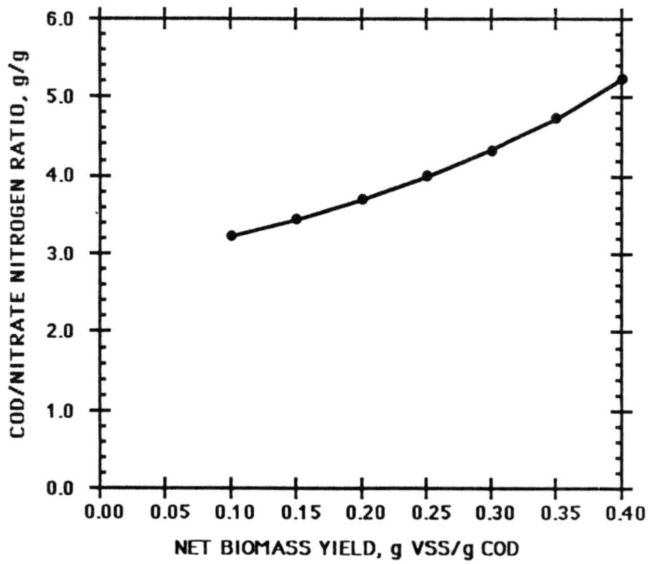

Figure 2.3 Effect of net biomass yield on COD consumed for nitrate reduction.

2.86 = Oxygen equivalent of nitrate nitrogen in oxidation-reduction reactions, g/g
m = Volatile suspended solids (VSS) synthesized from COD used, g
1.42 = Assumed COD of VSS produced, g/g

The net yield (Y_N) times the COD used equals the VSS produced:

$$m = Y_N \text{COD} \tag{2.32}$$

The net yield will be a function of the synthesis yield, endogenous decay coefficient, and SRT:

$$Y_N = \frac{Y}{1 + K_d \, \text{SRT}} \tag{2.33}$$

where:

Y = Synthesis yield coefficient, g VSS/g COD used
K_d = Endogenous decay coefficient, d^{-1}

Combining Equations (2.29) through (2.32) yields:

$$\text{COD} = 1.42\,(Y_N)\,\text{COD} + 2.86\,N_E \tag{2.34}$$

The COD/N ratio relates the COD used to the nitrate nitrogen used for energy-producing oxidation-reduction reactions and to that used to supply nitrogen for cell synthesis. Assuming that the VSS produced contains 10% nitrogen, the COD/N ratio is as follows:

$$\text{COD/N} = \text{COD}/(N_E + N_{sy}) \tag{2.35}$$

$$\text{COD/N} = \text{COD}/(N_E + 0.10\,Y_N\,\text{COD}) \tag{2.36}$$

Substituting N_E from Equation (2.34) into Equation (2.36) yields the COD/N ratio for denitrification.

$$\frac{\text{COD}}{\text{N}} = \frac{2.86}{-1.134 Y_N} \tag{2.37}$$

BIOLOGICAL KINETICS

The interest in biological kinetic rates for denitrification usually applies to three common types of nitrogen removal system designs:

(1) Systems with anoxic zones preceding aeration, where the influent wastewater provides the carbon source to create a biological demand for nitrate.
(2) A post-aeration reactor where the endogenous respiration oxygen demand of the bacteria is used for nitrate reduction.
(3) A post-nitrification reactor system with a separate clarifier where methanol is added to grow bacteria and to create a demand for the nitrate electron acceptor.

Other systems, such as intermittent aeration or systems that add methanol to a post-anoxic zone, may be used, and modifications of the denitrification rate expressions to follow will be required. Biological kinetics for pre-anoxic zones using domestic wastewater for the carbon source are presented first.

Pre-Anoxic Denitrification Kinetics

A preferred design approach is to use Equation (2.23), presented earlier, with proper mass balances to describe nitrate removal. This can be done for a single-stage or multiple-stage pre-anoxic zone. A mass balance on the organic substrate should be done first to determine the organic substrate removal rate. A relationship between organic substrate used and nitrate nitrogen reduction can then be used to determine the nitrate reduction rate due to cell synthesis and substrate removal and endogenous decay. This approach is analogous to that used to determine oxygen requirements in a biological system as a function of substrate removal and endogenous decay (Metcalf and Eddy, 1990) and is shown by Equation (2.38) for nitrate reduction. As in Equation (2.23), the fraction of heterotrophic organisms consuming nitrate (F_{DN}) is also an important factor affecting the endogenous decay nitrate requirement.

$$R_{no} = AR_{su} + \frac{1.42}{2.86} K_d X (F_{DN}) \qquad (2.38)$$

where:

R_{no} = Nitrate reduction rate, mg/L-d
R_{su} = Substrate removal rate, mg/L-d
A = Nitrate nitrogen used for organic substrate removal for cell synthesis, g/g

The value for A can be determined by rearranging Equation (2.34) and using Y instead of Y_N since this portion of Equation (2.38) represents an immediate substrate and nitrate use for synthesis.

$$A = \frac{1 - 1.42Y}{2.86} \tag{2.39}$$

The steady-state substrate mass balance is shown, for example, for the first anoxic zone stage for the anoxic-aerobic system in Figure 2.4.

$$O = QS_o + rQSe + RQSe - (Q + rQ + RQ)S_1 - R_{su}V \tag{2.40}$$

The substrate utilization rate, R_{su}, is related to R_{xn} in Equation (2.23).

$$R_{su} = R_{xn}/Y \tag{2.41}$$

$$R_{su} = \frac{\mu_m}{Y}(F_{DN}) \left[\frac{S_1}{K + S_1}\right] \left[\frac{K_o}{K_d + O_1}\right] \left[\frac{NO_1}{NO_s + NO_1}\right] \tag{2.42}$$

Equations (2.40) and (2.42) can be used to solve for S_1 and R_{su} and then Equation (2.38) can be used to solve for R_{no}. The value for R_{no} can be used in the following steady-state nitrate mass balance for stage 1 to determine the nitrate nitrogen concentration in the flow entering stage 2.

$$O = RQNO + rQNO - (Q + rQ + RQ)NO_1 - R_{no}V \tag{2.43}$$

The accuracy of using this rational modeling approach is dependent on the values of the various coefficients and parameters in the equations. Reasonable values for K_o, K_s, and Y can be obtained. The half-velocity constant for nitrate reduction (NO_s) has been shown to be 0.5-1.0 mg/L

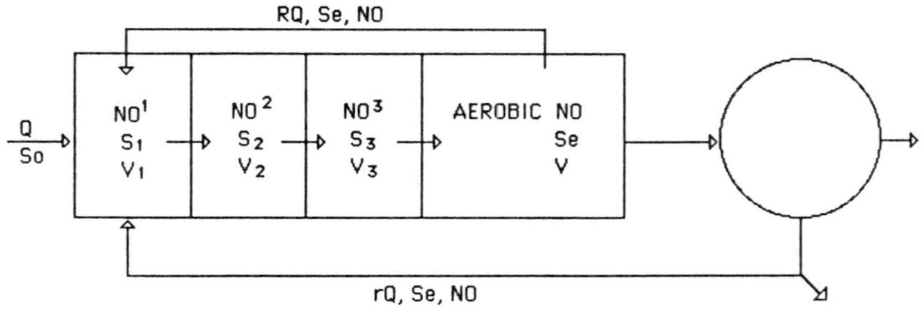

Figure 2.4 Substrate changes in a staged anoxic-aerobic activated sludge system.

(Jones et al., 1990; Requa and Schroeder, 1973). However, the major problem and unknown is the value for F_{DN}, and there is limited data available for μ_m, Y, and K_s under nitrate reduction. McClintock et al. (1988) found Y to be 53%, K_d to be 51% and the μ_m to be about the same for nitrate reduction versus oxygen respiration cultures treating a synthetic wastewater.

Another problem is the effect of rapid substrate uptake and storage, which can occur in contrast to Equation (2.23) showing substrate utilization as a function of the substrate concentration in the bulk liquid. Estimating the F_{DN} value is compounded by the fact that facultative, strictly aerobic, and perhaps strictly anoxic populations can co-exist and vary in the activated sludge system, and the substrate utilization rate may also vary for a facultative organism when nitrate is used as the electron acceptor instead of oxygen. Due to the difficulty of separating these critical kinetic coefficients, empirical relationships have been developed to describe a specific denitrification rate in pre-anoxic systems. Such a rate can theoretically be illustrated using R_{no} from Equation (2.38).

$$\text{SDNR} = R_{no}/X_H \tag{2.44}$$

or

$$\text{SDNR} = \frac{A(R_{su})}{X_H} + \frac{1.42K_d F_{DN}}{2.86}$$

where:

SDNR = Specific denitrification rate, mg NO_3–N/mg VSS-d

Alternately SDNR has also been based on the MLSS concentration instead of the MLVSS concentration from laboratory and field data.

The SDNR can be expected to be proportional to the food to MLSS (F/M) ratio for the anoxic stage, provided that most of the substrate is used since the specific substrate utilization rate (SSUR) is a function of the F/M ratio:

$$\text{SSUR} = \frac{R_{su}}{X_H} = \frac{Q\,(S_o - S)}{V\,(X_H)} \cong F/M \tag{2.45}$$

This approach was used to model observed denitrification rates at about 20°C in a single-stage pre-anoxic zone in a pilot-scale study and in a full-scale Bardenpho plant (Stensel, 1981; Burdick et al., 1982) with the resultant data fit as shown in Figure 2.5.

$$\text{SDNR} = 0.03\,(F/M_A) + 0.029 \qquad (2.46)$$

where:

$F/M_A = F/M$ ratio loading to the anoxic zone, g BOD/g TSS-d

Higher SDNR values than predicted by Equation (2.46) may be expected for anoxic zones with multiple stages and high BOD loadings. In addition, SDNR values can vary as a function of different mixed liquor denitrifying active mass fractions for different systems. Nevertheless, the model does relate the SDNR to an important fundamental that controls the denitrification rate, which is the substrate utilization rate. Panzer et al. (1980) also related specific denitrification rates that ranged from 0.4 g/g-d for the first stage of a high-loaded four-stage pre-anoxic zone, to 0.05 g/g-d for a single-stage anoxic zone, to the COD loading.

Other investigators have developed denitrification rate expressions as a function of temperature for pre-anoxic zones, but these have limited usefulness in design since the rates are not related to BOD loadings and substrate

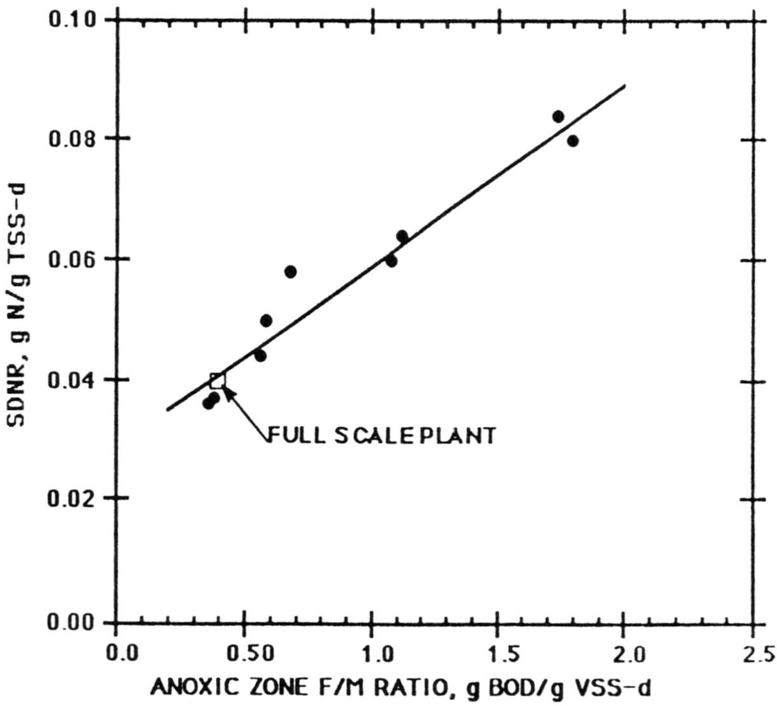

Figure 2.5 Specific denitrification rate function of anoxic zone *F/M* ratio.

TABLE 2.13. Denitrification Rate Temperature Coefficients for Pre-Anoxic Zone (Domestic Wastewater Carbon Source).

Value of Temperature Correction Coefficients	Reference
1.10	Mulburger (1971)
1.06	Dawson and Murphy (1972)
1.09	Barnard (1973)
1.20	Ekama and Marais (1984)
1.07	Kaas and Anderson (1989)

utilization rates. A wide scatter of SDNRs have been shown in the *U.S. EPA Nitrogen Control Manual* (1975) ranging from 0.03 to 0.15 over a temperature range of 12-27°C. Again these data were not related to the organic loadings. Ekama and Marais (1984) reported SDNR values ranging from a maximum rate of 0.72 g/g-d down to 0.072 g/g-d as substrate was depleted in batch denitrification rate tests.

Temperature effects on pre-anoxic zone denitrification rates have been proposed from observed data. The accuracy of these rates is not known since the data correlations did not consider the effect of substrate loadings. Table 2.13 summarizes temperature correction values for (*j*) in the following equation:

$$SDNR_T = (SDNR_{20})\, \theta^{T-20} \tag{2.47}$$

Post-Anoxic Denitrification Rates—Endogenous Respiration

Following nitrification, a denitrification zone can be incorporated into the activated sludge system before the secondary clarifier. After nitrification, the organic substrate concentration is at a minimal level and the denitrification rate is dependent on the respiration rate of the bacteria using stored food reserves or substrate released from endogenous decay. The *U.S. EPA Nitrogen Control Manual* (1975) showed endogenous SDNRs generally ranging from 0.015 to 0.06 g/g-d. Stensel (1981) found that the wide variation in these SDNRs could be related to the fact that the mixed liquor would have different respiration rates as a function of the system operating SRT. Laboratory experiments were performed with an alternating aerobic-anoxic activated sludge system, and SDNRs were related to the operating SRT and Specific Carbonaceous Oxygen Uptake Rate (SCOUR). The equivalent oxygen utilization rate under nitrate respiration was approximately equal to one-half the oxygen utilization rate under aerobic respiration conditions. Wuhrman (1968) also found a similar relationship between oxygen and nitrate endogenous respiration rates. A data fit of the SDNR versus SRT data at 20°C resulted in the following equation (Burdick et al., 1982).

$$\text{SDNR} = 0.12(\text{SRT})^{-0.706} \tag{2.48}$$

Steady-state reactor conditions were also reached at 10°C, and the temperature correction factor (θ) per Equation (2.47) was 1.02.

Equation (2.48) was developed for a specific wastewater condition, and the SDNR rate could change with the wastewater characteristics. Thus it is more appropriate to derive the SDNR from equations that can be used to predict the SCOUR.

$$\text{SCOUR} = \frac{An}{Y_{NT}}\left(\frac{1}{\text{SRT}}\right) \tag{2.49}$$

where:

SCOUR = Specific carbonaceous oxygen uptake rate, g O_2/g TSS-d
Y_{NT} = Net TSS production including inert solids and biomass, g TSS/g BOD_5 removed
An = Net carbonaceous oxygen requirement, g O_2/g BOD_5 removed

The SDNR can be derived from the SCOUR as follows:

$$\text{SDNR} = \left(\frac{F}{2.86}\right)\left(\frac{An}{Y_N}\right)\left(\frac{1}{\text{SRT}}\right) \tag{2.50}$$

where:

F = Fraction of the carbonaceous oxygen respiration rate when using nitrate instead of molecular oxygen, g/g
2.86 = Oxygen equivalent in nitrate, g/g

Values for An and Y_{NT} can be calculated as follows for the wastewater:

$$An = 1.5 - 1.42Y + \frac{1.42K_dY\,\text{SRT}}{1 + K_d\,\text{SRT}} \tag{2.51}$$

$$Y_{NT} = \frac{Y}{1 + K_d\,\text{SRT}} + \frac{X_I}{S_o} \tag{2.52}$$

TABLE 2.14. Denitrification Kinetic Coefficient Using Methanol for Organic Substrate.

Parameter	20°C	10°C
Synthesis yield coefficient, Y gr VSS/grCOD used	0.18	0.17
Endogenous decay coefficient, K_2, g/g-d	0.04	0.05
Maximum specific substrate utilization rate, K g COD/g VSS-d	10.3	3.1
Half-velocity coefficient, K_S, mg/L	9.1	12.6

where:

Y = Synthesis yield coefficients, g TSS/g BOD_5 removed
X_I = Influent inert, nonbiodegradable solids concentrations, mg/L
S_o = Influent BOD_5 concentration, mg/L

Denitrification with External Carbon Source

Methanol has been used in the past as an external carbon source for denitrification in a separate activated sludge system following nitrification. Methanol can also be added to a post-anoxic zone following nitrification to provide necessary organic substrate or to increase the denitrification rate. A wide range of SDNRs have been reported for methanol addition systems [*U.S. EPA Nitrogen Control Manual* (1975)]; 0.10-1.2 g NO_3-N/g TSS-d. This range of observed rates was due to different organic substrate levels during the denitrification reactions, and application of basic kinetic information can provide a more rational design approach for these systems. Table 2.14 summarizes basic denitrification kinetic coefficients available for systems using methanol (Stensel et al., 1973). These coefficients can be used in an organic substrate mass balance for methanol addition to determine the organic substrate utilization rate that can then be related to the denitrification rate as shown earlier with Equations (2.39) and (2.44).

The steady-state design approach for a completely mixed denitrification system using methanol is illustrated here. The first step is to determine the methanol COD concentration needed in the influent flow. This can be done using Equation (2.37) and determining the effluent soluble COD concentration:

$$S_o - S = \frac{2.86}{1 - 1.134Y_N}(N_o - N) \qquad (2.53)$$

where:

S_o = Influent methanol COD concentration, mg/L
S = Effluent methanol COD concentration, mg/L
N_o = Influent nitrate nitrogen concentration, mg/L
N = Effluent nitrate nitrogen concentration, mg/L

For a completely mixed reactor, S can be determined as a function of the system design SRT and kinetic coefficients (Lawrence and McCarty, 1970):

$$S = \frac{K_s\,[1 + K_d\,\mathrm{SRT}]}{\mathrm{SRT}\,[YK - K_d] - 1} \qquad (2.54)$$

Thus the SRT is selected first, which will determine the value for S and Y_N using Equation (2.33). The value for $N_o - N$ is based on the influent nitrate and desired effluent nitrate nitrogen concentrations. Once these are known, the value for S_o can be calculated. The value for the product of the denitrification biomass concentration and hydraulic retention time can then be determined from a mass balance on solids (Lawrence and McCarty, 1970).

$$(X)\mathrm{HRT} = \frac{Y(S_o - S)\,\mathrm{SRT}}{(1 + K_d\,\mathrm{SRT})} \qquad (2.55)$$

where:

HRT = Hydraulic detention time, d
X = Denitrification biomass concentration, mg/L

Based on the expected sludge thickening characteristics and clarifier recycle sludge rates, X can be selected, and the denitrification detention time determined.

The specific denitrification rate can be calculated for the resultant design as follows:

$$\mathrm{SDNR} = \frac{N_o - N}{(X)(\mathrm{HRT})} \qquad (2.56)$$

TABLE 2.15. Effect of SRT on Denitrification Reactor Design ($T = 20°C$) (Nitrate Nitrogen Removal = 29.0 mg/L).

SRT, d	2	3	4	5	6	8
S, mg/L	3.7	2.3	1.7	1.4	1.1	0.9
S_o, mg/L	105.8	103.6	102.3	101.3	100.0	99.0
X HRT, mg/L-d	35.3	50.0	63.5	74.9	87.1	108.0
Detention time*, hrs.	0.4	0.6	0.8	0.9	1.0	1.3
SDNR, g/g-d	0.82	0.58	0.46	0.40	0.33	0.27

*MLVSS = 2000 mg/L.

The equations show that if more methanol is added than is needed, the value for X(HRT) will be greater and the resultant SDNR will be lower.

Table 2.15 shows the effect of design SRT on the resultant SDNR and required detention time to reduce nitrate nitrogen in the influent to a separate stage denitrification system using methanol, from 30.0 mg/L to 1.0 mg/L.

PRINCIPLES OF BIOLOGICAL PHOSPHATE REMOVAL

DEFINITIONS

In conventional wastewater treatment, it is sufficient to define the organic carbon content in terms of BOD, COD, or TOC. For the purpose of this discussion, COD has been selected as a measure of the total organic carbon content. However, it is necessary to make further distinctions between the different fractions making up the overall COD since these play an important part in the practice of biological phosphate removal (BPR).

The nondegradable fraction of the COD plays no role whatsoever, and for this reason we need to refer only to the biodegradable fraction (DCOD). This fraction varies but is normally about 93% of the total, and if this percentage is used as a default value, the error in using COD to indicate biodegradable carbon is very small. The overall COD could further be split into the particulate fraction, consisting of settleable solids and colloidal solids, and the COD in solution. The COD in solution could be divided into the more complex molecules and the volatile fatty acids (VFA). The VFA are again divided between the long-chain VFA and the short-chain VFA (SCVFA).

In fresh sewage, a small fraction of the COD is soluble and an even smaller fraction consists of VFA. At the moment that the sewage is discharged to the sewer, acid fermentation begins, caused by attached growth in the sewers under the liquid level, especially in slow-flowing sewers. Fermentation is especially pronounced when the wastewater is strong, ambient temperatures are high, easily degradable material such as brewery wastes are added, there are many pump stations and force mains from which oxygen is excluded, or the wastewater has to be conveyed over long distances in slow-flowing sewers.

CHEMICAL PRECIPITATION OF PHOSPHATES

It was once believed that the removal of phosphates observed in some activated sludge plants was entirely due to chemical precipitation of forms of calcium phosphate formed towards the end of the aeration basin when excessive aeration drives off the carbon dioxide and raises the pH fractionally. It was not possible, however, to identify any crystals resulting from precipitation. It is today recognized that the principle mechanism in the removal of phosphates in BNR plants is biological. Hoffman (1977) showed that in the low-alkalinity wastewaters around Capetown, about 16% of the phosphates are removed by chemical precipitation. Lan et al. (1983) found that, at most, 15 to 27% of the phosphates removed during anaerobic-aerobic treatment could have been precipitated. This figure may be higher for harder waters, but will not constitute the major portion of the phosphates removed.

BIOLOGICAL REMOVAL OF PHOSPHATES

While experimenting with biological nitrogen removal, Barnard (1974) noticed the correlation between biological P removal and nitrates in the effluent. Considering also the previous observation in full-scale plants and in the PhoStrip process, he postulated that for biological P removal it was necessary that the organisms pass through an anaerobic stage in the absence of both nitrates and dissolved oxygen. Fuhs and Chen (1975) identified *Acinetobacter* ssp and especially the strain Lwoffi as the organism responsible for accumulating excess phosphates in their cells when they have short-chain volatile fatty acids (SCFA), especially acetates, as feed stock. Nicholls and Osborn (1978) suggested that the anaerobic stage, free of nitrates, proposed by Barnard was necessary to allow the *Acinetobacter* ssp to selectively take up acetates into the cells, using stored polyphosphates as the energy source and releasing phosphates to the liquid phase. The acetates taken up were stored as poly-β-hydroxybuterate (PHB) inside the cells until they reached the aeration zone where the PHB was metabolized under aerobic conditions, providing energy for the uptake of all available orthophosphate.

BIOLOGICAL MODEL

Comeau et al. (1986) produced the first comprehensive model for the biological removal of phosphates by *Acinetobacter*. Comeau proposed that under anaerobic conditions, undisassociated acetic acid is transported across the membrane with a decrease of one H^+ in the pH gradient for every acetate transported. Polyphosphate is used for either ATP synthesis or for the excretion of protons across the membrane. Degradation of polyphosphate leads to accumulation of intracellular orthophosphate, which is released by the action of a pH-sensitive carrier. Energy for PHB synthesis is provided by polyphosphate.

Acetyl-CoA is partially metabolized by the tricarbolic acid (TCA) cycle to provide NAOH for the reduction of acetoacetyl-CoA in PHB synthesis. Under aerobic conditions, the PHB carbon reserves are used to take up phosphates and metabolize them to polyphosphate (see Figures 2.6 and 2.7).

This model was modified by Wentzel et al. (1986) who postulated that ATP/ADP and NADH/NAD ratios control polyphosphate and PHB synthesis and degradation. According to this postulation, the NADH/NAD ratio rises under anaerobic conditions as a result of the lack of a terminal electron acceptor, while repression of the oxidative phosphorylation leads to a decrease in ATP/ADP. This controls the TCA cycle and PHB synthesis and ensures that acetyl-CoA is shared between the two pathways as shown in Figure 2.8.

The proton motive force that is dissipated by the uptake of acetate is reinstated by the uptake of hydroxyl ions (see Figure 2.9).

Under aerobic conditions in the absence of exogenous carbon compounds, PHB degradation is controlled by the decrease in the NADH/NAD ratio, which also activates the tricarbolic and glyoxylic acid cycles. PHB degradation provides the cell with energy, and the resultant increase in the ATP/ADP ratio enhances polyphosphate synthesis.

The presence of nitrates in the anaerobic zone would provide certain organisms with a terminal electron acceptor. This will prevent fermentation of easily degradable compounds to form new SCVFA such as acetates. Also, acetate used for energy through the anaerobic respiration chain, using nitrate as electron acceptor, will not be converted to polyhydroxybuterate. Synthesis of this product requires both acetyl-CoA and high NADH levels, which will not be available when sufficient nitrates are present. The lack of accumulated carbon will reduce the ability of the organisms to use this energy source to take up phosphates under aerobic conditions.

According to Lotter (1989), polyphosphate accumulation is dependent on the stimulation of polyphosphate kinase, which occurs by simultaneous

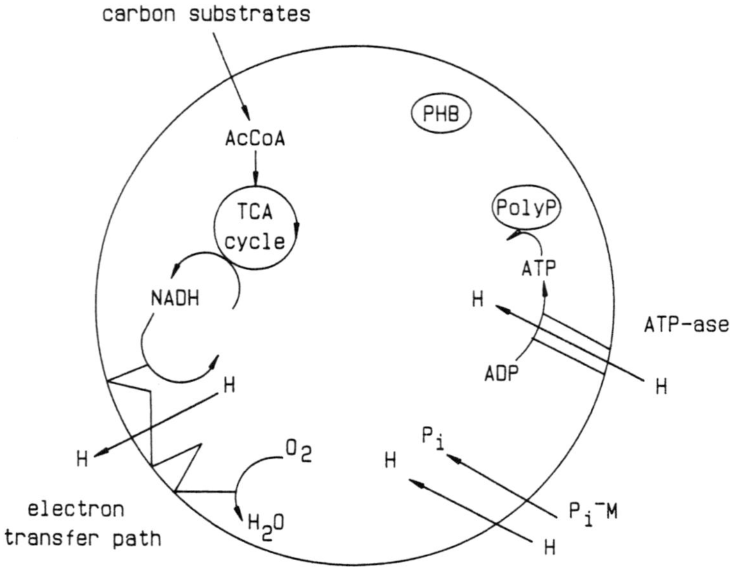

Figure 2.6 Comeau model for anaerobic conditions.

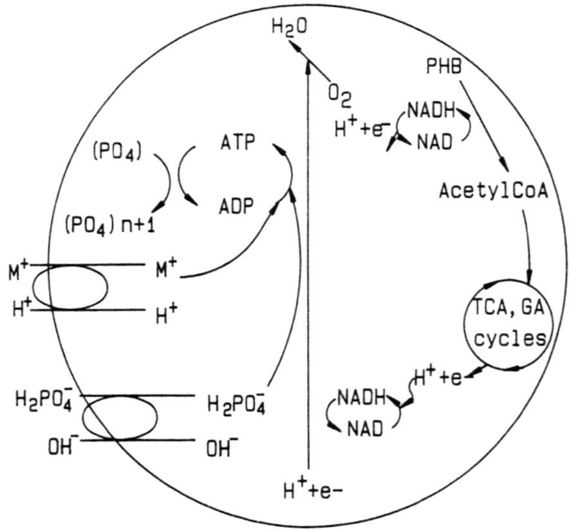

Figure 2.7 Comeau model for aerobic conditions.

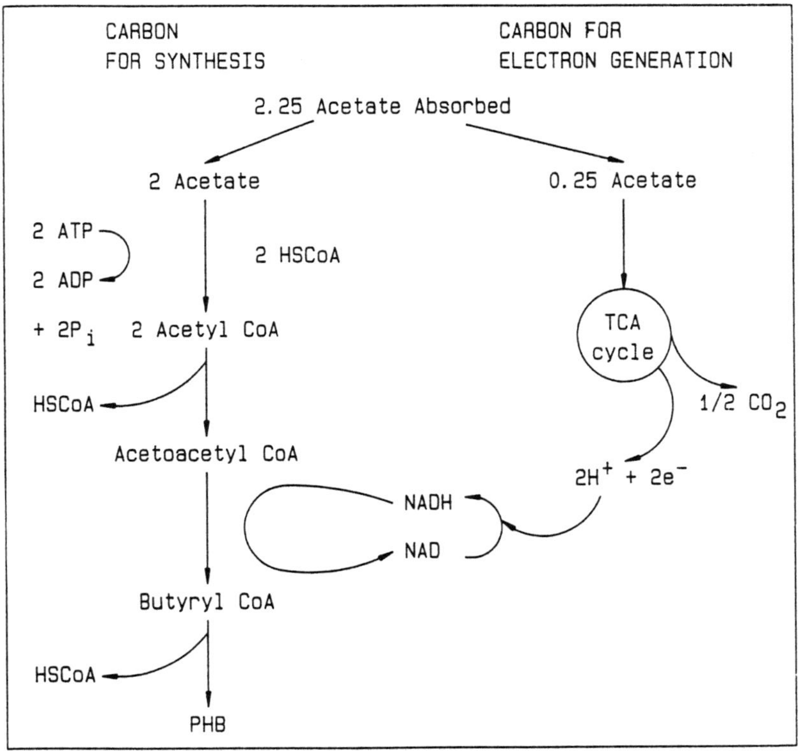

Figure 2.8 Division of acetate between TCA and PHB.

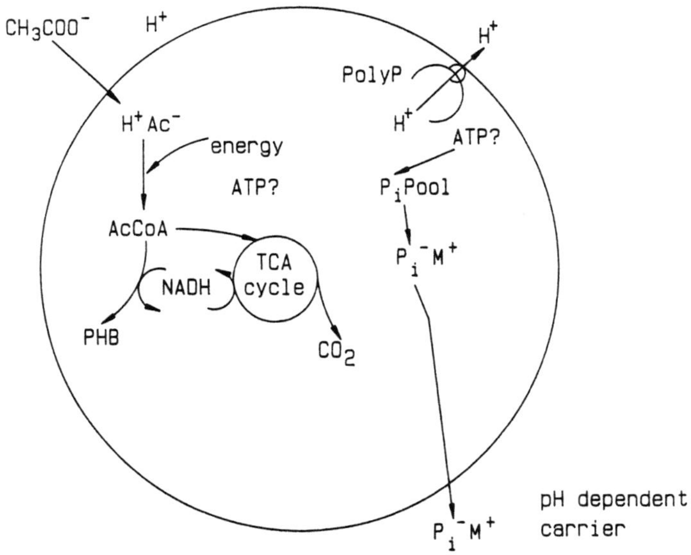

Figure 2.9 Wentzel anaerobic metaboles.

phosphate release and substrate uptake under anaerobic conditions. Without sufficient oxygen in the aeration zone, the electron transfer path and oxidative phosphorylation to form ATP will not function satisfactorily. Poly-β-hydroxybuterate will not be degraded and the anaerobic uptake of acetate will subsequently be reduced. Excess ATP conversion to polyphosphate will not occur.

With the ability to adapt to the various environmental conditions in the BNR plant, plus the well-established ability to form dense flocs and thus favor the selection process of sedimentation, the BioP organisms will grow selectively and become predominant, providing a convenient practical unit process for the removal of phosphates.

SECONDARY RELEASE OF PHOSPHATES

Fuhs and Chen (1975) noted in their historic paper that phosphates are released by the uptake of actetates as well as under other conditions unrelated to the uptake of acetates. When bubbling carbon dioxide through the BioP organisms obtained from a BioP removal plant, or when adding acid, they found that a substantial release of phosphates took place. Since this release was not accompanied by the uptake of energy, none was available for the uptake of phosphates in the aeration basin. Not much attention was paid to this aspect of phosphate removal until Barnard (1984) noticed that a consistent release of phosphates took place in the second anoxic zone of a five-stage Bardenpho process, to which no nitrates were discharged (see Figure 2.10).

This release disappeared when the second anoxic zone was aerated. He called this "secondary release," being P release not associated with acetate uptake. A similar release was noted in Kelowna (Barnard 1984), when in

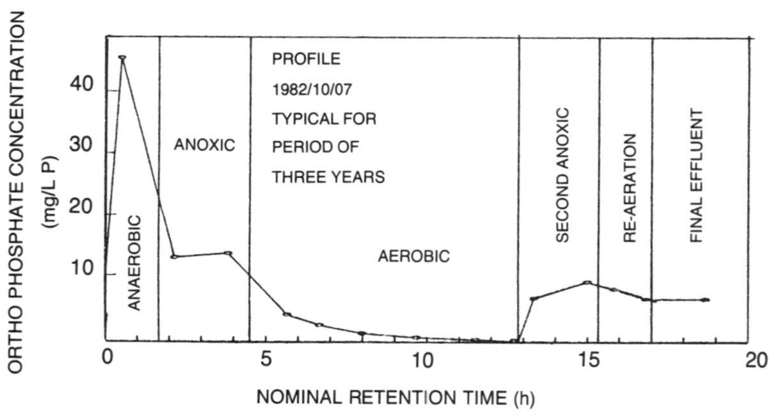

Figure 2.10 Phosphate release and uptake Randfontein.

an effort to improve P removal the anaerobic zone was operated with an upflow sludge blanket. The release of phosphates doubled but there was a marked deterioration in the overall removal. This indicated that in the anaerobic basin both primary and secondary release took place, but when the secondary release predominated, there was not sufficient PHB in the cells to effect the uptake of all the released phosphates.

This theory would also explain the need to have a sidestream chemical step for phosphate precipitation in the patented PhoStrip process. In this case, all of the return activated sludge is passed through a long anaerobic phase (up to thirty hours), while relying on fermentation of the activated sludge for the production of VFA. This leads to some primary, but also a large secondary, release. Provided the released phosphates could be stripped out in a sidestream, the VFA produced is sufficient for uptake of the phosphates in the aeration basin. This does not mean that the process cannot perform its intended function of removing phosphates to very low levels in the effluent. The release of phosphates under various conditions was demonstrated by Gerber (1986) when feeding both nitrates and various carbon substrates to the activated sludge obtained from a BNR plant. His findings are presented in Figure 2.11.

By feeding nitrates with the various substrates, he was effectively blocking any fermentation in the anaerobic zone. During the period when nitrates were being reduced to zero, there was sufficient substrate available for uptake by the BioP organisms, resulting in the release of phosphates. When feeding acetate, there was a rapid release of phosphates until the substrate was depleted, which started a slower rate of release. This slow rate could not be due to slow fermentation since only a single substrate was added. This is clearly a release not associated with energy uptake and would thus qualify as a secondary release. The results indicated that propionic acid could also be taken up by the cells, but other acids such as butyric showed no release of phosphates until the nitrates were depleted. Then a slow release started, which may have been partially caused by slow fermentation of the butyric acid to acetic acid and a secondary release.

DESIGN IMPLICATIONS

In the design of a BNR plant, the secondary release must be avoided as much as possible in order to reserve all the available energy for removal of the incoming and primary-released phosphates. This requires that prefermentation of the incoming COD be optimized to provide all the necessary acetates and minimize fermentation in the anaerobic zone of the BNR plant.

Before the development of the technology of prefermentation, Brodisch (1983) found that only BNR plants where *Aeromonas punctata* were present in addition to *Acinetobacter* were efficient in P removal. *A. punctata* are

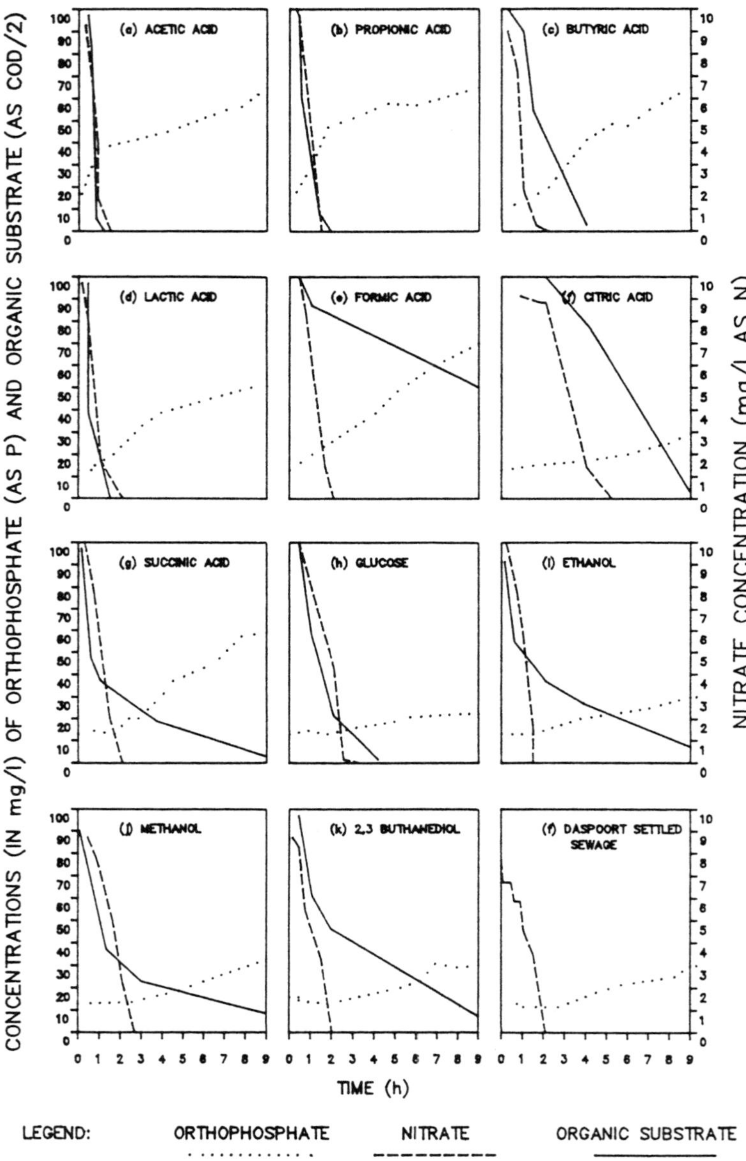

Figure 2.11 Effect of various easily degradable substrates.

heterotrophic organisms capable of acid fermentation. The accent is now on limiting fermentation in the anaerobic basin to prevent secondary release. Stevens (1985) found that in Kelowna, Canada, reducing the anaerobic stage from a nominal HRT of three hours, to two hours, and then to one hour, improved the plant performance considerably. It would appear that especially when treating fairly fresh, weak sewage, any effort at acid fermentation in the anaerobic basin is counterproductive, although this may not necessarily be the case with stronger wastewaters in warmer climates.

On the other hand, when treating stronger or partially fermented waste, a longer anaerobic retention time may be useful, especially in the absence of prefermentation or sufficient prefermentation. This was demonstrated on two occasions in South Africa. Gerber et al. (1982) found that for the Daspoort Pretoria waste without prefermentation, the five-stage Bardenpho performed reasonably, but when the plant was switched to the UCT configuration, phosphate removal was lost until the size of the anaerobic basin was doubled. For the same plant, the anaerobic mass fraction of the UCT configuration is only half of that of the Bardenpho configuration. The longer anaerobic mass fraction in this case increased the VFA production. The same phenomenon was observed for a full-scale plant near Bloemfontein. Contrary to this, phosphate removal was improved in the new Westbank plant in British Columbia, Canada by switching to the UCT mode even though nitrates (the main reason for the UCT flow sheet) were not present in the RAS underflow. As in Kelowna, reducing the anaerobic mass fraction improved the plant performance.

These anecdotes demonstrate the need for caution when relying on the anaerobic basin for fermentation. With prefermentation the designer can make sure that conditions will be optimal for the growth of *Acinetobacter* even with weak wastewater and low temperatures.

DENITRIFICATION

The above discussions stressed the importance of limiting the nitrates in the feed or in the return activated sludge in order to preserve all the available VFA for phosphate removal. Nitrates constitute the single most important factor that must be controlled to ensure good performance by BNR plants.

This does not mean that the presence of nitrates in the effluent will always result in loss of phosphate removal. Some wastes contain sufficient VFA in the influent that irrespective of the nitrates in the RAS, P removal will result. This is especially important in retrofitting or experimenting with existing full-scale plants. In situations where most of the wastewater is pumped to the plant and the temperatures are favorable, P removal will take place without a serious attempt at reducing nitrates. One example is the plant at Walt Disney World, Florida (Knight et al., 1987) where switching off a few aerators near the inlet zone resulted in good P removal, as can be seen in Figure 2.12, in spite of the fact that full nitrification was also experienced. In this case there were sufficient VFA generated in the force mains to remove the nitrates while providing sufficient food for the BioP organisms. When experimenting with existing plants, one should not assume automatically that it would not be possible. Also when performing pilot-plant studies, the operator should make sure that the characteristics of the waste feed match those of the full-scale situation entirely.

The importance of avoiding nitrate interference in BNR plants has led to the invention of a dozen or so variations of the original Phoredox principle, as indicated in Chapter 1. Unfortunately, many "inventors" became so biased towards their own configuration that only favorable results were reported, to the point where process novelty was even claimed for a round tank as opposed to a rectangular one. It is the duty of designers to design according to the effluent requirements and the special characteristics of the wastewater, using the best process configuration for those specific conditions.

Varied configurations have been developed in order to ensure that nitrates will not interfere with the phosphate removal mechanism. To this effect

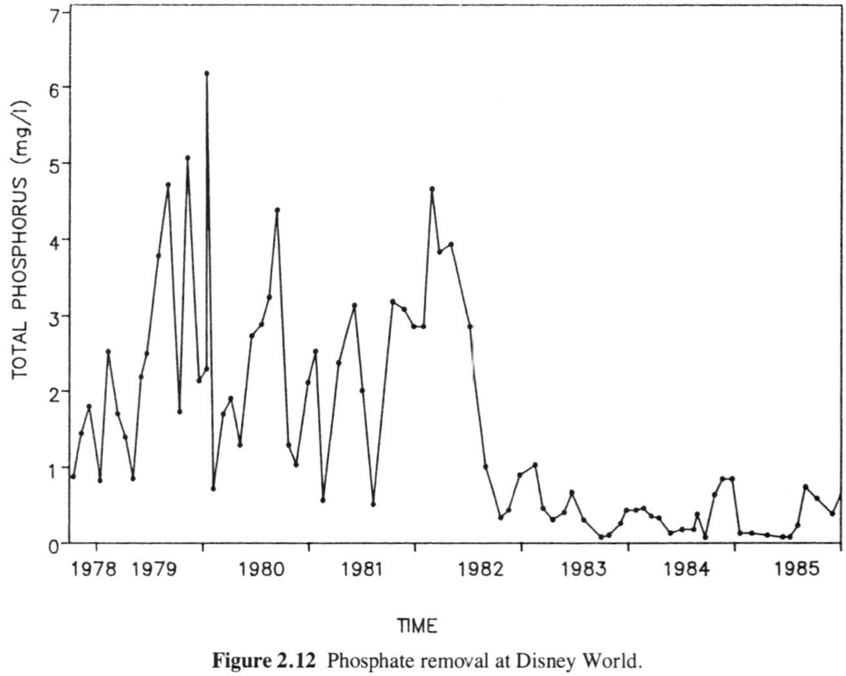

Figure 2.12 Phosphate removal at Disney World.

there are a number of design considerations that will be fully discussed in Chapter 4. Suffice it to say at this stage that nitrates will interfere even in two-stage Phoredox (AO) plants, designed with a short sludge age to avoid nitrification, since nitrification cannot be avoided when mixed liquor temperatures exceed 24°C.

There is natural competition between the denitrifying organisms and the BioP organisms. When high percentage removal of both nitrogen and phosphate is required, the plant should be biased towards nitrogen removal, since augmenting P removal with chemicals is much less costly and easier to control. Volatile fatty acids (VFA) as well as other easily degradable materials are responsible for the first high rate of denitrification demonstrated in Chapter 2. But the idea is to reserve the SCVFA for the BioP organisms. Whatever remains must then suffice for nitrogen removal. If the available carbon is not sufficient to satisfy both needs, the carbon can be augmented by prefermentation, which "steals" some of the carbon that would normally pass to the anaerobic digesters. This does not place an energy burden on the BNR plant since nitrates in the effluent are a source of energy already stored from a previous use. In plants where good nitrogen removal is achieved, good P removal is normally also relatively simple.

While some subspecies of *Acinetobacter* have the ability to remove nitrates through denitrification, the major portion of the organisms found in a typical BNR plant removes little or no nitrates (Lotter, 1989). Thus the food they remove preferentially is not available for denitrification. Fortunately, there is a side effect to biological phosphate removal that seems to benefit denitrification. This relates to the storage mechanism triggered in heterotrophic facultative organisms when mixing influent wastewater containing a high concentration of easily degradable COD with the RAS in the selector. Thus, while some of the feed needed for the rapid denitrification of nitrates in the anoxic zone has been sequestered by the BioP organisms, the stored COD (Van Huyssteen et al., 1988) more than makes up for the shortfall and allows a high percentage denitrification not previously thought possible.

The Modified University of Capetown (MUCT) process (Figure 2.13) and the Queensland process (first mentioned in UOQ notes) (Figure 2.14) assumes that nitrate removal is only necessary to protect the anaerobic basin and allows for good P removal while still having a high concentration of nitrates in the effluent. Both processes subject the RAS to anoxic conditions to remove the nitrates before passing the RAS to the anaerobic zone. In the Modified UCT process and its offshoot (called the New Johannesburg Process), the RAS is contacted with the effluent stream of the anaerobic contact unit to rid it of nitrates, while in the Queensland process endogenous respiration of the sludge with some assist from primary settling tank effluent without the VFA from the fermenter is used for this purpose.

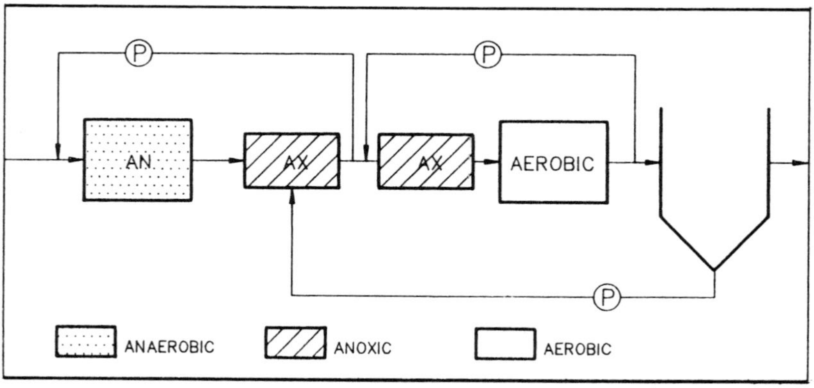

Figure 2.13 Modified UCT process.

pH VALUE OF THE MIXED LIQUOR

Not much work has been done specifically on the effect of pH on the functioning of the *Acinetobacter* organisms. From the proposed mechanisms, it is evident that more energy will be required to maintain the proton balance and thus to take up acetates against a higher H^+ concentration in the anaerobic basin. Thus, overall phosphate removal may be reduced with decrease in pH value, while the opposite may be true under conditions of higher pH values.

The significance of these findings relates to low-alkalinity wastewaters where nitrification may lower the pH value of the mixed liquor well below the neutral value. In such cases, optimum denitrification to restore pH value

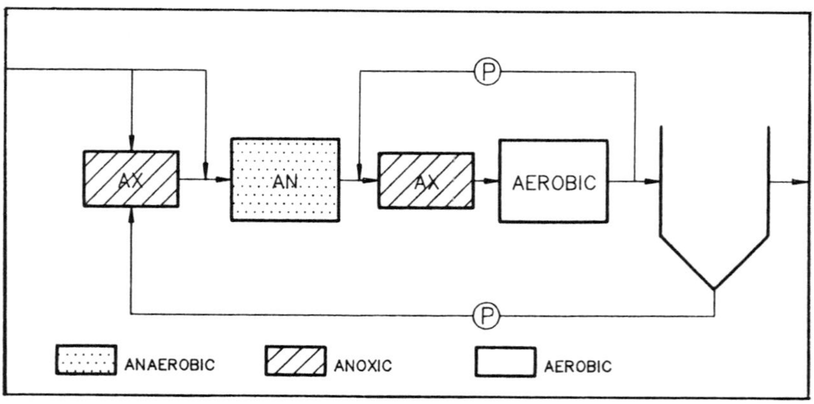

Figure 2.14 Flow diagram for the removal of nitrogen from RAS.

may be required. Even this may not be sufficient and some lime addition may have to be provided to ensure a sufficient pH value for optimum P removal.

MAINTAINING DISSOLVED OXYGEN CONCENTRATION

While some anaerobic and anoxic zones may be required for the reasons quoted above, a sufficient DO level is required somewhere in the aeration basin for phosphate uptake by the BioP organisms. This is not normally a problem with most flow patterns, but it has proven to be a problem with channel systems such as the Pasveer and Orbal systems. The latter consists normally of multiple channels, allowing a high DO concentration in the last channel. The Pasveer plant, on the other hand, may consist of a single or double channel with a large number of smaller aerators such that there is no zone in the plant where the DO is sufficiently high to ensure good P removal.

The contact of the RAS with VFA-containing liquor seems to change the behavior of the organisms in terms of storage of food in the cells. This changes the oxygen utilization rate (OUR) patterns in the aeration basin (Van Huyssteen et al., 1989) which may require careful pilot studies to ensure that the oxygen supply along the aeration basin matches the requirements, and that deficits are not experienced in the full-scale plant.

TEMPERATURE

Temperature does not seem to influence P removal as much as COD removal, but may play a role in nitrification and denitrification, which can affect the process. For the two-stage Phoredox process, lower temperatures may be beneficial in that it would be easier to design for P removal at a practical MCRT without fear of nitrates' interference.

MEAN CELL RESIDENCE TIME (MCRT)

The mean cell residence time seems to play a lesser role in practical terms than expected. This is attributed by Wentzel et al., (1988) to the low observed endogenous decay rate of the BioP organisms, which is about 0.05 per day, as opposed to 0.24/d for normal heterotrophic organisms. What this means is that at longer MCRTs, a proportionally larger part of the active mass will be taken up by *Acinetobacter*, with a resultant higher phosphate content.

It has been shown in practice that good P removal is possible at MCRTs ranging from three to sixty days. It is not excluded that some form of crystallization may take place at higher MCRTs (Kerdachi and Roberts, 1982) thus accounting for the stability of the process.

At the other end of the scale Wentzel et al. (1988) showed that while BPR is possible at MCRTs of less than three days, the process is not as stable and the effluent is not clear. Since the sludge particles contain phosphates, it is essential to have a clear effluent if the purpose is to obtain a low effluent P value without subsequent or chemical treatment.

At an MCRT of about six days the process is more stable, but at higher temperatures nitrification may interfere.

TOXIC INTERFERENCE

Little is known about inhibition of the BioP removal process, mostly because incidences of inhibition are rare under normal operating conditions. In only one instance out of more than fifty full-scale plants was the process inhibited by some substance that did not also inhibit the nitrification process. This happened in a plant treating only slaughterhouse waste. The offending stream has been identified as originating from the lairages, but the offending compound has not been identified. It was sufficiently inhibitory to block any acetate uptake in the anaerobic zone of the pilot plant. In the absence of the lairage washwater, there was no inhibition. No inhibition has been observed in domestic wastewater unless the nitrification process was also inhibited. More research is needed in this area.

LEACHING FROM WASTE SLUDGES

Since phosphates are released from the sludge under anaerobic conditions, it follows that anaerobic treatment will release phosphates to the liquid phase. Dewatering of sludge for disposal must be done under aerobic conditions to avoid too much leaching back. (See further in Chapter 8.)

REFERENCES

Anthonisen, A. C. et al. 1976. "Inhibition of Nitrification by Ammonia and Nitrous Acid," *J. Wat. Poll. Con. Fed.*, 48:835.

Antoniou, P. et al. 1990. "Effect of Temperature and pH on the Effective Maximum Specific Growth Rate of Nitrifying Bacteria," *Wat. Res.*, 24:97.

Barnard, J. L. 1973. "Biological Denitrification," *J. Int. Wat. Poll. Con. Fed.*, 72:6.

Barnard, J. L. 1974. "Cut P and N Without Chemicals," *Water and Wastes Eng.*, 11(7):33-36.

Barnard, J. L. 1975. "Biological Nutrient Removal Without the Addition of Chemicals," *Wat. Res.*, 9:485.

Barnard, J. L. 1984. "Activated Primary Tanks for Phosphate Removal," *Water S.A.*, 10:121-126.

Barth, E. F. and H. D. Stensel. 1981. "International Nutrient Control Technology for Municipal Effluents," *J. Wat. Poll. Con. Fed.*, 53:1691

Barth, E. F., R. C. Brenner and R. F. Lewis. 1968. "Chemical Biological Control of Nitrogen and Phosphorus in Wastewater Effluent," *J. Wat. Poll. Con. Fed.*, 40:2040.

Baas-Becking, L. G. M. and G. S. Parks. 1927. "Energy Relations in the Metabolism of Autotrophic Bacteria," *Physiol. Revs.*, 7:85.

Batchelor, B. 1982. "Kinetic Analysis of Alternative Configurations for Single-Sludge Nitrification/Denitrification," *J. Wat. Poll. Con. Fed.*, 54:1493.

Beccari, M., D. Marani and R. Ramadori. 1979. "A Critical Analysis of Nitrification Alternatives," *Wat. Res.*, 13:185.

Beckman, W. J. et al. 1972. "Design and Operation of Combined Carbon Oxidation-Nitrification Activated Sludge Plant," *J. Wat. Poll. Con. Fed.*, 44:1916.

Benefield, L. D. and C. W. Randall. 1980. *Biological Process Design for Wastewater Treatment*, Prentice-Hall, Inc.

Bidstrup, S. M. and C. P. L. Grady, Jr. 1988. *J. Wat. Poll. Cont. Fed.*, 60:351.

Bliss, P. J. and D. Barnes. 1981. "Biological Nitrogen Control in Wastewaters," *Effl. and Wat. Treatment J.*, Feb. p. 65.

Blum, D. J. W. and R. E. Speece. 1991. "A Database of Chemical Toxicity to Environmental Bacteria and Its Use in Interspecies Comparisons and Correlations," *Res. J. Wat. Poll. Con. Fed.*, 63:198.

Bock, E., P. A. Wilderer and A. Freitag. 1988. "Growth of *Nitrobacter* in the Absence of Dissolved Oxygen," *Wat. Res.*, 22:245.

Boon, B. and H. Laudelout H. 1962. "Kinetics of Nitrate Oxidation by *Nitrobacter winogradsyi*," *Biochem. J.*, 85:440.

Bremner, J. M. and K. Shaw. 1958. "Denitrification in Soil II. Factors Affecting Denitrification," *J. Agric. Sci.*, 51:39.

Brodish, K. E. U. and S. J. Joyner. 1983. "The Role of Microorganisms Other Than *Acinetobacter* in Biological Phosphate Removal in Activated Sludge Process," *Wat. Sci. Tech.*, 15:87-103.

Burdick, C. R., D. R. Refling and H. D. Stensel. 1982. "Advanced Biological Treatment to Achieve Nutrient Removal," *J. Wat. Poll. Con. Fed.*, 54:1078.

Charley, R. C., D. G. Hooper and A. G. McLee. 1980. "Nitrification Kinetics in Activated Sludge at Various Temperatures and Dissolved Oxygen Concentrations," *Wat. Res.*, 14:1387.

Chudoba, J., J. S. Cech and P. Chudoba. 1985. "The Effect of Aeration Tank Configuration on Nitrification Kinetics." *J. Wat. Poll. Con. Fed.*, 57:1078.

Comeau, Y., K. H. Hall, R. E. W. Hancock and W. K. Oldham. 1986. "Biochemical Model for Enhanced Biological Phosphate Removal," *Wat. Res.*, 20(12).

Dawson, R. N. and K. L. Murphy. 1972. "The Temperature Dependency of Biological Denitrification," *Wat. Res.*, 6:71.

Downing, A. L. and A. P. Hopwood. 1964. "Some Observations on the Kinetics of Nitrifying Activated Sludge Plants," *Schweizerische Aeitschrift fur Hydrologie.*, 26:271.

Downing, A. L., H. A. Painter and G. Knowles. 1964. "Nitrification in the Activated Sludge Process," *J. Proc. Inst. Sew. Purif.*, p. 130.

Downing, A. L. and L. J. Scagg. 1958. "The Effect of Synthetic Detergents on the Rate of Aeration in Diffused Air Activated Sludge Plants," *Water Waste Treatment J.*, 7:102.

Eckenfelder, W. W., Jr. 1990. "*Principles of Biological and Physical/Chemical Nitrogen Removal: Principles and Practice of Nutrient Removal from Municipal Wastewater*," Soap and Detergent Association, New York, NY.

Ekama, G. A., G. v. R. Marais and I. P. Siebritz. 1984. "Biological Excess Phosphorus Removal," *Design and Operation of Nutrient Removal Activated Sludge Processes*, Water Research Commission, Pretoria S. A.

Engle, M. S. and M. Alexander. 1958. "Growth and Autotrophic Metabolism of *Nitrosomonas europaea*," *J. Bact.*, 76:217.

Eylar, O. R. and E. L. Schmidt. 1959. "A Survey of Heterotrophic Microorganisms From Soil for Ability to Form Nitrite and Nitrate," *J. Gen. Microbiol.*, 20:473.

Focht, D. D. and A. C. Chang. 1975. "Nitrification and Denitrification Processes Related to Waste Water Treatment," *Adv. Appl. Microbiol.*, 20:631.

Fuhs, G. W. and M. Chen. 1975. "Phosphate Removal in the Activated Sludge Process," *Microbial Ecology*, 2:119.

Gayle, B. P. et al. 1989. "Biological Denitrification of Water," *J. of Env. Engr. Amer. Soc. of Civil Engr.*, 115:930.

Gerber, A., E. S. Mostert, C. T. Winer and R. H. DeVilliers. 1986. "The Effect of Acetate and Other Short Chain Carbon Compounds on the Kinetics of Biological Nutrient Removal," *Water SA*, 12:7-12.

Gerber, A., E. S. Mostert, C. T. Winter and R. H. DeVilliers. 1982. "Biological Nutrient Removal in the UCT Modification of the Activated Sludge Process," *IMSEA*, Dec. 1982/Jan. 1983.

Gould, G. W. and Lees H. 1960. "The Isolation and Culture of the Nitrifying Organisms. Part I. *Nitrobacter*," *Can. J. Microbiol.*, 6:299.

Hall, I. R. 1974. "Some Studies on Nitrification in the Activated Sludge Process," *J. Wat. Poll. Con. Fed.*, 73:538.

Hall, E. R. and K. L. Murphy. 1980. "Estimation of Nitrifying Biomass and Kinetics in Wastewater," *Wat. Res.*, 14:297.

Hanaki, K., C. Wantawin and S. Ohgaki. 1990a. "Effects of the Activity of Heterotrophs on Nitrification in a Suspended-Growth Reactor," *Wat. Res.*, 24:(3):289-296. Printed in Great Britain.

Hanaki, K., C. Wantawin and S. Ohgaki. 1990b. "Nitrification at Low Levels of Dissolved Oxygen with and without Organic Loading in a Suspended-Growth Reactor," *Wat. Res.*, 24(3):297-302. Printed in Great Britain.

Haug, R. T. and P. L. McCarty. 1971. *Nitrification with the Submerged Filter; Technical Report No. 149*, Dept. of Civil Engr. Stanford University, Palo Alto, Calif.

Haug, R. T. and P. L. McCarty. 1972. "Nitrification with the Submerged Filter," *J. Wat. Poll. Con. Fed.*, 44:2086.

Hockenbury, M. R. and C. P. L. Grady, Jr. 1977. "Inhibition of Nitrification – Effects of Selected Organic Compounds," *J. Wat. Poll. Con. Fed.*, 49:768.

Hoffman, J. R. 1977. "Phosphorus Removal in the Modified Activated Sludge Process," M. S. Thesis, University of Cape Town, Rondebosch, Republic of South Africa.

Hultman, B. 1971. "Kinetics of Biological Nitrogen Removal." *Inst. Vattenforsorjmingsoch Avloppsteknik samt Vattenkemi, KTH, Pub.*, 71:5, Stockholm.

IAWPRC Task Group on Mathematical Modeling for Design and Operation of Biological Wastewater Treatment. 1986. "Final Report: Activated Sludge Model," *IAWPRC Scientific and Technical Reports No. 1.*

Johnson, W. K. and G. J. Schroepfer. 1964. "Nitrogen Removal by Nitrification and Denitrification," *J. Wat. Poll. Con. Fed.*, 36:1015.

Jones, W. L., E. D. Schroeder and P. A. Wilderer. 1990. "Denitrification in a Batch Wastewater Treatment System Using Sequestered Organic Substances," *J. Wat. Poll. Con. Fed.*, 62:259.

Kaas, J. C. and K. K. Andersen. 1989. "Design for Nitrogen and Phosphorus Removal at Low Temperatures," Presented at the *62nd Annual Water Poll. Control Fed. Conf., San Francisco, CA, October 17, 1989.*

Kerdachi, D. A. and M. R. Roberts. 1982. "Full-scale Phosphate Removal Experience in the Umhlatuzana Works at a Different Sludge Ages," *IAWPR Post Conference Seminar on Phosphate Removal in Biological Treatment Processes, Pretoria, Republic of South Africa, 5-6 April.*

Knight, R. L. et al. 1987. "Performance of a Natural Wetland Treatment System for Wastewater Treatment," *J. Wat. Pol. Con. Fed.*, 59:9.

Knowles, G., A. L. Downing and M. J. Barrett. 1965. "Determination of Kinetic Constants for Nitrifying Bacteria in Mixed Culture with the Aid of a Computer," *J. Gen Microbiol.*, 38:263.

Kugelman, I. J. et al. 1991. "Aerobic Denitrification in Activated Sludge," *Proceedings of the 1991 Envir. Engr. Spec. Conf., Amer. Soc. of Civil Engrs.*, Reno, Nev.

Lan, J. D., L. D. Benefield and C. W. Randall. 1983. "Phosphorus Removal in the Activated Sludge Process," *Wat. Res.*, 17(9):1193-1200. Pergamon Press (Great Britain).

Laudelout, H. et al. 1974. "Effect of Temperature on the Velocity of Oxidation of Ammonia to Nitrate in Mixed Nitrifier Culture," *Ann. Microbiol.*, 125B:75.

Laudelout, H., R. Lambert and M. L. Pham. 1976. "The Effect of pH and Partial Pressure of Oxygen on Nitrification," *Ann. Microbiol.*, 127A:367.

Lawrence, A. W. and C. G. Brown. 1976. "Design and Control of Nitrifying Activated Sludge Systems," *J. Wat. Poll. Con. Fed.*, 48:1779.

Lawrence, A. W. and P. L. McCarty. 1970. "Unified Basis for Biological Treatment Design and Operation," *Jour. San. Eng. Div., Proc. Amer. Soc. Civil Engr.*, 96:757.

Lees, H. and J. R. Simpson. 1957. "The Biochemistry of the Nitrifying Organisms, 5. The Nitrite Oxidation by Nitrobacter," *Biochem. J.*, 65:297.

Lotter, L. 1989. "Two-year Study on the Enhancement of Biological Phosphate Removal by Altering Process Feed Composition (Metabolic Control Mechanisms)," WRC Rep. No. 137/3/89, Wat. Res. Com., Pretoria, Republic of South Africa.

Loveless, J. E. and H. A. Painter. 1968. "The Influence of Metal Ion Concentration and pH Value on the Growth of a *Nitrosomonas* Strain Isolated from Activated Sludge," *J. Gen. Microbiol.*, 52:1.

McCarty, P. L., L. Beck and P. St. Amant. 1969. "Biological Denitrification of Wastewaters by Addition of Organic Materials," Proceedings of the *24th Purdue Ind. Waste Conf., Purdue University, Lafayette, Ind.*

McClintock, S. A. et al. 1988. "Nitrate versus Oxygen Respiration in the Activated Sludge Process," *J. Wat. Poll. Con. Fed.*, 60:342.

Metcalf and Eddy. 1991. "Wastewater Engineering; Treatment, Disposal, Reuse," McGraw-Hill, Inc.

Mulbarger, M. C. 1971. ''Nitrification and Denitrification in Activated Sludge Systems,'' *J. Wat. Poll. Con. Fed.*, 43:2059.

Nagel, C. A. and J. G. Haworth. 1969. ''Operation Factors Affecting Nitrification in the Activated Sludge Process,'' paper presented at the *42nd Annual Conference of the Wat. Pollut. Control Fed., Dallas, TX.*

Nelson, D. H. 1931. ''Isolation and Characterization of *Nitrosomonas* and *Nitrobacter*. Zentr.,'' *Bakt. Parasitenk*, 83:280.

Nelson, L. M. and R. Knowles. 1978. ''Effect of Oxygen and Nitrate on Nitrogen Fixation and Denitrification by *Azospirillum brasilense* Grown in Continuous Culture,'' *Can. J. Microbiol.*, 24:1395.

Neufeld, R. D., A. J. Hill and D. O. Adekoya. 1980. ''Phenol and Free Ammonia Inhibition to *Nitrosomonas* Activity,'' *Wat. Res.*, 14:1695.

Nicholls, H. A. and D. W. Osborn. 1978. ''Optimization of the Activated Sludge Process for Biological Removal of Phosphorus,'' *Prog. in Water Tech.*, 10(1):2.

Nommik, H. 1956. ''Investigations on Denitrification in Soil,'' *Acta Agric. Scand.*, 6:195.

Okun, D. A. 1949. ''System of Bio-precipitation of Organic Matter from Sewage,'' *Sew. Works J.*, 21:763.

Painter, H. A. 1970. ''A Review of Literature on Inorganic Nitrogen Metabolism in Microorganisms,'' *Wat. Res.*, 4:393.

Painter, H. A. and J. E. Loveless. 1983. ''Effect of Temperature and pH Value on the Growth-rate Constants of Nitrifying Bacteria in the Activated-Sludge Process,'' *Wat. Res.*, 7:237.

Panzer, C. C., M. Komanowsky and G. E. Senske. 1980. ''Improved Performance in Combined Nitrification/Denitrification,'' Presented at the *35th Annual Purdue Ind. Waste Conf., Lafayette, Ind. May 1980.*

Payne, W. J. 1981. *Denitrification*, John Wiley & Sons, N.Y.

Payne, W. J. and P. S. Riley. 1969. ''Suppression by Nitrate of Enzymatic Reduction of Nitric Oxide,'' *Proc. Soc. Exp. Biol. Med.*, 132:258.

Peeters, T. L., A. D. Van Goal and H. Laudelout. 1969. ''Kinetic Study of Oxygen Limited Respiration in Nitrifying Bacteria,'' *Bact. Proc.*, 141.

Poduska, R. A. 1973. *A Dynamic Model of Nitrification For the Activated Sludge Process*, Ph. D. Thesis, Clemson University.

Prakasam, T. B. S. et al. 1979. ''Pilot-Scale Studies of Single-stage Nitrification,'' *J. Wat. Poll. Con. Fed.*, 51:1904.

Randall, C., V. Pattarkine and S. McClintock. 1991. ''Nitrification Kinetics in Single-Sludge Biological Nutrient Removal Activated Sludge Systems,'' *Proc. IAWPRC Seminar, Interactions of Wastewater, Biomass and Reactor Configurations in Biological Treatment Plants*, Paper 3.8, 19 pages, Copenhagen, Denmark.

Requa, D. A. and E. D. Schroeder. 1973. ''Kinetics of Packed Bed Denitrification,'' *J. Wat. Poll. Con. Fed.*, 45:1696.

Rittman, B. E. and W. E. Langeland. 1985. ''Simultaneous Denitrification with Nitrification in Single-channel Oxidation Ditches,'' *J. Wat. Poll. Con. Fed.*, 57:300.

Robertson, L. A. et al. 1988. ''Simultaneous Nitrification and Denitrification in Aerobic Chemostat Cultures of *Thiosphaera pantotropha*,'' *Appl and Env. Microbiol.*, 54:2812.

Rozich, A. F. and D. J. Castens. 1986. "Inhibition Kinetics of Nitrification in Continuous-Flow Reactors," *J. Wat. Poll. Con. Fed.*, 58:220.

Schloesing, T. and Muntz A. 1877. "Sur las nitrification par les ferments organises," *Compt. Rend. Acad. Sc.*, 84:301.

Schoberl, P. and Engel H. 1964. "Dar verhalten der nitrification kakterien gegenuber gelostem," *Archiv Fin Mikrobiolgie.*, p. 393.

Sharma, B. and Ahlert R. C. 1977. "Nitrification and Nitrogen Removal," *Wat. Res.*, 11:897.

Skerman, V. B. D. and I. C. MacRae. 1957. "The Influence of Oxygen Availability on the Degree of Nitrate Reduction by *Pseudomonas denitrifications*," *Can. J. of Microbiol.*, 3:505.

Skinner, F. A. and N. Walker. 1961. "Growth of *Nitrosomonas europaea* in Batch and Continuous Culture," *Arch. Mikrobiol.*, 38:339.

Stankewich, M. J. Jr. 1972. "Biological Nitrification with the High Purity Oxygenation Process," *Proceeding, 27th Annual Purdue Industrial Waste Conference*, Purdue University, West Lafayette, Ind.

Stensel, H. D. 1981. "Biological Nitrogen Removal System Design," *Water Amer. Inst. of Chem. Engrs.*, 237.

Stensel, H. D., R. C. Loehr and A. W. Lawrence. 1973. "Biological Kinetics of Suspended-Growth Denitrification," *J. Wat. Poll. Con. Fed.*, 45:249.

Stensel, H. D., C. S. McDowell and E. D. Ritter. 1976. "An Automated Biological Nitrification Toxicity Test," *J. Wat. Poll. Con. Fed.*, 48:2343.

Stenstrom, M. K. and R. A. Poduska. 1980. "The Effect of Dissolved Oxygen Concentration on Nitrification," *Wat. Res.*, 14:643.

Stenstrom, M. K. and S. S. Song. 1991. "Effects of Oxygen Transport Limitation on Nitrification in the Activated Sludge Process," *Res. J. Wat. Poll. Con. Fed.*, 63:208.

Stevens, J. 1985. Private communication.

Terai, H. and Mori T. 1975. "Studies on Phosphorylation Coupled with Denitrification and Aerobic Respiration in *Pseudomonas denitrificans*," *Bot. Mag.*, 38:231.

University of Capetown. 1984. *Theory, Design and Operation of Nutrient Removal Activated Sludge Processes*, Water Research Commission, Pretoria S. A. 5-4.

U. S. Environmental Protection Agency. 1975. Process Design Manual for Nitrogen Control, Office of Technology Transfer, Washington, DC. October.

Van Huyssteen, J. A., J. L. Barnard and J. Hendriksz. 1988. The Olifantsfontein Nutrient Removal Plant. *Proceedings International Specialized Conference on Upgrading of Wastewater Treatment Plants, IAWPRC*, Munich Germany. Sept. 3-7, pp. 1-8.

Wentzel, M. C., L. H. Lotter, R. E. Loewenthal and G. v. R. Marais. 1986. "Metabolic Behavior of *Acinetobacter ssp.* in Enhanced Biological Phosphorus Removal—A Biochemical Model," *Water SA*, 12(4):209-224.

Wentzel, M. C., G. A. Ekama, P. L. Dold, R. E. Lowenthal and G. v. R. Marais. 1988. "Biological Excess Phosphate Removal," WRC Rep. No. 148/1/88, Wat. Res. Com., Pretoria, Republic of South Africa.

Werzernak, C. T. and Gannon J. J. 1967. "Oxygen-Nitrogen Relationships in Autotrophic Nitrification," *Appl. Microbiol.*, 15:211.

Wheatland, A. B., M. J. Barnett and A. M. Bruce. 1959. "Some Observations on Denitrification in Rivers and Estuaries," *Inst. Sewage Purif. J. Proc.*, p. 149.

Wild, H. E., C. N. Sawyer and T. D. McMahon. 1971. "Factors Affecting Nitrification Kinetics," *J. Wat. Poll. Con. Fed.*, 43:1845.

Wiljer, J. and C. C. Delwiche. 1954. "Investigations on the Denitrifying Process in Soil," *Plant Soil*, 5:155.

Winogradsky, M. S. 1891. "Recherches sur les organismes de la nitrification," *Annals Institute Pasteur*, 5:92.

Wuhrman, K. 1963. "Effect of Oxygen Tension on Biochemical Reactions in Sewage Purification Plants," Proc. 3rd Manhattan Conf., *Advances in Biological Waste Treatment*, McMillan, N. Y.

Wuhrman, K. 1968. "Research Developments in Regard to Concept and Base Values of the Activated Sludge System in Advances in Water Quality Improvement," Ed. Gloyna, E. F. and Eckenfelder, W. W. Jr., University of Texas Press, 143.

Young, J. C., L. O. Thompson and D. R. Curtis. 1977. "Control Strategy for Biological Nitrification Systems," *J. Water Pollut. Control Fed.*, 51:1824.

CHAPTER 3

Design of Prefermentation Processes

INTRODUCTION

A number of conventional activated sludge plants in the U.S.A., notably the Rilling Road plant in San Antonio, the Backwater plant in Baltimore, and the Hyperion plant in Los Angeles, were reported to remove phosphates down to levels below 1 mg/L. These plants had several characteristics in common in that they were all high-rate plants designed in a plugflow mode with semi-coarse bubble aeration. Milbury (1971) reported that all plants in the U.S.A. showed a marked release of phosphates in the initial aeration zone with a slow uptake through the entire length of the aeration basin. Milbury reported that in his study of the Backwater plant the oxygen in the initial zone was "near zero."

The patented PhoStrip process (Levin, 1970) passed the return activated sludge (RAS) through a thickener where the microorganisms used up all the oxygen through endogenous respiration and then actively fermented. Phosphates were released to the liquid, some of which were separated via the overflow to form a sidestream rich in phosphates which could be treated with lime for precipitation of the phosphates. The bottom sludge withdrawal of the thickener was mixed with the influent wastewater and passed to the aeration basin for normal aeration followed by sedimentation in the final clarifier. Good removals of phosphates were reported.

While experimenting with nitrogen removal through anoxic zones in the activated sludge process, Barnard (1974) discovered good P removal in the system when at the same time the nitrates in the effluent were low. Barnard postulated that the common factor to all the observations was that the active organisms passed through an anaerobic stage in which no nitrates or dissolved oxygen were present. Such a zone might be the first zone as Milbury described, the fermenter/thickener of the PhoStrip process, or the second anoxic zone in the case of nitrogen removal plants.

When Fuhs and Chen (1975) identified the organisms responsible for excess P uptake as *Acinetobacter*, which feeds preferably on acetates, it was assumed that, as was evident from the PhoStrip plant, the fermentation products were formed in the anaerobic zones and that the products so formed would be sufficient for sustaining the biological removal of phosphates. While Fuhs and Chen discovered that phosphates were released to the liquid stage during the uptake of acetates, as described in Chapter 2, they also discovered substantial additional P release when the mixed liquor was subjected to anaerobic conditions. This "secondary release" as it was referred to later (Barnard 1984), posed a problem in that using an anaerobic zone in the activated sludge process for acid fermentation has two opposing results. The longer the retention time in the zone in order to get more uptake of energy, the more secondary release and the greater the energy required for uptake of the released phosphates. The best example of this dilemma is the original PhoStrip process, in which all the required short-chain Volatile Fatty Acids (VFA) must be generated in the fermentation zone, resulting in such massive releases of phosphates that some of the phosphates must be precipitated by chemicals.

In the Phoredox-type systems where the incoming wastewater is contacted with the RAS, the short-chain volatile fatty acids (SCVFA) in the influent are not destroyed through prior aeration and the anaerobic retention time can be much shorter, leading to much lower secondary release which eliminates the need to add chemicals for meeting the standards of less than 1 mg/L of phosphates in the effluent.

The alternative to the generation of VFA in the activated sludge plant is to use a prefermenter, in which the fermentation step is independent of the activated sludge system, and carbon available in the incoming raw sewage is fermented to generate short-chain VFAs. In many instances sufficient VFA is present in the raw sewage to obviate any special process for their generation. This may have resulted from fermentation in slow-flowing outfall sewers or in pump stations and force mains. Where a substantial portion of the waste is pumped to the plant through long force mains, excellent P removal may result without any real effort. This may be the reason for the P removal observed in many of the conventional plants in the U.S.A.

Generally, it has been found that without any incidental fermentation in outfalls of force mains or deliberate fermentation in, for instance, primary sedimentation tanks, it is very difficult to produce sufficient VFA in the anaerobic zone of a BNR activated sludge plant when the influent is weak (influent BOD less than 200 mg/L) and winter mixed liquor temperatures are below 17°C. Somewhere below this point the additional secondary release is more than the uptake possible through the production of SCVFA, and enlargement of the anaerobic zone becomes counterproductive.

Thus, while it has been found that with stronger wastes and high winter temperatures there may be merit in enlarging the anaerobic zone, for weaker wastes and lower winter temperatures, the smaller the anaerobic basin the better, provided that sufficient VFA are produced in a prefermentation step. The anaerobic zone then becomes merely an anaerobic contact zone to which air need not be totally excluded and where strict anaerobic conditions may be counterproductive. Gerber et al. (1986) have shown that when acetates were fed to sludge from a nutrient removal plant, primary release of phosphates commenced immediately even in the presence of nitrates and dissolved oxygen. At this point it is not yet clear which is the more harmful under these conditions – excessive anaerobiosis or some oxygen or nitrate. Since no more fermentation is required, oxygen or nitrate will only allow some organisms to utilize some of the substrate while secondary release under anaerobic conditions would require more SCVFA for the uptake of such released phosphates.

MECHANISM OF PREFERMENTATION

In the normal anaerobic fermentation of primary sludge, the larger proteins and carbohydrate compounds are reduced by acid fermentation to a variety of simple products. Typical fermentation products are:

- acetates
- propionates
- butyrates
- hydrogen

A typical biological reaction is shown in Figure 3.1.

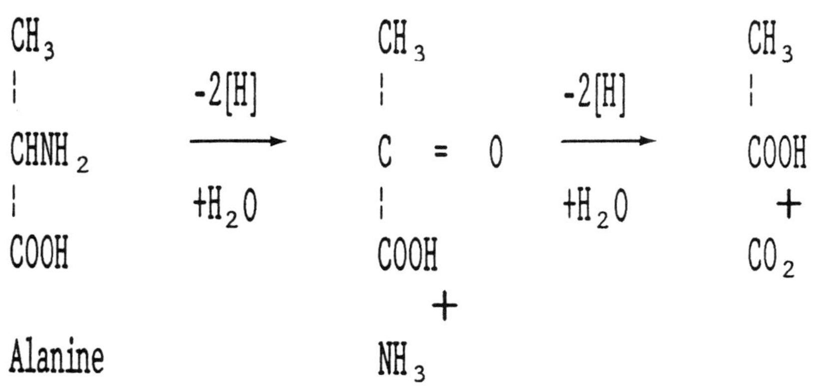

Figure 3.1 Biological reaction of acid fermentation.

Under normal operating conditions at pH near neutral, 70 to 85% of the by-products formed are acetates and 10 to 20% could be in the form of propionates.

In anaerobic digesters, methane fermentation is established spontaneously, following the biological pathways shown in Figure 3.2.

While acid fermentation can take place at a redox potential of more than −300 mV, methane fermentation (and sulfate reduction) takes place at a redox potential of less than −550 mV. In the natural sequence, when starting up a fermenter, acid fermentation will take off, and as the redox potential drops, methane organisms will start growing and will soon reach an equilibrium wherein the acetates are consumed at the rate at which they are produced. The growth of methane organisms must be avoided in order to obtain optimal acetate production.

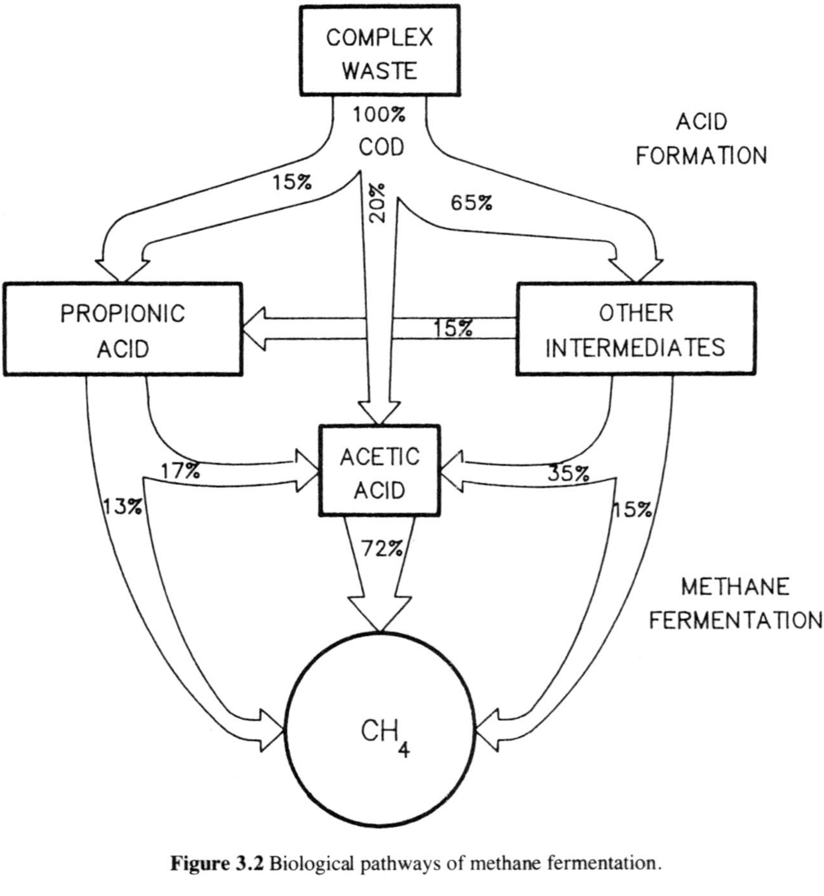

Figure 3.2 Biological pathways of methane fermentation.

DESIGN OF SYSTEMS FOR OPTIMIZING SCVFA PRODUCTION

The acetates and propionates produced are very soluble and can be elutriated from the sludge. The following methods have been proposed for achieving this.

DEWATERING THE UNDERFLOW FROM GRAVITY THICKENERS

Initial attempts for elutriating the SCVFA from the thickener underflow at Kelowna B.C., Canada consisted of passing the underflow over a static screen for the removal of the coarser material (Barnard et al., 1984). This will only work when spraying water on the screen to continuously wash the sludge, but this resulted in too much solid material passing through the screen. An alternative to this method could be the use of some dewatering device such as a centrifuge or belt press, and with return of the liquid to the anaerobic basin. Note that this refers to the dewatering of thickened primary sludge and not of digested sludge.

RETURN THE THICKENER OVERFLOW DIRECTLY TO THE ANAEROBIC BASIN

This would be normal practice in combination with the BNR plants but does not optimize the elutriation of VFA. In the thickening process, liquid containing VFA is expelled continuously, but as the sludge thickens and the optimal VFA production is reached, less liquid is expelled and much of the VFA is not passed to the AS unit unless the next step is sludge dewatering.

RECYCLE ON THE PST OR GRAVITY THICKENER

One of the simplest ways of producing VFA is to allow a sludge blanket to form within the PST and to slowly recycle the sludge to the inlet. This concept is the basis of the "activated primary sedimentation tanks" or a-PST (Barnard, 1984). The constant recycling process inoculates the incoming solids with actively fermenting organisms, elutriates the VFA formed in the sludge blanket, and prevents the formation of methane and hydrogen sulfide through the constant exposure to air with every recycle. Such recycle may also be applied to the gravity thickener or from the thickener back to the PST.

The equivalent of the daily production of primary sludge must be removed to the anaerobic digestion system. The amount may be difficult to determine since one is not dealing with a completely mixed system. It becomes essential that the depth of the sludge blanket be determined each day or that automatic sludge blanket control be installed for determining the rate of primary sludge wasting.

Since any continuous system will eventually lead to a slow buildup of methane organisms, the sludge should be wasted completely at regular intervals. These intervals vary from one location to the next and even from season to season, and must be optimized for each individual plant.

One of the problems experienced with a recycle on the PST is the additional load on the scraper mechanism. In most instances the mechanisms were not designed for the heavy sludge load and could not cope with the deeper sludge blanket. In new plants the mechanisms can be designed to deal with this problem.

COMPLETELY MIXED FERMENTER

The sludge removed daily from the PST can be discharged to a completely mixed fermentation unit and from there to a thickener for removal of the liquid containing the VFA. The SRT of the sludge can be controlled easily by the volume of sludge wasted per day from the PST. For example, if the minimum volume of sludge wasted per day from the PST is x m^3 (i.e. the maximum thickening possible in the PST), the volume of the fermenter may be $8x$, which will result in the maximum SRT of eight days.

By wasting for longer periods from each PST and thus wasting a larger volume per day, the retention time in the fermenter is reduced and the SRT can be controlled. Even though this may mean that more liquid is wasted, the liquid will be removed and discharged to the BNR plant. Provision should be made for emptying the fermenter at regular intervals in order to block the development of methane organisms, which tend to develop and build up slowly in spite of the adverse conditions. Alternatively, one may provide for regular aeration of the fermenter over short periods to counteract the development of methane organisms. This has been successfully employed at the city of Johannesburg plants.

BATCH SYSTEMS

At the city of Pietermaritzburg in Natal, South Africa, four small redundant digesters are used for batch treatment of the primary sludge. The PST underflow is pumped for two days to one batch unit, and then the pumping is automatically switched to the second, and so forth. At the other end of the batch units, pumps automatically empty the batch units to the thickener on a constant basis, the thickened sludge being discharged to the digesters and the VFA-rich liquid to the BNR plant. This has the advantage of full automation and achieves optimization of the process in an odorless situation.

SOME EXPERIMENTAL INFORMATION

Laboratory experiments at the University of Capetown produced the following results (Wentzel et al., 1988).

Batch Tests

From a number of tests run over periods of nine to nineteen days with initial VSS concentrations ranging from 3000 to 40,000 mg/L, they made the following conclusions:

- The lag phase before acid production started varied from zero to seven days.
- Peak acid concentration was achieved six to nine days after the start of fermentation, which seemingly followed a first-order reaction.
- Maximum yield of about 0.125 kg VFA as COD/kg initial VSS as COD. At three to four days, the yield was 0.075 kg/kg.
- Sixty to 90% of COD filtered through a 0.45 μm filter was VFA in proportions of 1:1:0.08:0.07 acetic:propionic:butyric and valeric acid.
- As batch tests progressed, more long-chain acids seemed to be produced.

Semi-Continuous Series Reactors

Four reactors in series were used, each initially filled with primary sludge. Fresh sludge was added to the first one, which replaced sludge to the second, the third, and the fourth reactors in line. Four tests were operated for twenty days after steady state, using influent VSS concentrations between 37,000 and 57,000 mg/L. The following observations were made:

- There was no lag period. Between 6 and 10% of the VSS (as COD) was solubilized to VFA (as COD).
- About 50 to 70% of the 0.45 μm filtered COD was VFA.
- The acetic:propionic:butyric:valeric acid ratio of the VFA was 1:1:0.3:0.1.
- Of the total VFA generated, 43% was acetic, 41% propionic, and 16% butyric and valeric.
- The series system conferred little advantage over a single reactor with equal retention time.

RETURN OF ORGANIC CARBON

In the design of the BNR plant, cognizance should be taken of the return of organic carbon in the form of BOD or COD to the secondary plant. This is not more than would normally be expected when thickening primary sludge, but in the event that extensive fermentation is required, the return of COD may be substantial and must be reflected in the calculations of the BNR plant. This may be especially true where the COD:TKN ratio of the primary effluent is too low to ensure good denitrification, such as when approaching a value of eight.

MONITORING OF THE PROCESS

The main purpose of acid fermentation is to maximize the production of *a* acetates. However, the analysis for acetate is difficult to perform, needs special equipment, and is not within reach of many of the smaller plant operators. The result is that most of the work is done on a "may-it-hit" basis, which is not entirely unrewarding in many instances. The main item to watch is the performance of the plant in terms of P removal. In one case while experimenting with the a-PST, the primary sludge was recycled for one day, then wasted to the digesters down to a certain sludge density judged entirely by the operator. When the primary tank was taken out of service, the plant soon stopped removing phosphates. When the recycling was restarted, the P removal was soon restored to the previous level. Ideally about 50 mg/L HAc as COD in the primary effluent is the aim. Figure 6.21 illustrates the importance of an activated primary at a South African plant.

It has been demonstrated in practice that it is possible to remove phosphates biologically to levels below 0.1 mg/L as P (after filtration). If this level of P removal is desired, a more scientific method for the monitoring of the prefermenter is required and investment in a gas chromatograph is recommended. With the assistance of good information, a seasonal history of the performance of the plant could be built up. If this is not possible, a check should be run on the performance of the fermenter by a commercial laboratory.

The operator without proper instrumentation can generally judge from the appearance of the fermenter when it is going into the methane fermentation mode because sulfide development will impart a black color to the mixture. A grayish color tending to black may be the best indicator of proper operation while the pH drops to about 5.5. As in so many instances in the field of wastewater treatment, the judgment of the experienced operator is invaluable.

The literature is very confusing on the ideal solids retention time for optimal results. This may be caused by the type of sludge arriving at the plant, as well as by its condition. Some sludge may contain easily degradable solids, others more complex proteins. The solids or the liquid arriving at the plant may be very fresh or may have undergone a long period of prefermentation. The wastewater may be strong or quite weak, the temperatures may range over 20°C. Attempts have been made to produce models, but in the end the matter needs to be solved by experimentation on the plant level. No attempt has yet been made to heat the fermenters, and seasonal variation must then also be taken into account.

SOME PRACTICAL POINTS ABOUT PREFERMENTERS

The less mechanical equipment used the better, especially since the unit must be enclosed for odor control. This creates a very corrosive atmosphere that attacks not only metal but also the concrete used for the cover. Since a high degree of stirring is not desired, external pumps may be considered while the cover may be aluminum or may be in touch with the liquid, which would prevent concrete corrosion. Provision should be made for venting the air space to an odor control unit. The effluent from a prefermenter should be handled with care so as to prevent odor problems.

The recycling of sludge in the fermenter leads to the buildup of a fibrous goo that is very difficult to handle, blocks pumps, and will accumulate in the digesters. While a fine screen in the sludge line to the fermenter will help, the best location for this screen is on the recycle line, since most of the fibrous mass seems to reform in the fermenter. Since the goo is odorous and difficult to handle, a press is required to squeeze the liquid out and return it to the fermenter or the PST. The remainder of the fibers could be wasted to the screenings. Since these fibers mostly end up in the digesters, the cost of removing them is worthwhile because of the additional benefit.

The importance of the fermenter for reliable phosphate removal would require that sufficient redundancy be provided to ensure that the production and supply of VFA is not ever interrupted. A number of fermenters and at least two thickeners would be required. Alternatively, provision could be made for recycling around the primary sedimentation tanks, of which there should be, for obvious reasons, more than one.

There are still a number of unknowns and contradictions in the art of prefermenting. For some reason or other, digester supernatant, even though one can measure what appears to be adequate VFA, seems to inhibit rather than help the process. This has been attributed solely to the presence of hydrogen sulfide but one of the most reliable plants for phosphate removal had an influent with a very high hydrogen sulfide concentration. Rabinowitz et al. (1985) observed that while the SCVFA kept increasing as the solids

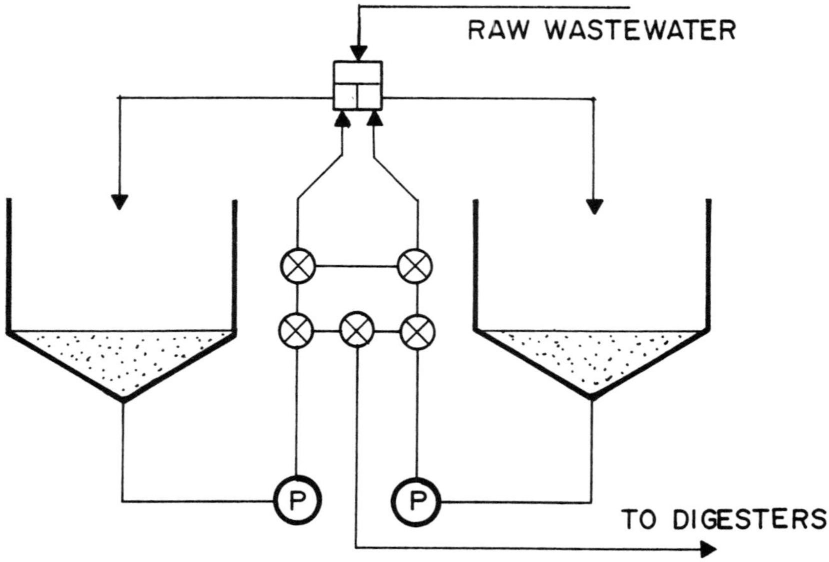

Figure 3.3 Arrangement of two activated primary tanks.

retention time of the fermenter was increased up to ten days, optimal phosphate removal was observed at an SRT in the fermenter of between six and eight days.

While most of the experience from the literature mentions the need for regular removal of the sludge from the prefermenter to avoid the deterioration that comes from either aging, the onset of methane production, or the onset of hydrogen sulfide production, some fermenters work continuously without any ill effects. The thickener/fermenter at Kelowna is a case in point, but in this instance it is not a completely mixed unit or one with recycling. New sludge is added on top and the sludge thickens gradually, squeezing the liquid upwards. Sludge removed from the bottom is not recycled and the process acts more like a batch unit.

When using two primary sedimentation tanks, as shown in Figure 3.3, there are many possible ways in which the plant could be operated. In this case the sludge is recycled back to the influent of each tank separately. The pumps are connected directly to the underflow of each tank and the lines are interconnected by means of two-way valves. Thus, through the pressing of a button on a panel, any of about twelve combinations of permutations is possible. For example, the underflow of tank no. 1 can be pumped to tank no. 2 while the underflow of tank no. 2 is pumped to the digesters or recycled on itself and then pumped to the digesters. It is thus possible to keep fermentation going in one tank while cleaning the sludge out of the other.

REFERENCES

Barnard, J. L. 1974. ''Cut P and N Without Chemicals,'' *Water and Wastes Eng.*, 11(7):33-36.

Barnard, J. L. 1984. ''Activated Primary Tanks for Phosphate Removal,'' *Water S.A.*, 10:121-126.

Barnard, J. L. et al. 1984. ''Design Strategies for Nutrient Removal Plant,'' *Wat. Sci. Tech.*, 11/12:133.

Fuhs, G. W. and M. Chen. 1975. ''Phosphate Removal in the Activated Sludge Process,'' *Microbial Ecology*, 2:119.

Gerber, A., E. S. Mostert, C. T. Winer and R. H. De Villiers. 1986. ''The Effect of Acetate and Other Short Chain Carbon Compounds on the Kinetics of Biological Nutrient Removal,'' *Water SA*, 12:7-12.

Levin, G. V. 1970. U. S. Patent No. 3,654,147. U. S. Patent Office, Washington, D.C.

Milbury, W. F., D. McCauley and G. Hawthorne. 1971. ''Operation of Conventional Activated Sludge for Maximum Phosphorus Removal,'' *J. Wat. Pol. Cont. Fed.*, 43(9):1890-1901.

Rabinowitz, B. and W. K. Oldham. 1985. ''The Use of Primary Sludge Fermentation in the Enhanced Biological Phosphorus Removal Process,'' *Proceedings of University of British Columbia Conference on New Directions and Research in Waste Treatment and Management, Vancouver, Canada.*

Wentzel, M. C., G. A. Ekama, P. L. Dold, R. E. Lowenthal and G. v. R. Marais. 1988 and 1989. ''Enhanced Polyphosphate Organism Cultures in Activated Sludge Systems. Part 1: Enhanced Culture Development,'' *Water SA*, 14:81-92. ''Part 2: Experimental Behavior,'' *Water SA*, 15:71-88. ''Part 3: Kinetic Model,'' *Water SA*, 15:89-102.

CHAPTER 4

Design of Activated Sludge Biological Nutrient Removal Plants

METHODS for the design of activated sludge biological nutrient removal (BNR) processes are presented in this chapter. The material is subdivided in accordance with the complexity and type of BNR system. The considerations and procedures for the design of two-stage BPR processes are presented first. This is followed by methods for design of three-stage, four-stage, and five-stage systems. Emphasis is placed on zone sizing, mixing/aeration requirements, and temperature effects. Methods for the design of Carrousel-type and sequencing batch reactor (SBR) type BNR systems are presented next, and the chapter concludes with a presentation of designer plant case histories.

SECTION I: TWO-STAGE BIOLOGICAL PHOSPHORUS REMOVAL SYSTEMS

The basic principle of biological phosphorus removal (BPR) design is very simple. An anaerobic zone is placed ahead of an aerobic zone in a single-sludge activated sludge system, then the influent wastewater is fed directly into the anaerobic zone. The primary BPR design consideration is that the bacterial biomass must cycle through anaerobic and aerobic environments, with sufficient available organics present in the anaerobic stage for intracellular storage and subsequent utilization in the aerobic stage. The sizing of the anaerobic zone is also relatively simple, although factors such as pH and temperature effects must be considered, but the design considerations required to ensure true anaerobic conditions in the zone during operation are relatively complex. For example, care must be taken to prevent significant entrainment of dissolved oxygen (DO) into the anaerobic mixed liquor during mixing, and the recycle of DO and nitrates with the return sludge needs to be minimized. Resolubilization of phosphorus during clarification and recycle of phosphorus from sludge processing should also be minimized. The number one consideration, however, is that the designer must incorporate sufficient flexibility for the operator to adjust the system to varying conditions if good performance and consistent results are to be achieved.

The treatment objectives of a BPR activated sludge system are to remove organic matter and phosphorus from the mixed liquor to required levels, and to develop a sludge that settles well enough to produce an effluent low in suspended solids after standard clarification. The process configuration is simple and requires only a small change from conventional activated sludge design. The influent wastewater should flow into an anaerobic stage, followed by an aerobic stage, which is followed by clarification and recycle of the settled sludge back to the anaerobic stage where it mixes with the influent wastewater. The basic configuration for a continuous flow system is shown in Figure 4.1.

It is generally assumed that nitrification is neither required nor desired when a two-stage BPR system is selected. Thus, the primary design considerations are: how big should the anaerobic zone be relative to the aerobic zone? is prefermentation required or desired? and what operating SRT should be used? Secondary considerations are: should the anaerobic zone be subdivided by baffles? how much mixing is required? what must be done to minimize DO and NO_X inputs into the anaerobic zone? and how should the secondary clarifiers and waste sludge processing train be designed to minimize solubilization and recycle of phosphorus to the biological reactor?

For the most precise BPR design, the wastewater must be characterized by the method of Marais and his co-workers (Ekama et al., 1982; Siebritz, 1983; Ekama and Marais, 1984a), as recently described by Gibson and Dold (1991). The empirical design method they developed may then be used (Ekama and Marais, 1984b), or the designer may use the modification of the IAWPRC model proposed by Dold (1990). These methods and models are presented in Chapter 7. This level of wastewater characterization is

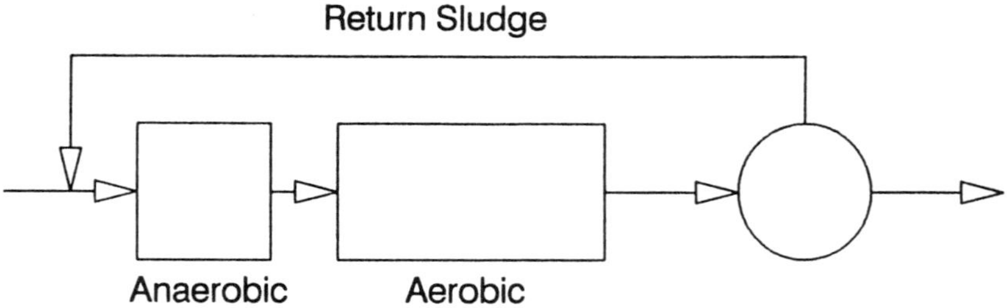

Figure 4.1 Two-stage biological phosphorus removal system.

seldom available to the designer, however, and may be impossible to obtain for many situations. Experience has shown that acceptable BPR designs can be developed without extensive wastewater characterization by incorporating operating flexibility into the treatment system. This latter approach is presented herein.

PHOSPHORUS LIMITATION VERSUS COD LIMITATION

As stated in Chapter 1, the components of the wastewater influent to the anaerobic zone determine the rates of reaction in that zone and the performance of the biological phosphorus removal (BPR) process. If a substantial concentration of readily available organics such as acetic, propionic, and butyric acid are in the influent, the rates of reaction are very fast and very little volume or HRT is needed for the polyP bacteria to accomplish storage of organics with release of orthophosphates. The extent of this reaction will be limited by either the phosphorus available, or the storable organics available. That is to say, all wastewaters are either phosphorus limited or storable organics (COD) limited with respect to the BPR process. If phosphorus is limiting, the available organics will not be completely removed in the anaerobic stage and soluble organics will enter the aerobic stage. If available COD is limiting, the phosphorus removal will be limited and the desired effluent phosphorus concentration may not be achievable without either prefermentation to increase the available organics or supplemental chemical addition to precipitate the excess phosphorus.

It should be recognized that available organics can be limiting for two reasons. They may be limiting because the influent wastewater does not contain enough total organics, as measured by BOD or COD, compared to the influent TP for removal of the phosphorus down to the desired concentration. This can be determined by the COD:TP ratio, or, alternately, the BOD:TP ratio, as discussed in Chapter 1. The available organics may also be limiting because the wastewater organic compounds have not undergone sufficient fermentation to generate the short-chain volatile fatty acids (SCVFAs) needed by the polyP bacteria to form stored organics such as PHB. This can be determined for most municipal wastewaters by comparing the biodegradable COD:TP ratio to the BOD_5:TP ratio. If the biodegradable COD:TP ratio is considerably higher than 40:1 whereas the BOD_5:TP ratio is considerably lower than 20:1, then the wastewater has not undergone substantial fermentation and design of the anaerobic stage should be modified to accomplish the desired level of fermentation. This can be achieved by simply making the anaerobic stage larger, thus giving the fermenting bacteria a longer time to generate the desired SCVFAs. Alternatively, a prefermentation unit can be used.

ANAEROBIC HRT VERSUS ANAEROBIC MASS FRACTION

The effect of increasing the anaerobic zone size, i.e., nominal hydraulic retention time (HRT), is shown in Figure 4.2(a) using data from the VIP pilot plant study (Daigger et al., 1987). Data points from the 72 ML/d Mauldin Road wastewater treatment plant operated in the three-stage BNR mode by the Western Carolina Regional Sewer Authority, Greenville, SC, and the 100 ML/d two-stage BPR module of the Bonnybrook WWTP, Calgary, Canada, are included for comparison. The average effluent TP concentrations of all three systems were less than 1.0 mg/L for the data shown, and the line of best fit has a low slope because the TCOD:TP ratio was greater than 40:1, i.e., the systems were phosphorus rather than COD limited. Even though a line with a slope has been fitted to the data in Figure 4.2(a), the data points indicate that there was no significant difference in phosphorus removal when the anaerobic HRT was increased from 2 to 2.5 hours. It is recommended that the data not be extrapolated beyond 2.5 hours. In fact, research indicates that an increase in influent phosphorus increases the fraction of polyP in the activated sludge when the system is not COD limited, and this results in an increase in the phosphorus removal capacity for the same anaerobic HRT, or anaerobic mass fraction.

When the TCOD:TP ratio of the wastewater influent to the VIP biological process was changed to 21 by the addition of phosphorus, the data point in the upper right hand corner of the graph resulted. Clearly the process was capable of removing considerably more phosphorus than the removals observed before the wastewater phosphorus concentration was increased. While the amount of phosphorus removed by the process increased almost threefold, the effluent SP concentration increased to 1.59 mg/L, and the effluent TP was 1.98 mg/L. This illustrates the basic relationships of BPR. If the available COD:TP ratio of the wastewater influent to the process is high and the plant is properly operated, low effluent phosphorus concentrations will result, the percent phosphorus in the MLVSS will be relatively low, and a change in the anaerobic HRT will have only a small effect on process performance. The results will be considerably different when the available COD:TP ratio is low.

The effect of the anaerobic HRT on phosphorus removal when the TCOD:TP ratio is low is illustrated by Figure 4.2(b). This data was obtained from operation of the 26 ML/d York River, Virginia WWTP of the Hampton Roads Sanitation District (HRSD) when it was being operated in the two-stage, three-stage, and VIP BNR configurations. The data indicates a much steeper slope than that seen in Figure 4.2(a) but also indicates that increasing the HRT above 2 hours provided little additional benefit. Another interesting aspect of the results is that the MLVSS concentrations in the VIP process were only about 50% of those in the A/O and A2/O

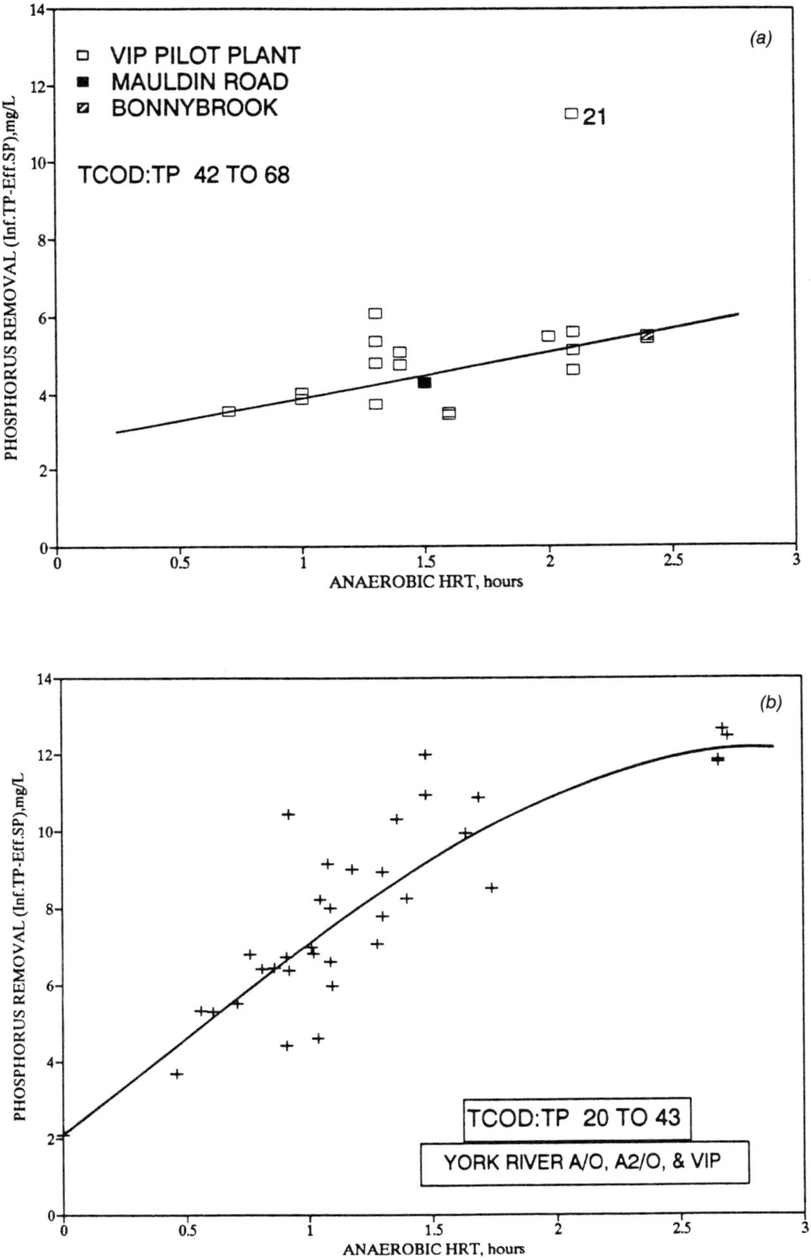

Figure 4.2 Effect of anaerobic HRT on biological phosphorus removal.

processes, yet the data fit reasonably into the same plot. This indicates that the fermentation time in the anaerobic zone was more important than the anaerobic mass fraction, i.e., the anaerobic HRT may be more important than the anaerobic mass fraction for design purposes. It should be noted, however, that the influent TP concentration to the process was considerably less during the VIP demonstration because of a phosphate detergent ban that was implemented between the A2/O experiments and the VIP experiments. Consequently, there was less phosphorus to remove and the process was borderline phosphorus limited during the VIP experiments while it was clearly COD limited during the A/O and A2/O experiments.

Phosphorus removal as a function of the anaerobic mass fraction is shown in Figure 4.3 for the same data set as used for Figure 4.2(b). It can be seen that there is a greater scatter in the data when plotted with mass fraction than when plotted with HRT. Similar results were obtained when the VIP pilot plant data were plotted with anaerobic mass fraction. The anaerobic mass fraction obviously correlates strongly with phosphorus removal at steady state, but it is believed this is because it is inseparably related to the anaerobic HRT and the amount of fermentation that takes place. In other words, the mass of short chain volatile fatty acids (SCVAs) available is the primary factor that determines phosphorus removal. When the SCFAs are available determines the appropriate design HRT and volume of the anaerobic zone, i.e., the anaerobic HRT must be long enough to achieve

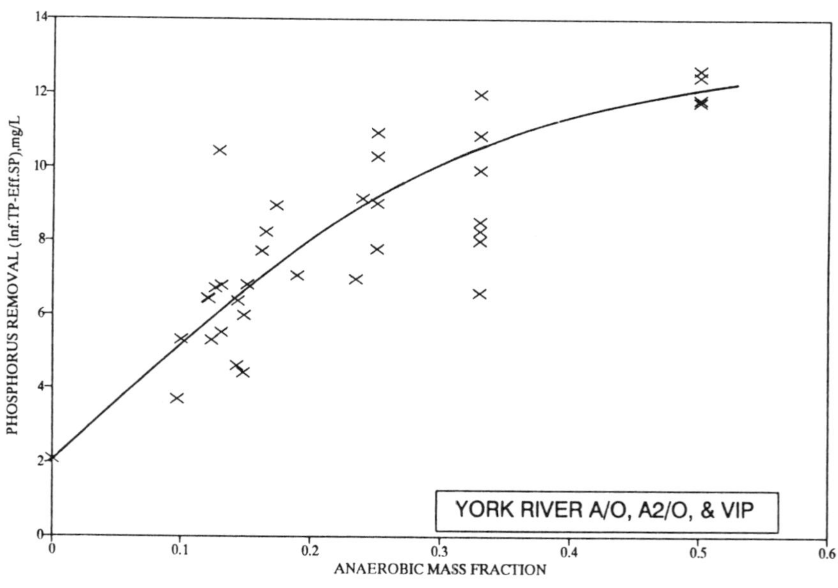

Figure 4.3 Effect of anaerobic mass fraction on biological phosphorus removal.

substantial fermentation if the influent does not contain sufficient quantities of SCFAs. If the influent already contains sufficient SCFAs, then the anaerobic zone only needs to be large enough to store the SCFAs as PHB or similar polymers.

FACTORS AFFECTING PHOSPHORUS REMOVAL

Phosphates can only be removed from the treatment system through the waste sludge. Therefore, phosphorus (P) removal is a function of the percent P in the MLSS, and the operating SRT. Because the growth advantage is given to the polyP bacteria in BPR systems by putting all of the wastewater into the anaerobic zone, they will grow to reach an equilibrium with the limiting factor, i.e., either phosphorus or available COD, for nearly all municipal wastewaters. Thus, either phosphorus or COD will determine the fraction of polyP bacteria in the activated sludge population, and this will determine the percent P in the MLVSS. If COD is limiting, the activated sludge will be dominated by the polyP bacteria and the percent P in the MLVSS will be high. However, the excess P remaining after the COD has been fully utilized will be discharged in the effluent, and the effluent P will tend to be high (greater than 1.5 mg/L). Conversely, if phosphorus is limiting, the soluble P in the effluent will be very low (less than 0.5 mg/L) unless secondary release occurs in either the aerobic zone or the secondary clarifier. The excess COD may be enough to permit significant growth of non-polyP aerobic bacteria, however, and the polyP bacteria may not be dominant. If so, the percent P in the MLVSS will be relatively low, i.e., 4 to 6%.

Paradoxically, then, if the percent P in the MLVSS is high, the effluent P concentration will probably be high also. And, if the percent P is low, the effluent P will be low for a properly designed and operated BPR system. Thus, the available COD to P ratio in the wastewater influent to the anaerobic zone determines both the percent P in the MLVSS and the effluent P concentration, and both parameters are inversely related to the COD:P ratio. These relationships are shown in Figures 4.4 and 4.5.

Because they are both functions of the COD:P ratio, there also tends to be a correlation between MLVSS percent P and effluent soluble phosphorus (SP), but it may be weak at times. For example, during the York River VIP demonstration, the percent P varied from less than 7% to more than 14%, yet the effluent SP monthly averages varied only from 0.49 to 1.16 mg/L when the process was not affected by excessively high flows or incomplete nitrification conditions. This occurred because the system was phosphorus rather than COD limited. In contrast, when the same plant was operated as a two-stage BPR plant and with an excess of phosphorus, there was an

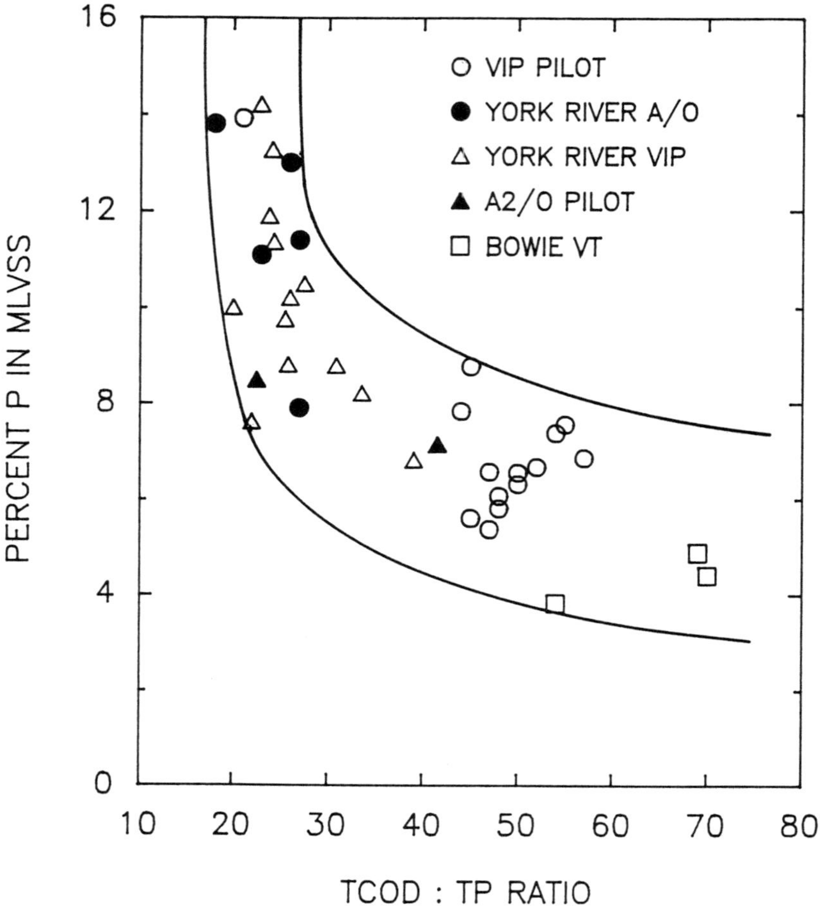

Figure 4.4 Effect of influent TCOD:TP ratio on percent P in MLVSS.

increase of effluent SP from about 2 to almost 4 mg/L when the percent P in the MLVSS increased from 10 to 13%.

Effect of SRT

Because of the changes in the polyP bacterial population as a function of the COD:P ratio, described above, BPR systems are virtually independent of SRT over the range from two to forty days (Barnard, 1988). Nevertheless, it is possible to develop a graph relating percent P in the MLVSS to the observed sludge yield and, therefore, the operating SRT, all as a function of the COD:P ratio, for any given wastewater and BPR system. Figure 4.6 is

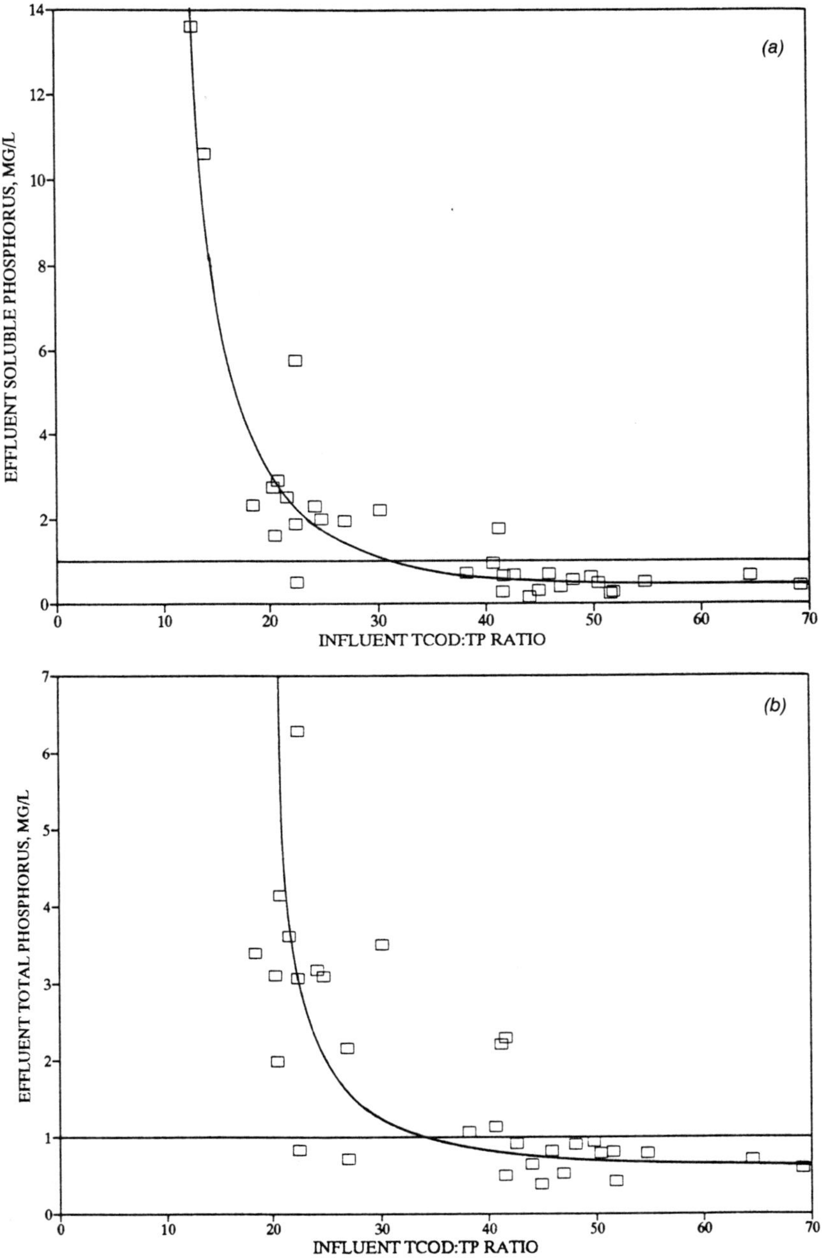

Figure 4.5 Effluent phosphorus as a function of the influent TCOD:TP ratio.

a graph of this type constructed from data obtained during the full-scale demonstration of the UCT/VIP process at the HRSD York River plant, located near Yorktown, Virginia. A theoretical graph similar to Figure 4.6 has been presented by Reddy (1990), and compared with data from the VIP pilot plant experiments (Daigger et al., 1987). Figure 4.7 is a graph of the VIP pilot plant data without the theoretical curves. Comparison of the data from the two plants shows that the actual values of the relationships are not transferable between the two systems.

Reddy (1990) also proposed that at fixed influent organic loading and process operating conditions, a BPR process will have a fixed maximum phosphorus storage capability. This concept is true for instantaneous conditions, but is not important over the normal range of conditions for BPR processes at equilibrium. That is to say, if the phosphorus loading is suddenly increased for a given COD loading condition, the P removal capacity of the polyP population may be exceeded temporarily, but the polyP population will increase with time and the P removal capacity will increase until either available P or available COD becomes limiting. Research performed by Wentzel et al. (1988) showed that the actual maximum P storage capability is 32% of the MLVSS, which is much greater than that needed to remove P from municipal wastewaters. For example, if the percent P of the MLVSS is 8%, and the MLVSS concentration is 3000 mg/L,

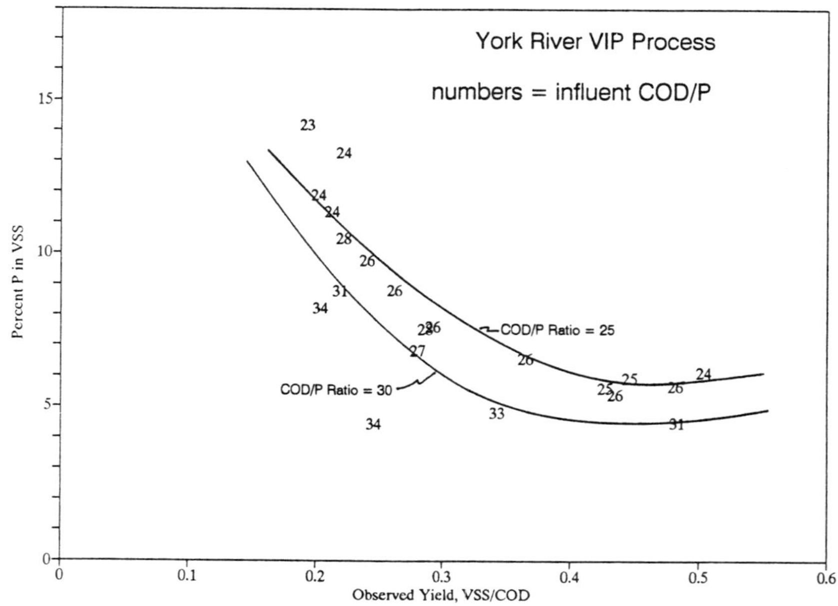

Figure 4.6 Percent P in the VSS as a function of the observed yield.

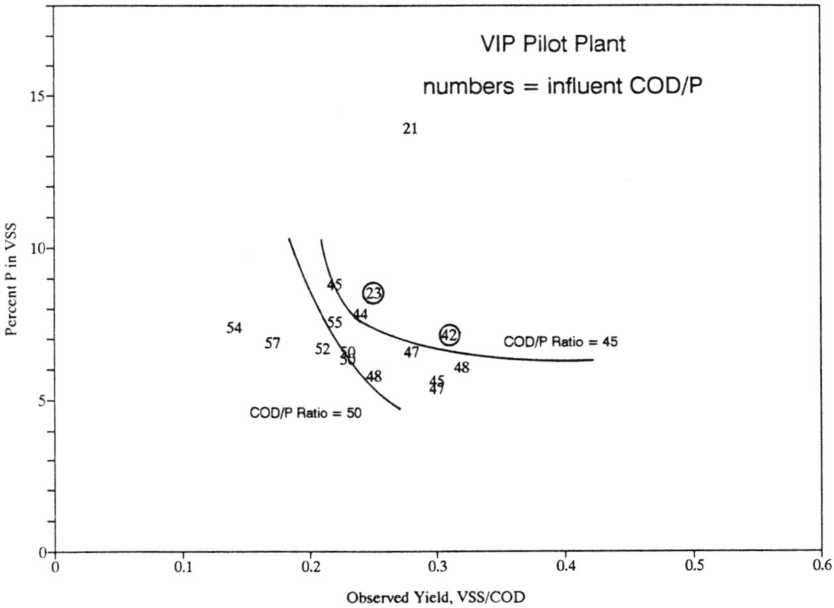

Figure 4.7 Percent P in the VSS as a function of the observed yield.

operation at an SRT of forty days would result in the removal of 6 mg/L TP. This is a typical amount that needs to be removed from U.S.A. wastewaters to produce an effluent SP concentration of less than 1.0 mg/L.

The maximum P storage capacity can be used to explain why BPR performance is independent of the operating SRT. When the SRT is increased (sludge wasting rate is decreased), the percent P increases up to the maximum value, and P removal changes little or not at all. The percent P increases because the fraction of the polyP bacteria in the MLVSS increases, as shown by Wentzel et al. (1988). Therefore, the operating SRT for two-stage BPR processes usually will be selected to prevent or minimize nitrification, typically four to five days. Of course, SRT values this low maximize waste sludge production, and this may not be desired. However, if SRT values high enough to permit partial or complete nitrification are selected, steps to prevent NO_X recycle to the anaerobic zone must be taken. The UCT/VIP configuration without nitrate recycle is a potentially suitable selection, as is any configuration that removes the nitrates from the RAS before it enters the anaerobic zone.

The above indicates that the best way to size the anaerobic zone of a BPR system is to make the size variable. That is, sufficient baffling plus mixing and aeration equipment should be installed to expand or reduce the size of the anaerobic zone in order to obtain the desired effluent SP concentration over the likely range of influent wastewater characteristics and operating conditions. Of course, if prefermentation is used, then the influent wastewater characteristics can be controlled and a very small anaerobic zone can be used, i.e., on the order of 5% of the total activated sludge basin volume. This is because the uptake and storage of SCFAs is a very rapid reaction when not limited by fermentation.

Factors Affecting Substrate Storage

Because of the interrelationships described above, the available COD/available P ratio of the wastewater influent to the biological process is the most important factor affecting the performance of BPR systems, and should be used for design. As Figures 4.4, 4.5, 4.6 and 4.7 show, the total COD:total P ratio approximates the available COD:available P ratio sufficiently for design purposes. The P removal actually obtained, however, is determined by the amount of stored substrate contained by the polyP bacteria when they enter the aerobic zone, and other factors affect the efficiency of storage in the anaerobic zone.

The amount of substrate stored by polyP bacteria in the anaerobic zone is a function of:

(1) The quantity of VFAs in the anaerobic zone
(2) The type(s) of VFAs in the anaerobic zone
(3) The mass of polyP bacteria in the anaerobic zone
(4) The type(s) of polyP bacteria in the anaerobic zone
(5) The quantity of stored phosphorus contained by the polyP bacteria when they enter the anaerobic zone
(6) The mass of the electron acceptors, DO and NO_X, entering the anaerobic zone per unit time
(7) The cations available in the anaerobic zone
(8) The HRT of the anaerobic zone
(9) The temperature of the anaerobic zone mixed liquor
(10) The pH of the anaerobic zone mixed liquor

The last three factors may prevent completion of the organic storage reaction for any given operating conditions and, consequently, limit phosphorus removal.

The reactions in the aerobic zone also determine system phosphorus removal. There must be sufficient DO for the polyP bacteria to completely metabolize the stored organics, and the DO concentration should not limit the rate of oxygen transfer to the cells. There should also be enough DO in the bioreactor effluent to prevent phosphorus release in the secondary clarifier, but the amount contained in the return activated sludge (RAS) should be near zero. DO concentrations ranging from 0.7 mg/L at the influent end of the aerobic zone to 3.0 mg/L at the effluent end appear to be satisfactory for high rate, plug-flow systems (Barnard, 1988). Experience with high internal recycle rate, high SRT oxidation ditch systems has shown that performance is determined by the mass of oxygen transferred rather than the concentration maintained, and it may be necessary to maintain the DO concentration below 0.3 mg/l to prevent recycle of excessive oxygen (Sen et al., 1990a). When this is the case, redox monitoring should be used instead of DO monitoring for aeration control purposes, or some other surrogate parameter such as effluent turbidity or alkalinity can be used (Sen et al., 1990b).

There is also evidence that phosphorus removal in the aerobic zone does not begin until all of the VFAs have been removed from solution (Pattarkine, 1991). Therefore, if the VFAs are not completely assimilated in the anaerobic zone, the size of the aerobic zone should be enlarged to ensure completeness of the phosphorus removal reaction after soluble substrate removal. The breakthrough of readily available organics to the aerobic zone is also likely to stimulate filamentous growth, particularly if the aerobic zone is completely mixed rather than plug-flow. Organic breakthrough is unlikely to be significant when treating North American municipal sewages, and can easily be avoided by including an anoxic zone ahead of the aerobic zone if conditions are such that it would otherwise occur.

Cation Requirements

Results reported by Pattarkine et al. (1991) have shown that the cations, magnesium and potassium, are both essential for BPR, i.e., they must both be available to the polyP bacteria for P uptake and release. The minimum amounts required for each mole of P removed is 0.25 moles of magnesium and 0.23 moles of potassium. Fortunately, the requirements are considerably less than the quantities of these two cations found in most wastewaters, and they are seldom limiting. The requirements should be considered, however, when treating unusual wastewaters, such as those from industries.

Temperature Effects

Biological phosphorus removal processes are relatively insensitive to temperature changes compared to other biological processes. Air Products and Chemicals, Inc. (1981) concluded that the responsible bacteria are psychrophiles which grow better at low temperatures than the typical activated sludge BOD removing bacteria which have their maximum growth rate at 20°C. Moreover, in subsequent laboratory-scale automated fill-and-draw studies conducted over the temperature range of 5 to 15°C (Sell et al., 1981), they found that there was no difference in BPR efficiency at temperatures of 15 and 10°C, but there was a substantial improvement in performance when the temperature was lowered to 5°C. At an *F:M* ratio of 0.25/d based on soluble BOD_5, the P removal rate constant increased from 0.74 to 0.79 L/gVSS-hr, a modest increase, but at an *F:M* of 0.16/d, the increase was from 0.47 to 0.68 L/gVSS-hr, a 50% increase. Similar results were obtained when the experiments were repeated.

Ekama et al. (1984), also based on laboratory-scale research, reported greater BPR at 14°C than at 22°C, which seemingly confirmed the Air Products results. Nevertheless, full-scale and pilot-scale studies have shown that BPR can be affected by temperature, and this can cause reduced removals and higher effluent P concentrations during cold weather. Waltrip (1991) reported a decrease in mg/L removal by BPR with temperature decrease during the York River plant VIP demonstration. The data in his report can be fit by a theta of 1.045 for temperatures below 20°C, i.e., 13 to 20°C, and by a theta of 1.0375 for temperatures from 20 to 27°C. The reference phosphorus removal value at 20°C was 4.8 mg/L and the average SRT for all of the data was 5.8 days. The reference phosphorus removal value at 15°C in the Air Products study was 3.30 mg/L.

Pilot-scale studies by McClintock et al. (1991) over the temperature range from 10 to 20°C also have shown that BPR is affected by temperature, but the temperature coefficient was a function of the operating SRT. A theta of 1.038 was calculated for a fifteen-day SRT whereas a value of 1.137 was determined for a five-day SRT, but the BPR process essentially failed at 10°C and five-day SRT. The reference P removal values at 20°C were somewhat higher than those previously reported at 6.7 and 8.1 mg/L, respectively, at the fifteen- and five-day SRTs. Interestingly, the P removal rate constants, L/g VSS-hr, were smaller at 15°C than at 10°C for both SRTs, although the percent P concentrations in the MLVSS were higher at 15°C.

The seemingly contradictory results of the studies can probably be put into perspective by considering the diversity of the bacteria that can accomplish BPR, and the scale of the studies. Research by Kavanaugh (1991) and several other researchers have shown that the ability to remove excess P is common to a wide variety of bacteria. It is likely that some polyP bacteria are psychrophiles, while others are mesophiles, and still others may be thermophiles. The group and type of bacteria that dominates will depend on the initial culture and the operating conditions. Regardless of the bacteria present, however, the potential to recycle and entrain the electron acceptors, DO and NO_X, into the anaerobic mixed liquor is much greater at low temperatures than at high temperatures in systems that are not sealed. This would reduce the amount of substrate stored by the MLVSS in the anaerobic reactor and, consequently, would reduce phosphorus removal by the system. For this latter reason, a temperature coefficient should be used to estimate BPR at low temperatures during process design. Based on the limited information available, a theta value of 1.045 is recommended. It is also recommended that operation at low SRTs (five days) and low temperatures ($<12°C$) be avoided unless previously successfully demonstrated by pilot plant studies. More research on the effects of temperature on BPR is needed.

Process Effects

Data from the VIP pilot plant study indicated that the UCT/VIP process had a much higher phosphorus storage capacity than the A/O or A2/O processes. The graphs reported by Daigger et al. (1987) are shown in Figure 4.8. However, referral back to the York River data in Figure 4.4 shows that the pilot plant data was anomalous, and that the A/O and A2/O processes are capable of storing as much phosphorus as the UCT/VIP process. Nevertheless, the data in Figure 4.6 shows that the highest effluent P concentrations were observed with the A/O and A2/O configurations. The results actually emphasize the importance of preventing the entrance of NO_X or DO into the anaerobic zone. The UCT/VIP configuration provides more positive protection of the anaerobic zone, and, therefore, less likelihood of reduced performance caused by introduction of excessive aerobic electron acceptors into the zone.

pH Effects

The effects of pH on the performance of BPR were studied by Tracy and Flammino (1985), and their results are shown in Figure 4.9. The graph shows that BPR is optimum between pH 6.6 to 7.4, whereas lower pH values are detrimental. Higher pH values have not been adequately studied.

Fortunately, most municipal BNR processes operate in the 6.6 to 7.4 pH range. The most likely exceptions are when the plant is fully nitrifying without denitrification while treating a low-alkalinity wastewater. The pH can usually be adjusted back to the optimum range by including denitrification in the single sludge system.

Low pH can be a problem when treating industrial wastewaters containing high concentrations of volatile fatty acids such as acetic. Pully (1991) found that the anaerobic zones of both two-stage and three-stage BPR systems were capable of adjusting the pH of a wastewater high in acetic acid from 4.5 to only 5.5, even though COD removal was complete in the aerobic

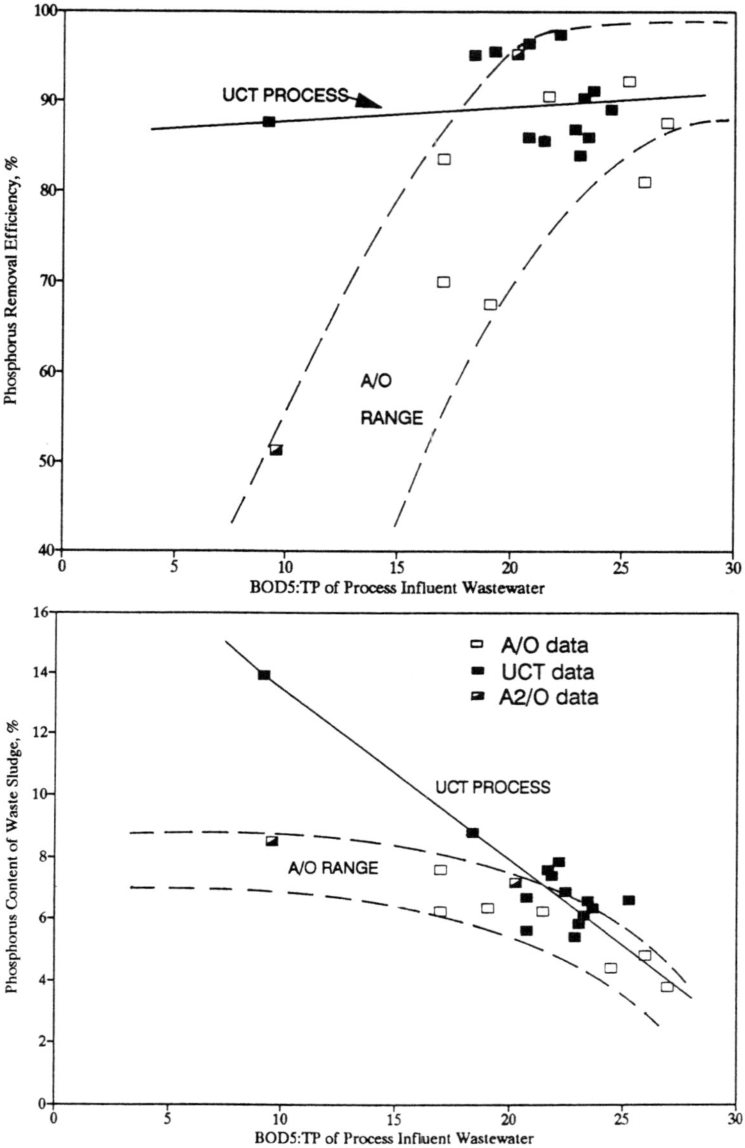

Figure 4.8 Comparison of UCT and A/O process performance.

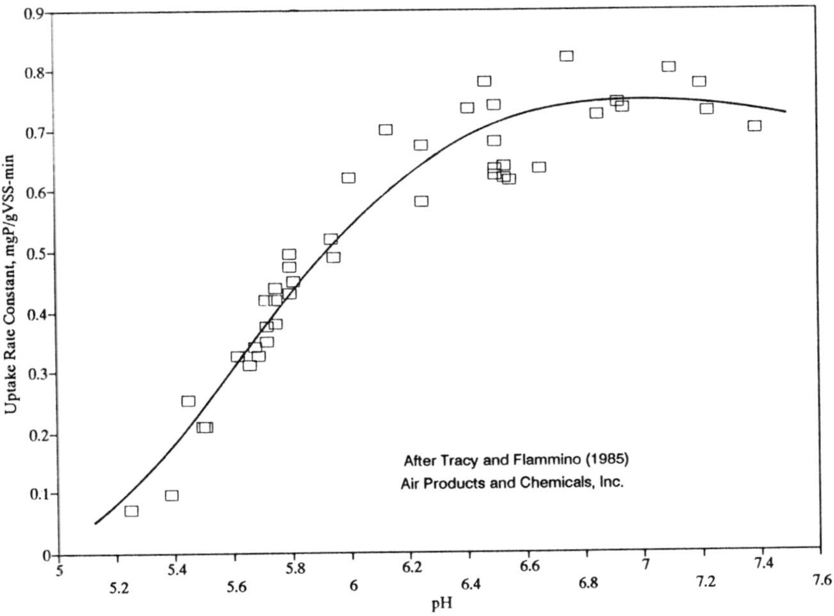

Figure 4.9 Effect of pH on the phosphate-uptake rate constant.

zone. The aerobic zone pH was 7.5, and the RAS rate was equal to the influent flow. Consequently, BPR could not be established without pH adjustment of the wastewater, even though neutralization was not necessary for COD removal.

SIZING THE ANAEROBIC ZONE

Rapid uptake of BOD and near-simultaneous release of phosphate typically occurs in the anaerobic zone of a BNR process, and the rate of the uptake reaction can be determined by measuring the rate of phosphorus release. Two phosphorus release and uptake curves are compared in Figure 4.10. The phosphorus profile for December 1986-January 16, 1987 is a six-week average profile for daily observations, and it shows that the release of phosphorus was virtually complete in section C3 of the anaerobic zone, which was equivalent to an anaerobic mass fraction of 25% and represented an anaerobic HRT of 1.5 hours. These data are from the 24.6 Megaliter per day (ML/d) York River, Virginia plant while it was being operated as a two-stage, i.e., anaerobic-aerobic, BPR system with RAS recycle to the anaerobic zone. The total nominal HRT for the biological process was 6.0 hours and the anaerobic zone was 33% of the total volume. Therefore, the anaerobic mass fraction was also 33% for this configuration and the nominal

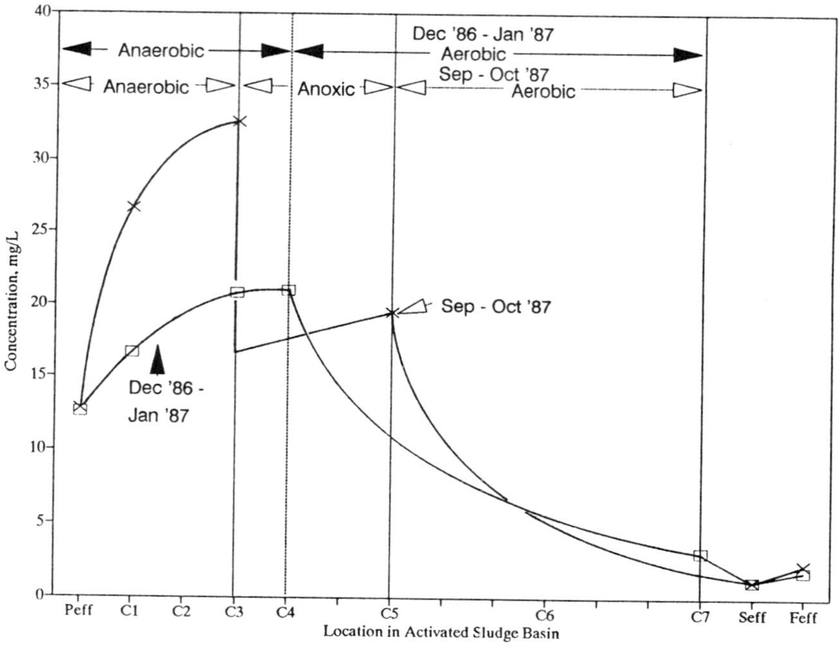

Figure 4.10 Phosphorus profiles at York River during two-stage and three-stage biological phosphorus removal.

HRT for the anaerobic stage was 120 minutes. The process was COD limited with a COD:TP ratio of 22:1, and the COD:BOD_5 ratio of 1.94:1 indicated the wastewater was only partially fermented. Nevertheless, the influent TP concentration of 11.76 mg/L was reduced to an effluent SP concentration of 2.74 mg/L, for a reduction of 9 mg/L. Clearly, the anaerobic HRT was long enough for near-complete fermentation within the zone. Furthermore, phosphorus release, which occurs simultaneously with substrate storage, was complete after only 90 minutes even though extensive fermentation was required in the anaerobic zone. The average raw influent temperature during the profile period was 13.6°C.

Later, during September and October, 1987, more complete fermentation of the influent wastewater, which occurred in the sewers, improved phosphorus removal, but only by 1.58 mg/L or about 18%. The process was being operated in the three-stage (A2/O) configuration and the anaerobic zone was 25% of the total volume. The average COD:TP ratio during this period was 25.6:1, whereas the BOD_5:TP ratio was 23.5:1. The raw influent temperature averaged 20°C over the eight-week term of the observation period. The influent wastewater was so well fermented that nearly 95% of the COD influent to the biological process measured as BOD_5, compared to only 51% during the previously described profile observation period. As a result, a lot more phosphorus was released, and at a faster rate than observed during December and January. Nevertheless, phosphorus removal increased by only a small fraction and the effluent quality did not improve at all. The secondary effluent soluble phosphorus averaged 1.16 mg/L for this profile compared to 1.17 mg/L for the December-January profile. Clearly, the BPR process was not very sensitive to temperature over the range of the two periods compared. The results also indicate that the COD:TP ratio of 40:1 is more reliable than the BOD_5:TP ratio of 20:1 for predicting an effluent TP concentration of less than 1.0 mg/L by BPR treatment, and that total phosphorus removal is not very sensitive to prefermentation for the anaerobic volume/mass fraction and HRT values used at York River.

The effect of the anaerobic sludge mass fraction on phosphorus removal by the York River BNR plant was shown previously by Figure 4.3. From the graph, it appears that phosphorus removal was strongly affected by the anaerobic mass fraction, however, the low mass fraction data were obtained when the potential for phosphorus removal was limited, i.e., the effluent phosphorus concentration was consistently less than 1 mg/L, which limited the phosphorus removal possible. The scatter of the data for the same anaerobic mass fraction indicates that other factors had a greater effect on phosphorus removal than did anaerobic mass fraction, and that greater phosphorus removal would have occurred if it had been possible. This appears to be confirmed by Figure 4.11, which shows that the effluent soluble phosphorus concentration was not a strong function of the anaerobic mass fraction when phosphorus was not limiting. Data from the VIP pilot plant study was also analyzed and it, too, indicated that phosphorus removal was a function of the anaerobic mass fraction. However, all of the sets of data indicate that, while the anaerobic mass fraction does affect performance, other factors appear to affect phosphorus removal even more.

Excellent performance has been obtained at Virginia Tech with a UCT pilot plant system operated with an SRT of fifteen days and a total HRT of 2.37 hours. The anaerobic volume was 31% of the total volume and the system had an anaerobic mass fraction of 16%. Thus, the anaerobic HRT was only 44 minutes and the mass fraction was relatively small, yet the system removed 21 mg/L of phosphorus from an influent concentration of 25 mg/L, producing an effluent of 4 mg/L. The influent COD concentration was 400 mg/L, the effluent COD was 30 mg/L, and the operating temperature was 20°C. The wastewater was typical municipal sewage except that phosphorus had been added to make certain that the system would not be phosphorus limited.

It appears, based on the wide variation of observed results, that the dominant factor that determines the appropriate size of the anaerobic zone is the composition of the wastewater entering the anaerobic zone. If the

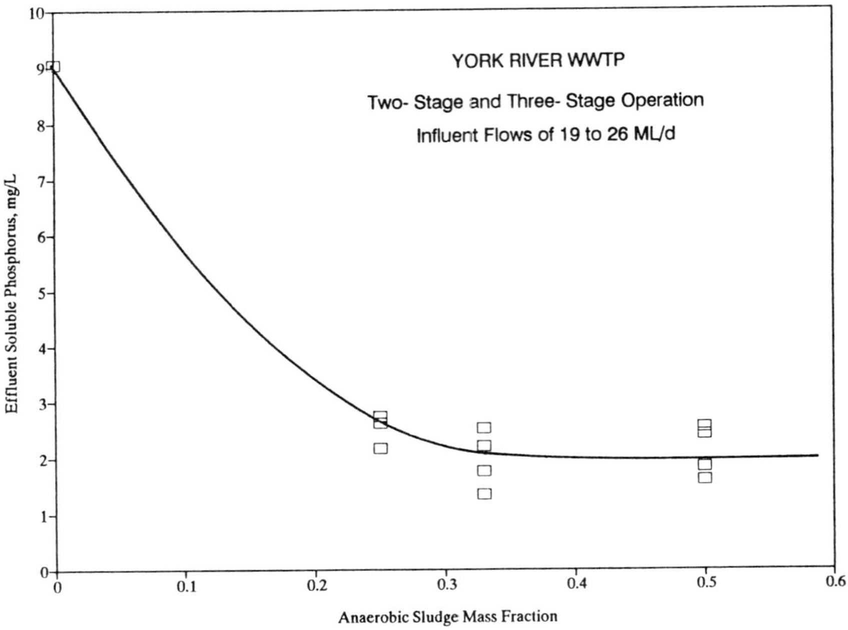

Figure 4.11 Effect of anaerobic sludge mass fraction on effluent soluble phosphorus.

wastewater is highly fermented, i.e., a large fraction of the organics consist of short chain volatile fatty acids (SCVFAs), substrate storage occurs very rapidly and a small anaerobic zone, on the order of 5% of the total basin volume, is sufficient for excess biological phosphorus removal. Such a volume is appropriate for systems that include prefermenters. On the other hand, if the wastewater is very fresh, i.e., contains virtually no SCFAs, then a much larger anaerobic zone should be used to promote fermentation. The data presented here indicate that an anaerobic mass fraction of 15% is sufficient for most municipal wastewater applications, and that 25% is sufficient even if extensive fermentation is required. The data also indicate that the anaerobic HRT is a more important parameter than the anaerobic mass fraction, but that 90 minutes is sufficient for extensive fermentation even at temperatures down to 13.6°C.

Pattarkine (1991) studied the kinetics of phosphorus release in the anaerobic zone of a pilot scale UCT/VIP BNR system treating municipal wastewater with supplemental phosphorus, and in batch experiments. He found that phosphorus release could be described by a first-order equation with time. His data support the selection of a relatively short anaerobic HRT/small anaerobic mass fraction for the treatment of municipal wastewaters.

AERATION REQUIREMENTS

The utilization of an anaerobic zone as the first treatment stage results in a reduction of the oxygen requirements for the treatment of wastewaters by BPR processes (Randall et al., 1992). This phenomenon has been termed "anaerobic stabilization," but the biochemical mechanisms involved are incompletely understood. Data from a pilot scale study designed to investigate the phenomenon are shown in Table 4.1.

Mass balance techniques were used to determine and compare the oxygen requirements of a conventional, fully aerobic activated sludge system and a UCT/VIP system, placed side by side and treating the same wastewater. Both units were operated at three temperatures and two SRTs. The results show that the BPR system always required less oxygen than the conventional system. The oxygen mass balance for the conventional system was within 4.5% of the theoretical amount for all experiments except IB. The oxygen mass balance for the BPR system was always less than the theoretical amount, by an average of 16.7%. Data from several other studies which support the existence of anaerobic stabilization are reported in the paper by Randall et al. (1992) and have been previously published (Randall et al., 1987; Randall et al., 1985; Daigger et al., 1987; Brannan, 1986).

Although a strong theoretical explanation for anaerobic stabilization has not been developed as yet, the existence of the phenomenon is considered to be conclusive enough to reduce the design oxygen requirements of activated sludge systems incorporating BPR processes. It is recommended that the theoretical requirements be reduced by 10% for design purposes.

MIXING REQUIREMENTS

The purpose of mixing in anaerobic and anoxic zones is simply to keep the mixed liquor solids suspended while minimizing surface turbulence to minimize transfer of oxygen from the atmosphere. Therefore, the minimum power input necessary to keep the solids suspended should be used, and the stirring mechanism should be designed to avoid vortexing.

An appropriate formula for determining the mixing requirements of activated sludge as a function of the MLSS concentration has been derived from data presented by Reynolds (1982), and is shown as follows:

$$P/V = 0.00094\ (\mu)^{0.3}\ (\text{MLSS})^{0.298} \qquad (4.1)$$

where

$\mu = 1.0087$ at 20°C
$P/V =$ kilowatts per 1000 liters
$\text{MLSS} =$ mg/L

TABLE 4.1. The Reduction of Oxygen Requirements by Biological Phosphorus Removal.

Experiment	Treatment System	Total MCRT (d)	Temp. (C)	Influent COD (mg/L)	Average MLVSS (mg/L)	O_2 Requirement Predicted (g/d)	O_2 Requirement Observed (g/d)	O_2 Reduction % of COD Stabilized
IA	Conv	5	10	262	1370	23.88	24.95	−4.5
	BPR	5	10	262	1180	20.16	16.47	18.3
IB	Conv	5	15	190	1050	31.93	27.41	14.2
	BPR	5	15	190	1080	19.91	18.23	8.4
IC	Conv	5	20	320	970	43.27	42.41	2.0
	BPR	5	20	320	1420	36.19	32.56	10.0
IIA	Conv	15	10	170	2090	26.12	25.63	1.9
	BPR	15	10	170	2290	21.96	19.42	11.6
IIB	Conv	15	15	228	2100	34.91	34.07	2.4
	BPR	15	15	228	2650	28.71	21.71	24.6
IIC	Conv	15	20	219	1270	33.82	34.20	1.1
	BPR	15	20	219	2300	29.48	21.44	27.3

Conv = Conventional, fully aerobic activated sludge system with a nominal HRT of 8 hours and an RAS of 100% of influent flow.
BPR = Biological nuutrient removal activated sludge system with anaerobic, anoxic and aerobic zones with nominal HRTs of 2/2/8 hours, respectively, and anoxic, nitrate and RAS recycle rates of 100% of influent flow.
Influent flow was 151.2 liters per day of unsettled municipal sewage.

DESIGN STEPS

The following steps are recommended for BPR design using the graphs and techniques of this section:

(1) Select the desired effluent phosphorus concentration.

(2) Determine the COD:P and BOD_5:P ratios of the process influent.

(3) Use Figure 4.5 to determine the effluent phosphorus concentration possible with the COD:P ratio of the process influent.

- If the COD:P ratio is greater than 40:1, the effluent P concentration possible with optimum design and operation will be less than 1.0 mg/L.
- If the COD:P ratio is less than 40:1, an effluent P concentration of less than 1.0 mg/L will be difficult to achieve and maintain. If the required effluent concentration cannot be obtained by BPR with the process influent, then either prefermentation to supplement the influent SCFAs or effluent polishing by chemical precipitation is necessary.

(4) Compare the COD and BOD_5 values to estimate the degree of fermentation of the process influent.

- If the COD:BOD_5 ratio is >2, and the COD:P ratio is >40, use Figure 4.2(a) to select the appropriate anaerobic HRT, based on the phosphorus removal desired, if the removal required is less than 6 mg/L. If the removal required is greater than 6 mg/L, use an anaerobic mass fraction of 25% or greater.
- If the COD:BOD_5 ratio is >2, and the COD:P ratio is <40, use Figure 4.2(b) to select the required anaerobic HRT if it is possible to achieve the required effluent P concentration. If the effluent P cannot be achieved, select either prefermentation or chemical polishing. If chemical polishing is selected, use the anaerobic HRT obtained from Figure 4.2(b). If prefermentation is selected, use an anaerobic mass fraction of 5 to 10%.
- If the COD:BOD_5 ratio is <2, and the COD:P ratio is >40, use a minimum anaerobic HRT of 90 minutes or a minimum anaerobic mass fraction of 15%, whichever is more appropriate for the system being designed.
- If the COD:BOD_5 ratio is <2, and the COD:P ratio is <40, use a minimum anaerobic HRT of 90 minutes or a minimum anaerobic mass fraction of 15%, but use Figure 4.2(b) to estimate whether or not enough phosphorus removal can be obtained with an anaerobic HRT of two hours or less to achieve the desired effluent quality. If sufficient removal cannot be

predicted, then prefermentation or chemical polishing should be used to accomplish BPR.

NOTE: Chemical co-precipitation with either aluminum or iron salts, i.e., addition of the chemical directly to the activated sludge, can be used to obtain effluent phosphorus concentrations as low as 0.7 mg/L simultaneously with BPR. If lower concentrations are required, then effluent filtration and/or tertiary precipitation will probably be required.

(5) Adjust the selected design anaerobic HRT for temperature using a theta coefficient of 1.045 for temperatures less than 20°C, if the system is COD limited, i.e., TCOD:TP < 40. Assume that the design HRT times from Figure 4.2 are for 20°C. No adjustment is necessary if the system is phosphorus limited.

(6) Adjust the temperature-adjusted anaerobic HRT for pH using Figure 4.9.

(7) Determine the design aerobic HRT for the system using phosphorus uptake kinetics.

(8) For a two-stage BPR system, design the activated sludge basin as a plug-flow unit and divide the anaerobic zone into at least three sub-zones using baffles. Equip the last anaerobic sub-zone with both mixing and aeration equipment so that it can be operated as either an anaerobic or an aerobic section. If designing a three-stage BNR system, the three anaerobic sub-zones should be equipped with mixing equipment only, but part or all of the anoxic zone should be equipped for either anoxic or aerobic operation. The purpose is to provide the operator with as much flexibility as possible for unforeseen problems and changing circumstances.

DESIGN EXAMPLE

Given: a municipal flow of 10 ML/d with a COD of 380 and a BOD_5 of 160 mg/L, after primary settling. Other primary effluent concentrations are: 6.0 mg/L TP, 24 mg/L TKN, 140 mg/L alkalinity, and 150 mg/L TSS, with a pH of 7.2. Design a two-stage BPR activated sludge plant to meet effluent concentrations of 20 mg/L BOD_5, 20 mg/L TSS, and 1.0 mg/L TP. Assume no effluent ammonium limitations for the primary design but also develop an alternate design with an effluent ammonium limitation of 2.0 mg/L. The effluent pH must be between 6 and 9. Design for a temperature of 12°C.

(1) The COD:TP ratio of the primary effluent is 380/6.0 = 63. Figure

4.5 predicts that an effluent concentration of less than 1.0 mg/L can be easily accomplished with BPR.

(2) The COD:BOD_5 ratio of 2.4 indicates that the wastewater needs substantial fermentation to maximize BPR. Figure 4.2(a) predicts that an anaerobic HRT of 1.9 hours is needed to remove the required 5.0 mg/L of phosphorus. Select an anaerobic HRT of 2 hours for design. Use an anaerobic mass fraction of 0.33 for a total nominal HRT of 6 hours. Figure 4.3 indicates that the required phosphorus removal could be obtained with a much smaller anaerobic mass fraction, in fact, about 0.1. Therefore, the system could easily be modified for nitrogen removal at a later date without compromising BPR.

(3) The system can be designed with an operating SRT of 4 days to prevent nitrification under most conditions. For a total nominal HRT of 6 hours in the activated sludge basin, the aerobic SRT would be only 4/6 * 4 = 2.7 days. Commonly accepted nitrification kinetics predict that little or no nitrification would occur up to mixed liquor temperatures in excess of 20°C. Appropriate design and operation of the secondary clarifier, or utilization of a small anoxic zone near the basin effluent would prevent the recycle of nitrates formed during partial nitrification back to the anaerobic zone even at higher temperatures. Alternatively, the RAS recycle could be returned to an anoxic zone before entering the anaerobic zone, such as in a UCT-type modification or a preanoxic section without wastewater addition.

(4) An appropriate design MLSS concentration is 4000 mg/L or greater, to insure that the clarifiers are sized sufficiently to provide operational flexibility. Assume a percent MLVSS fraction of 85%, for a design MLVSS concentration of 3400 mg/L. However, operation at an SRT of four days would result in an MLVSS concentration of only 2000 mg/L, based on a COD yield coefficient of 0.4 and an endogenous decay coefficient of 0.1 day^{-1}, adjusted by a temperature coefficient (theta) of 1.05. The SRT at design MLSS concentration and temperature would be eight days, which would provide sufficient operating flexibility to maintain BPR at low temperatures.

NOTE: It has generally been accepted, based on laboratory research results reported by Sell et al. (1981) and Ekama and Marais (1985), that BPR is not detrimentally affected by low temperatures. In fact, both groups reported that BPR was superior at low temperatures compared to high temperatures. This may be true for systems that are phosphorus limited, but full-scale and laboratory results indicate that it is not true for COD limited systems, as shown by the results of Waltrip (1991) and McClintock et al. (1991). However, the system being designed is phosphorus limited, as indicated by the high COD:P ratio, i.e., >40:1, and a temperature adjustment for phosphorus removal is not necessary.

(5) There is an insignificant effect of BPR on alkalinity. In fact, it increases alkalinity slightly, and therefore there is no need for a pH adjustment of any of the design parameters for this example.

(6) Oxygen requirements for this example can be calculated by the following equation:

$$\text{Oxygen required} = Q(1 - 1.45Y)(S_o - S_e) + 1.45K_dXV_a$$

where

V_a = aerobic volume
X = MLVSS concentration
1.45 = the COD equivalence of the MLVSS

Therefore,

$$\text{kgs } O_2/\text{d} = 10(1 - 1.45 * 0.4)(380 - 20) + 1.45 * 0.1 * 4/6 * 2.5 = 2334$$

before the 10% credit for anaerobic COD reduction is taken.

$$\text{Net oxygen required} = 2334 - 0.1(2334) = 2100 \text{ kg/d}$$

(7) The mixing requirements for the anaerobic zone are calculated from Equation (4.1):

$$P/V = 0.00094\,(1.0087)^{0.3}\,(4000)^{0.238} = 0.00678 \text{ kW}/1000 \text{ L} \quad (4.1)$$

$$\text{Total power required} = 0.00678 * 2/6 * 2.5 \times 10^6 = 5650 \text{ kW}$$

Alternate Design

The alternate design would require an HRT sufficient to maintain nitrification. Nitrification would require increased amounts of oxygen and would consume considerable alkalinity. However if the nitrates were recycled to utilize influent BOD for denitrification, the increased oxygen requirements for nitrification would be substantially reduced, and a significant fraction of the alkalinity would be recovered. The details for nitrification and denitrification design are presented later in this chapter.

SECTION II: DESIGN OF THREE-STAGE, FOUR-STAGE, AND FIVE-STAGE BNR SYSTEMS

The actual performances of full-scale plants are much more complex than those of laboratory units, and even the performance of pilot plant units sometimes can make decisions, based purely on laboratory-generated mathematical models, irrelevant. Laboratory results should be used as guidelines for intelligent design, and should not dictate design. Even though a design procedure will be given below, this should not be followed strictly without consideration for special waste characteristics or the characteristics of the particular oxygenation device to be used.

Deciding on a three-, four-, or five-stage unit will depend on a number of factors, some of which are listed below:

- waste characteristics
- effluent requirements
- COD:TKN ratio
- COD:P ratio
- selection of aeration device
- type of clarifiers used
- minimum and maximum operating temperatures
- alkalinity of the wastewater

WASTE CHARACTERISTICS

Waste characteristics differ tremendously from place to place. In countries having a high standard of living and where water is abundant, the wastewaters tend to be weak while the COD:P ratio is low. See Table 4.2 later.

At the other end of the scale, third world countries and those with scarce water resources tend to have stronger wastes with a higher COD:P ratio.

Industrial effluent may have a marked effect on the ability to biologically remove nitrogen and phosphates during treatment. The discharge of fertilizer effluent may overcharge the wastewater with nitrogen, making it difficult to treat, while those of packing plants and breweries may lower the nitrogen content, making BNR that much easier. Slaughterhouse wastewater may contain as much as 30 to 40 mg P/L, but will normally have sufficient carbon to effect the removal of both nitrogen and phosphates, provided there are no toxic compounds present.

Breweries, canneries, and soft drink bottling plants usually discharge wastes that have sufficient short-chain carbon compounds to ensure good P removal, although the organics are not necessarily in the form of short chain volatile fatty acids (SCVFA). However, the organics are easily degradable material that can be converted to SCVFA either in the sewer lines or in the activated primary tanks or even in the anaerobic zone of the BNR plant. On the other hand, edible oil wastes may contain slowly biodegradable material that hinders SCVFA formation, but can be very useful for denitrification in the aerobic basin. Thus, the form of the organics can profoundly influence the choice of process, as will be explained later.

The waste characteristics can be changed by the topography of the urban area and the ambient temperature. Where the topography is flat and temperatures high, as is experienced along much of the south and southeast coasts of the United States, wastewaters are frequently pumped and the force mains, because they develop slime growths, serve as excellent acid fermenters, thus preparing the wastes for BNR plants. With steeper topography, many sewers have self-cleaning velocities and pumping is seldom necessary. These wastes arrive at the plant in a fresh condition and need to be prefermented.

At times, papers are written about the experience of a single plant without mentioning the specific conditions at that plant that favor or do not favor good BioP removal. As a result, these papers may be quite misleading to the uninitiated reader. For example, the area serviced may have flat topography and many lift stations that produce sufficient VFA to require only minimal denitrification and yet allow for good P removal. On the other hand, a waste with slowly degradable organics may need prefermentation but no special anoxic zones, especially when using point source aerators. The slowly degradable waste will exert an oxygen demand throughout the length of the aeration basin. Thus, the anoxic zones formed by point source aerators in the aeration basin will have sufficient carbon for the total denitrification of all the nitrates formed as opposed to a normal domestic wastewater in which the carbon is rapidly removed and only endogenous respiration is available for denitrification. In one such plant, no nitrate or ammonia could be detected at the end of the aeration zone. In another instance, the combination of a weak wastewater and bad clarifier design that required a high rate of RAS recycle was not recognized as the weakness of the specific plant, but was thought to be a problem with the process.

EFFLUENT REQUIREMENTS

The number of stages used in BNR plants will to a large extent be dictated by the effluent requirements. These requirements may entail:

(1) Phosphate removal with no nitrification
(2) Phosphate removal with nitrification but no denitrification
(3) Phosphate removal with only partial denitrification
(4) Phosphate removal with nitrification only in summer
(5) Year-round nitrogen and phosphate removal

The ways in which all these requirements can be accommodated will be discussed below.

COD:P RATIO

The COD:P ratios of sewages vary from one location to the next. Generally, higher ratios are found in developing countries and where the waste is influenced by industrial discharges. Table 4.2 lists the ratios for some locations around the world.

The ratios for cities in North America tend to approach the minimum value of about 36:1, although a lower ratio (27:1) occurs at York River, Virginia. Allowing for a COD:BOD ratio of approximately 1.8:1, this corresponds to a BOD:P ratio of 20:1. Excellent removals at lower values have been reported, but at this ratio the situation gets critical and extra care should be exercised to ensure that the use of the available carbon is optimized. Again, no nitrates should be allowed to compete for the avail-

TABLE 4.2. COD:P Ratios for Various Locations.

Plant	Location	COD:P
Northern Works	Johannesburg	40:1
Goudkoppies	Johannesburg	72:1
Olifantsvlei	Johannesburg	67:1
Mutare	Zimbabwe	91:1
Bulawayo	Zimbabwe	60:1
Palmetto	Florida	36:1
Kelowna, B.C.	Canada	41:1
Patapsco	Maryland	36:1
Brasilia	Brazil	81:1
York River	Virginia	27:1
Lamberts Point	Norfolk, Virginia	49:1

able SCVFA, while the remaining carbon should be used optimally for denitrification. This would require maximum denitrification by endogenous respiration and a selection of stages to accomplish this.

Most important for these situations is that industrial wastes not be pretreated in such a way that the carbon is removed and the nitrogen and phosphates passed on to the BNR plant. For example, the wastewater tariffs may be based on COD alone, prompting a slaughterhouse manager to install a high-rate trickling filter to remove the BOD, thus discharging only the ammonia compounds and phosphates to the sewer and distorting the balance to the point where BNR to the desired effluent concentrations becomes a problem.

COD:TKN RATIO

Most wastes have a COD:TKN ratio on the order of 8 to 10:1 after primary sedimentation. These ratios may also be influenced by pretreatment as indicated above. There are instances where the groundwater contains nitrates that infiltrate into the sewers and upset the ratios. Phosphate reduction is less of a problem at high COD:TKN ratios since most of the nitrates are removed during treatment and do not interfere with the process for P removal. At the lower ratios, greater reliance on endogenous respiration is necessary to optimize utilization of the available carbon. Where the carbon is not sufficient and good N and P removal is required, primary sludge may have to be added after fermentation, or the prefermented primary sludge can be centrifuged with the centrate being returned to the anaerobic basin. A note of caution, however – some polymers used for sludge conditioning are toxic to *Nitrosomonas*.

SELECTION OF AERATION DEVICE

Point source aerators allow for a much higher rate of denitrification in the aeration basin, a fact that will not be reflected in any of the current models for BNR plants. In full-scale plants, nitrogen losses in the aeration basin of between 20% and 100% of the total have been observed. In a five-stage Bardenpho plant with fine bubble aeration, treating mostly domestic sewage, about 20% of the total influent nitrogen was lost in the aeration basin. In a plant with slow speed surface aerators receiving a small fraction of industrial waste, up to 50% of the total influent nitrogen was lost in the aeration basin (Van Huyssteen et al., 1989), while a plant mentioned previously lost all nitrogen either in the waste activated sludge or in the aeration basin. This latter plant was a five-stage Bardenpho plant, but as a result of the total nitrogen removal in the aeration basin no nitrates entered the second anoxic zone. This resulted in secondary release of phosphates that could not be taken up in the final aeration basin, and necessitated the installation of aerators in the second anoxic zone, turning the plant into a three-stage Bardenpho plant.

At the Orange County plant in North Carolina (OWASA), it was possible to get 85% N removal in the plant without formal anoxic zones by cutting air to the jet aerators.

While this latter instance was a deliberate cutting back of air, the former instances happened in spite of aerators fitted with draught tubes running at maximum energy in 4.5 m deep aeration tanks. The degree of denitrification possible can in certain cases obviate the need for formal anoxic basins, thereby influencing the tank configuration.

CLARIFIER SELECTION

Clarifier design is important since it has an influence on the possible RAS rate. The more the RAS rate can be reduced while still maintaining sufficient MLSS, the better for plant performance. The clarifiers should be operated with some sludge blanket, but not so deep that release of phosphates to the liquid will result. When the sludge passes through the critical flux zone, little liquid is exchanged to the upper layers, and if some P release takes place in this layer it is not serious. Simultaneously, the remaining nitrate in the sludge may be reduced to zero before it can reach the anaerobic basin. In one such plant with conical center feed scraped clarifiers, the RAS rate is reduced to a point where none of the about 3 mg/L of nitrate in the effluent appears in the underflow. Excellent P removal is obtained such that the final effluent contains less than 0.1 mg/L after filtration. In another plant, badly designed suction lift clarifiers resulted in a 150% sludge recycle to the anaerobic basin, leading to fresh and nitrate-rich sludge being recycled plus a reduction of the actual retention time in the anaerobic basin.

Clarifier selection will be further discussed in Chapter 5.

SELECTION OF THREE-, FOUR-, OR FIVE-STAGE PLANTS

It is difficult to put plants in categories of stages. As mentioned before, a simple plug-flow aeration basin using surface aerators, turbine aerators, or jet aerators, could be classified as a multistage nitrification/denitrification plant similar to a multi-channel plant such as a Carrousel or, especially, an Orbal plant. The mixed liquor entering the aeration basin is subjected to multiple recycling through aerobic and anaerobic zones before passing through an aeration zone prior to the final clarifiers.

When using fine bubble aeration, formal stages are required because the aeration pattern is more uniform and the plant behavior is closer to that predicted by formal mathematical models.

All the situations discussed below assume that there is sufficient SCVFA in the feed, or the feed or primary sludge is passed through a prefermenter for the production of SCVFAs. With a prefermenter in the form of a thickener, or a fermenter followed by a thickener, a concentrated stream of SCVFAs can be directed to any part of the plant.

Phosphate Removal but No Nitrification

If no nitrification is required and the temperatures are not high, the simple two-stage high-rate Phoredox (AO) process may be sufficient. However, with higher temperatures some nitrate formation cannot be avoided, and the RAS should be subjected to an anoxic stage to rid it of nitrates before mixing it with the influent wastewater. The flow diagram in Figure 4.12 may be used, or a variation of the UCT process (Figure 4.13), to remove sufficient nitrates so as not to upset the phosphorus removal process. The short MCRT and high temperatures will ensure that the endogenous respiration rate is high, and the high concentration of solids in the anoxic stage will lead to a rapid removal of nitrates in the first flow diagram.

Phosphate Removal with Nitrification but No Denitrification

In this case it is possible to follow the Figure 4.12 and 4.13 flow sheets and provide for a nitrification stage to follow. More recent experience indicates better economy by providing for near total denitrification, which saves energy and restores alkalinity.

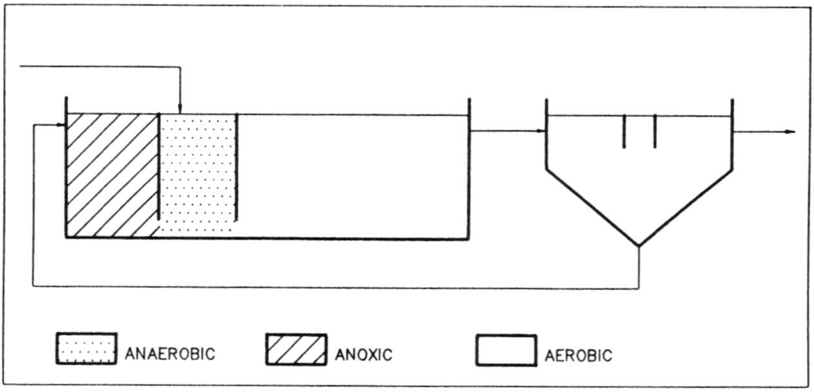

Figure 4.12 High-rate Phoredox with RAS denitrification.

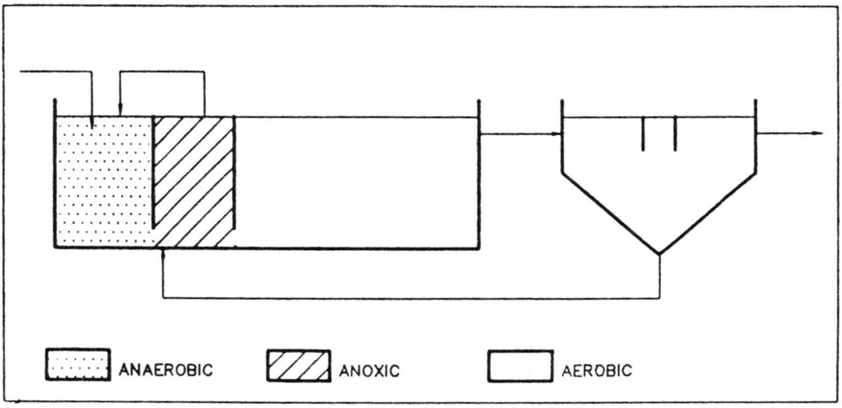

Figure 4.13 High-rate Phoredox with UCT recycle.

The MUCT process (Figure 4.14) is in theory the best flow sheet for this type of operation, moreso when using fine bubble aeration systems. Two mixed liquor recycle systems are required, as opposed to the option of using the three-stage Bardenpho option with nitrate reduction of the RAS by endogenous respiration and the addition of a small fraction of the incoming wastewater to the anoxic section (Figure 4.15).

When using surface aeration, the removal of nitrogen in the aeration basin may vary from 40 to 100%, and the effluent nitrate should be sufficiently low so as not to interfere with the anaerobic contact zone. By designing the aeration basin in a plug-flow configuration, some anoxic zones in the first section of the plant are inevitable, while the endogenous oxygen demand at

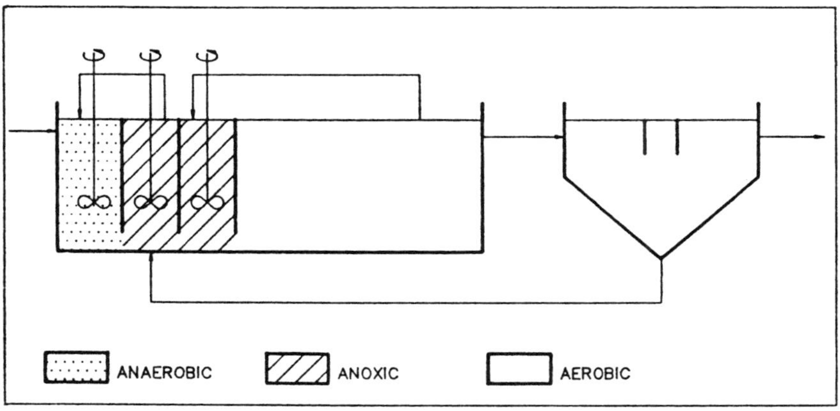

Figure 4.14 MUCT process.

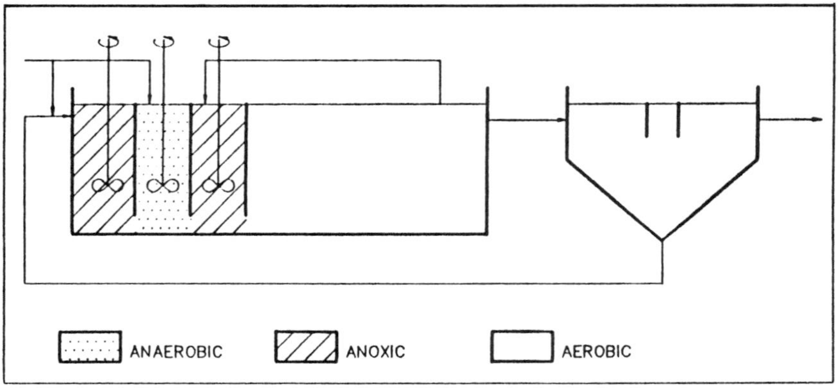

Figure 4.15 Three-stage Bardenpho with RAS denitrification.

the end of the aeration basin will be low and the D.O. level will increase to allow for good nitrification and phosphate uptake. Nitrates not removed in the aeration basin will be recycled to the anoxic zone. Thus, good overall nitrogen removal is achieved economically.

Phosphate Removal with Nitrification Only in Summer

When year-round nitrification is required, a generous safety factor is necessary to ensure nitrification at the coldest expected mixed liquor temperatures, allowing for some mishap. When nitrification in summer only is required, the safety factor can be reduced considerably while designing for summer temperatures only. This could lead to a great reduction in the overall cost of the plant, as will be shown later. The advantage is that the MCRT is short, the active fraction of the mixed liquor is higher, and the resultant denitrification rate is higher. This leads to shorter anoxic retention times and an overall saving in capital and operating costs.

The flow sheets in Figures 4.14 and 4.15, above, would be sufficient for achieving the desired results. In winter, when nitrification is lost, the anoxic zone could be aerated to avoid secondary release of phosphates within the zone. The phosphate removal will not suffer because of loss of nitrification. In fact, it typically improves.

High Percentage Nitrogen and Phosphate Removal

For high levels of nitrogen and phosphate removal, the five-stage process is required even though the removal of nitrates in the second anoxic zone is low. When a final effluent TN of less than 3 mg/L is required, the effluent nitrate must not exceed 1.5 mg/L in order to allow for some ammonia and nondegradable organic nitrogen that cannot be removed. Both ammonia removal and nitrate removal follow Monod kinetics, which makes it difficult to consistently meet low effluent standards.

Thus far, most instances of consistently good removals of both N and P have been performed by five-stage Bardenpho type plants having point source aerators. In some plants, initially good performances were recorded with subsequent disappointing results. This may have stemmed from a lack of understanding by the operators as to what was actually happening in the process. Actually, better than predicted removals are usually possible. The Kelowna plant in British Columbia, Canada will be used as an example. When the UCT model was applied to the plant, it predicted that the effluent nitrate concentration should be about 7 mg/L (Barnard et al., 1984). Yet during actual performance, less than 1 mg/L ammonia and nitrate could be maintained over months. This difference between theory and practice typically has been ignored, but if not taken into account one would have to come to a final conclusion that it is impossible to obtain a total N of less than 3 mg/L in spite of the fact that it has been produced consistently by a number of plants as can be seen from the plot in Figure 4.16.

All the plants that achieved such low TN values used either surface or turbine aerators. Surface aerators were mostly used in Carrousel plants forming the aeration portion of five-stage Bardenpho plants. The impact of

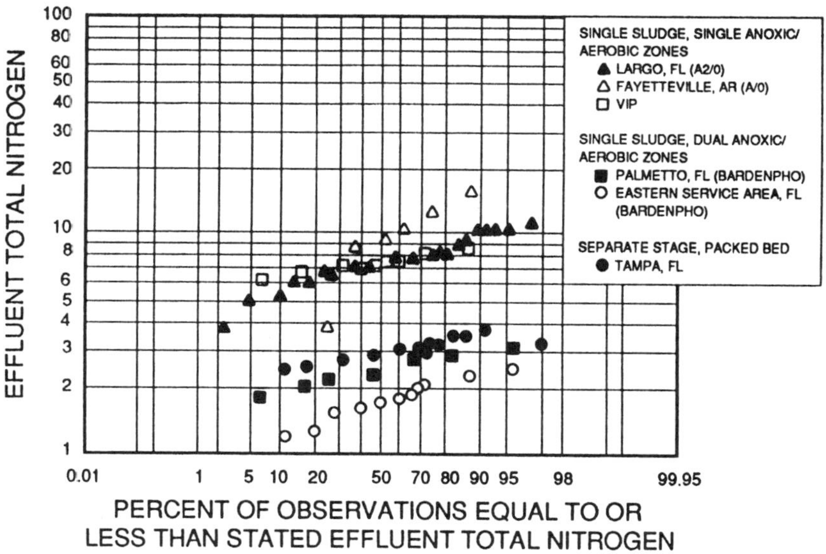

Figure 4.16 Performance of plants re nitrogen removal.

denitrification in the aeration basin will be demonstrated by using an example later in this chapter.

ZONE SIZING

Decisions with respect to zone sizing have to be made against the background of the design considerations outlined above. The simplest way to proceed with the design procedure is to consider option 4, i.e., high percentage nitrogen and phosphorus removal, and proceed from there.

The first assumption is that a total analysis of the wastewater is available and that the operating parameters have been determined. This may have to include judgement on the minimum mixed liquor temperatures likely to be experienced, as well as any toxic interferences for either biological nitrification or phosphate removal.

Information regarding the wastewater analyses would include

- chemical oxygen demand
- influent suspended solids
- settleable solids
- easily degradable COD
- acetate and propionate concentrations
- total influent phosphorus
- total Kjeldahl nitrogen
- total organic nitrogen
- alkalinity
- pH value
- diurnal variation of COD and TKN load
- rate of substance degradation (especially for a large industrial content)
- temperature range of mixed liquor

Zone sizing procedures will follow more or less the following steps:

(1) Determine if the acetate plus propionate concentration in the influent is sufficient for ensuring the required phosphate removal. Roughly five to six times the TP concentration is desired. Depending on choice of sludge handling processes, additional SCVFA may be produced by the primary sludge thickening process.

(2) If the SCVFAs are not sufficient, decide on some form of prefermentation for boosting the concentration. This may be an activated primary, a thickener with recycle, or a batch fermenter plus thickener.

(3) Determine the overall plant size, using any reliable model for the carbon and nitrification system, assuming that the anaerobic zone will

take up about 12% of the reactor capacity, the anoxic zones will take up about 38% and the aeration section will take up the remainder. Adjustments can be made later. This would involve making an estimate of the fraction of COD removed in the primary sedimentation tanks and the fraction returned as VFA to the biological reactor. Select the MCRT that will ensure nitrification at the lowest expected temperature, allowing for a safety factor that reflects the daily peak TKN values. Take into account that nitrification is only possible in the aerated sections of the plant and that some balancing will reduce the peaks and thus the necessary safety factor.

(4) Size the anaerobic basin for the uptake of acetate to a concentration of up to six times the TP in the influent at a rate of 6 mg COD/mg active VSS/d.

(5) Make an estimate of the remaining easily degradable COD on the basis that about 1.06 kg acetate as COD will be taken up by the *Acinetobacter* for every kg P released. Estimate the mass of nitrates that could be removed by the first high rate of denitrification.

(6) Determine the mass of nitrogen loss in the PST plus any feedback from the sludge treatment processes in order to calculate the mass of nitrogen to the BNR process. Estimate the loss of nitrogen in the Waste Activated Sludge (WAS) and in the effluent. Determine the mass of nitrogen that must be nitrified and denitrified.

(7) If possible, determine in a pilot plant the pattern of oxygen utilization in the aeration tank. This pattern may well be influenced by factors such as the *F/M* ratio of the first anaerobic stage. If the *F/M* ratio in the first section exceeds 10, there is likely to be a "selector effect," resulting in a change in the OUR patterns. If there is a consistently high rate of OUR through the aeration basin, the loss of nitrates through denitrification will be high depending on the type of aeration device used. This loss may range from 20% of the total nitrogen for a fine bubble aeration system to 100% removal for a surface aerator when more slowly degradable compounds such as edible oil are present. The degradation rate of the compounds could serve as an indicator as to the percent removal. If unable to determine the rate, allow for a maximum of 20% removal.

(8) Subtract the nitrates lost in the effluent, lost in the aeration basin, and lost through denitrification using the high rate of denitrification adjusted for minimum temperature, in order to determine the removal required through the lower rate and endogenous respiration.

(9) Assume a maximum mixed liquor recycle rate to the anoxic zone of $4Q$, based on an average wastewater flow rate of Q, and an RAS rate of $0.8Q$. Assuming that all the nitrates recycled to the anoxic basin

will be removed in the anoxic basin, calculate from a mass balance the nitrates that must be removed in the second anoxic zone to achieve the desired results. Size the second anoxic zone.

(10) Calculate the retention time required by the first high rate of denitrification, the time required for the removal of the remainder of the nitrates in the first anoxic zone by the second slower rate, and total these to give the retention time of the first anoxic zone.

(11) Check the sizes against the first assumptions and, if necessary, recalculate.

For less stringent nitrogen removal, use the same procedure but without the second anoxic zone. Calculate the nitrates in the effluent. Allow for some denitrification in the final clarifier and allow for some anoxic retention time for the removal of the remaining nitrates in the RAS. This anoxic zone could be ahead of the anaerobic zone with allowance for a small portion of the primary effluent to assist denitrification, as shown in Figure 4.15. Modifications of the UCT process could also be used.

DESIGN EXAMPLE

Assume that the plant is to be designed using the following information given below before primary sedimentation. Consider two situations, the first in which total effluent nitrogen must be less than 5 mg/L and the other where the effluent TN must be below 3 mg/L. Consider the flow sheet in Figure 4.17 with a mixed liquor recycle of aQ and an RAS recycle rate of bQ, where Q is the average influent wastewater flow rate.

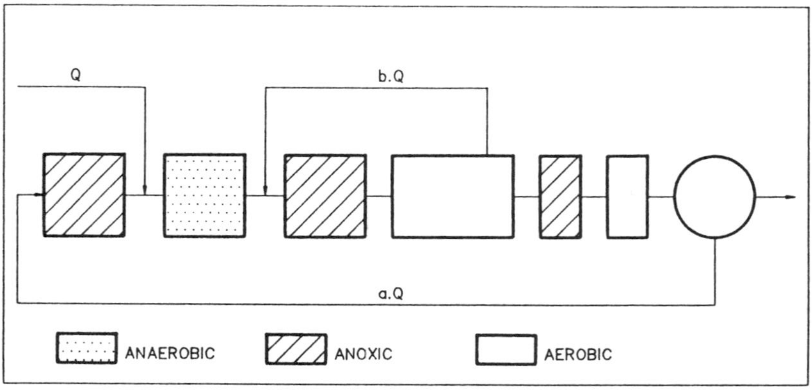

Figure 4.17 Flow diagram for design example.

Flow	$q = 80\ \text{ML/d}$
Influent COD	$S_o = 400\ \text{mg/L}$
Influent suspended solids	$X_o = 200\ \text{mg/L}$
Nondegradable COD	$S_{nd} = 30\ \text{mg/L}$
Total influent nitrogen (TKN)	$N_o = 40\ \text{mg/L}$
Total influent phosphorus (TP)	$\text{P} = 6\ \text{mg/L}$
Effluent suspended solids	$X_e = 10\ \text{mg/L}$
Specific growth rate of *Nitrosomonas*	$\mu_m = 0.35/\text{d}$
Minimum mixed liquor temperature	$T = 12°\text{C}$

Denitrification rates

$$D_1\,(T) = 0.72 * X_a * \theta_1{}^{(T-20)}\ \text{mg N/mg active MLSS/d}$$

$$D_2\,(T) = 0.16 * X_a * \theta_2{}^{(T-20)}\ \text{mg N/mg active MLSS/d}$$

$$D_3\,(T) = 0.072 * X_a * \theta_3{}^{(T-20)}\ \text{mg N/mg active MLSS/d}$$

Fraction of active mass nondegradable	$f = 0.2$
Yield constant for carbonaceous compounds	$Y = 0.48\ \text{g/gCOD}_r$

Endogenous decay rate

$$K_b\,(T) = \frac{0.24 * (1.029)^{(T-20)}}{\text{d}}$$

$$\theta_1 = 1.2$$
$$\theta_2 = 1.03$$
$$\theta_3 = 1.03$$

Calculate the values of the denitrification rates:

$$D_1\,(12) = 0.72 * 1.2^{(12\text{-}20)}$$

$$= 0.167\ \text{mg/mg active MLSS/d} \qquad (4.2)$$

$$D_2\,(12) = 0.160 * 1.03^{(12\text{-}20)}$$

$$= 0.126\ \text{mg/mg active MLSS/d} \qquad (4.3)$$

$$D_3\,(12) = 0.160 * 1.03^{(12\text{-}20)}$$

$$= 0.126\ \text{mg/mg active MLSS/d} \qquad (4.4)$$

Determine the minimum MCRT for nitrification using the nitrification model of the *EPA Manual*:

$$\mu_m = \mu_m * e^{0.098(T-15)} * \frac{\text{DO}}{K_o + \text{DO}} * [1 - 0.833\,(7.2 - \text{pH})] \quad (4.5)$$

where

μ_m = Actual growth rate at specific conditions
DO = Dissolved oxygen in the nitrification stage, mg/L
K_o = half saturation constant = 0.3
pH = pH value of the mixed liquor

Thus

$$\mu_m = 0.35\, e^{\,0.098(12-15)} * \frac{2}{0.3 + 2} * [1 - 0.833\,(7.2 - 7.0)]$$

$$= 0.189/\text{d}$$

The decay constant of the nitrifiers

$$K_d\,(T) = K_d\,(20)\,(1.022)^{(T-20)}$$

$$= 0.05 * 1.022^{(-8)}$$

$$= 0.042/\text{d}$$

The minimum MCRT for nitrification under aerobic conditions

$$\theta_m = \frac{1}{(0.189 - 0.042)}$$

$$= 6.8 \text{ days}$$

Assume that about 40% of the basin volume will be dedicated to anaerobic and anoxic volumes and apply a safety factor of 1.6.

Then, minimum safe MCRT for nitrification

$$\theta_m = 6.8 * 1.6/(1 - 0.4)$$

$$= 18.1 \text{ days}$$

Determine the COD, TKN, and suspended solids removal in the primary sedimentation tanks, taking into account any feedback from sludge treatment units. If actual figures are unavailable, assume the following removals

COD	35%
TKN	10%
SS	70%

Assume a 5% COD feedback from the activated primary or gravity thickener.

The masses of these parameters entering the BNR plant are as follows:

$$\text{COD}: MS_o = 0.7 * 400 * 80 = 22{,}400 \text{ kg/d}$$
$$\text{TKN}: MN_o = 0.9 * 40 * 80 = 2880 \text{ kg/d}$$
$$\text{SS}: MX_o = 0.3 * 200 * 80 = 4800 \text{ kg/d}$$

Since not all the COD is removed in the BNR plant, the effluent COD must be determined from a laboratory scale unit. In this case assume 30 mg/L is not degradable. Thus, the final effluent COD may be of the order of 35 mg/L, which would correspond to less than 5 mg/L BOD.

The COD removed in the BNR plant is

$$MS_{bi} = 22{,}400 - 35*80 = 19{,}600$$

Using the IAWPRC model, calculate the active mass, MX_a

$$MX_a = Y * MS_{bi} * \theta / [1 + K_b(T) * \theta]$$

$$= 0.48 * 19{,}600 * 18.1 / (1 + 0.191 * 18.1) \qquad (4.6)$$

$$X_a = 38{,}205 \text{ kg}$$

$$MX_v = YMS_{bi} * \theta * [1 + f * K_b(T) * \theta] / [1 + K_b(T) * \theta]$$

$$= 0.48 * 19{,}600 * 18.1 * (1 + 0.2 * 0.191 * 18.1)/(1 + 0.191 * 18.1)$$

$$= 64{,}620 \text{ kg/d} \qquad (4.7)$$

$$MX_t = MX_v / 0.83$$

$$= 64{,}620/0.83$$

$$= 77{,}860 \text{ kg/d} \qquad (4.8)$$

Calculate the nitrogen in the sludge assuming that the N content of the volatile solids is 10%. Thus the nitrogen loss, i.e., that which will not need to be denitrified, will be:

In primary sludge $0.10 * 40 * 80 = 360$ kg/d
In waste act. sludge $0.10 * 64{,}620/18.1 = 357$ kg/d
In effluent SS $0.10 * 80 * 10 = 80$ kg/d
As effluent ammonia $0.5 * 80 = 40$ kg/d
Eff. undegradable TKN $0.4 * 80 = 32$ kg/d
Remaining to be nitrified/denitrified $= 2331$ kg/d
In terms of concentration $= 2331/80 = 29.1$ mg/L

Use any convenient model to calculate the sludge recycle (RAS) rate required if

Maximum MLSS $= 3000$ mg/L
SVI $= 120$ mL/g
RAS rate $= 0.6 * Q$

Conservatively, assume that 15% of the total nitrogen nitrified will be lost through simultaneous denitrification in the aeration basin. This will leave $N_n = 24.7$ mg/L to be denitrified in the anoxic zone.

Assume that a mixed liquor recycle rate of $5Q$ is used and that full denitrification takes place in the first anoxic zone. Then from a mass balance around the first anoxic zone, taking the nitrates in the effluent of the nitrification zone as $N_i(1)$,

$$N_n * Q = N_i(1) * (0.6 + 5 + 1) * Q$$

$$N_i(1) = N_n/(0.6 + 5 + 1)$$

$$= 24.7/6.6$$

$$= 3.75 \text{ mg nitrate-N/L}$$

Assume firstly that this is sufficient for discharge under the requirement of less than 5 mg/L TN and that no further second anoxic zone will be needed.

For this design, it would be necessary to provide some means for removal of nitrate from the RAS before discharge into the anaerobic basin.

Calculate the size of the final clarifiers and assume that there will be some sludge blanket formed at the tank bottom.

If the peak flow equals $2.5Q$ and the MLSS equals 3000 mg/L, the sludge flux to the final clarifiers will be:

$$\text{Sludge flux} = (2.5 + 0.6) * (80{,}000/25)\ (\text{m}^3/\text{h}) * 3\ (\text{kg/m}^3)$$

$$= 31{,}000\ \text{kg/h}$$

Limiting sludge flux:

$$= 6\ \text{kg/m}^2/\text{h}$$

Surface area of final clarifiers:

$$= 31{,}000/6$$

$$= 5166\ \text{m}^2$$

Use four clarifiers, each with a diameter of 40 m (5,026 m^2).

If an anoxic sludge blanket of 200 mm formed in each clarifier then:

$$\text{Anoxic vol} = 5062 * 0.2$$

$$= 1005\ \text{m}^3$$

Active mass contained in this volume can be calculated from Equations (4.6) and (4.7):

$$\text{Active mass} = 1005 * 3 * 38.1/77.86$$

$$= 1479\ \text{kg}$$

Nitrates that can be removed through endogenous respiration in the final clarifiers can be calculated using Equation (4.4):

$$= D_3 * MX_a$$

$$= 0.057 * 1479\ \text{kg/day}$$

$$= 84.3\ \text{kg/d}$$

$$= 1.06\ \text{mg N/L}$$

Nitrates remaining to be removed in a pre-anoxic zone:

$$= 3.75 - 1.06$$

$$= 2.69 \text{ mg/L or } 215 \text{ kg/d}$$

Active mass required for endogenous respiration denitrification in the pre-anoxic zone:

$$MX_a = 215/0.057$$

$$X_a = 3772 \text{ kg}$$

$$\text{Required } MX_t = 3772 * 77.86/38.2$$

$$X_t = 7688 \text{ kg}$$

At a recycle rate of $0.6Q$, the nitrate concentration of the underflow of the final clarifiers will be 8 kg/m³.

Thus, volume of pre-anoxic basin will be:

$$= 7688/8$$

$$= 961 \text{ m}^3$$

Determine the volume of the first anoxic basin.

Assume that a P release of five times the P in the influent will be required. Thus a P release of up to 30 mg/L will be required in the anaerobic basin. This will take up about 30 mg/L of the easily degradable COD. If the easily degradable COD is 30% of the influent COD, i.e., 120 mg/L, then 90 mg/L will be left for denitrification using the high-rate form of the nitrification Equation (4.3), thereafter the second rate will be used. Assuming that about 7 mg COD is required for the removal of 1 mg nitrate nitrogen, then:

Mass of nitrogen removed per day by the high rate of denitrification

$$= 90/7 = 12.86 \text{ mg/L or}$$

$$= 1029 \text{ kg/d}$$

To be removed at the second rate

$$= 5 * 80 * 3.75 - 1029$$

$$= 471 \text{ kg/d}$$

The active fraction of the MLSS

$$= 3 * 38.2/77.86$$

$$= 1.47 \text{ kg/m}^3$$

Thus, the high rate of denitrification

$$= 0.167 * 1.47$$

$$= 0.246 \text{ kg/m}^3$$

Volume required for high rate

$$= 5 * 80 * 3.75/0.246$$

$$= 6097 \text{ m}^3$$

Volume for removing 472 kg by the second denitrification rate

$$= 472/(0.126 * 1.47)$$

$$= 2548 \text{ m}^3$$

Total first anoxic volume

$$= 6097 + 2548$$

$$= 8645 \text{ m}^3$$

Total basin volume

$$= MX_t/\text{MLSS}$$

$$= 77{,}860/3$$

$$= 25{,}953 \text{ m}^3$$

The anaerobic basin does not form part of the aerobic-anoxic reactor. Allow 1 h retention in the anaerobic basin. This will give the following volumes.

Pre-anoxic zone = 954
Anaerobic = 3333
First anoxic = 8645

Aerobic = 16,354
Total volume = 29,286 m^3

The anaerobic/anoxic fraction of the reactor is 37%, which is close to the assumed value.

If it is desired that the effluent total nitrogen be reduced to a value of less than 3 mg/L, the exercise could be repeated using a second anoxic zone with endogenous denitrification instead of a pre-anoxic zone. However, do not try to reduce the effluent nitrates below 1.5 mg/L since that would make the second anoxic zone unreasonably large. The TKN of the effluent should be below 1.2 mg/L giving a TN value of less than 3.

The mass of nitrates to be removed in the second anoxic zone will be

$$= 1.6 * 80 * (3.75 - 1.5)$$

$$= 288 \text{ kg/d}$$

Required active mass

$$= 288/0.057$$

$$= 5053 \text{ kg}$$

Required total mass

$$= 5053 * 77.86/38.2$$

$$= 10{,}298 \text{ kg}$$

Volume at 3 kg/m^3

$$= 3433 \text{ m}^3$$

The oxygen requirements could be determined using any reliable model.

MIXING/AERATION DESIGN

The mixing devices in the anaerobic and anoxic zones should impart sufficient energy to keep the mixed liquor in suspension. More energy is needed when designing for a plant without primary sedimentation. With primary sedimentation, about 4 W/m^3 will be sufficient to keep the solids in suspension without too much aeration of the basin contents. However, this energy is not sufficient for proper mixing of the incoming flow with the recycled flow and it is advisable to mix these streams upon entering the zone. When there is no primary sedimentation, there is a tendency for some sand and grit to accumulate in the anaerobic zone. Higher mixing intensity will be counterproductive and consideration should be given to a holding or balancing tank which could also serve for prefermentation and that could occasionally be emptied for cleaning.

Submerged mixers allow for more flexibility, but some bad experiences with famous trade names have swung favor back to the slow-speed vertical shaft mixers. Operators do not favor the regular removal of the submerged units.

The vertical shaft mixers are less flexible, and a breakdown may lead to stagnation and phosphate release. Provisions should be made for a stored standby unit for quick changeover or a standby submerged mixer.

MIXED LIQUOR RECYCLE

In smaller plants, low head centrifugal pumps may be used for this purpose, but in larger plants slow-speed axial flow pumps are essential. With good design the physical pumping head may be as little as 80 mm and all the energy will be consumed by friction losses. Slow-speed axial flow pumps reduce the energy for recycling mixed liquor to a few kW. Also, induced turbulence and the introduction of dissolved oxygen is reduced. This is important since the recycle is usually directed to anoxic or anaerobic zones.

The massive volume of mixed liquor to be recycled dictates the plant layout. If the plant layout could be arranged such that the end of the aeration zone and the start of the anoxic zone could be next to one another, no long large channels would be needed and the energy consumption could be kept to a minimum. This arrangement makes it difficult to measure the recycle flow. Accurate measurement is not that important, but at least at the startup of the plant the designer must check the recycle rate. This could be done by stream velocity meters when required.

AERATION DEVICES

The aeration device selected and the resultant geometry of the aeration basin play an important role in the design and operation of BNR plants.

Aeration devices could be broadly classified as diffused air aeration and point source aeration. Diffused air systems introduce the air over the entire aeration basin while point source aerators introduce air or oxygen at only a set number of points and use some form of mechanical pumping to disperse the oxygen through the tanks.

In BNR plants it is essential to have the ability to control the oxygen input.

This plays a role in the choice of aeration device. Overaeration may lead to the recycling of too much air to the anoxic zone, resulting in reduced denitrification and nitrates in the effluent and the RAS. When, for instance, the plant is underloaded it must still be possible to turn down the oxygen supply when the demand is low, e.g., early in the morning. Under these conditions sludge may settle out in the aeration basin or diffusers may clog, if the aeration system is not carefully designed.

Diffused air aeration could be in the form of

- fine bubble diffused air systems
- medium bubble systems
- course bubble systems

Energy considerations have resulted in the current popularity of fine bubble diffuser systems. New membrane fine bubble systems allow greater flexibility than ceramic domes in that they can be switched off without severe blockages occurring. During low air demand periods, banks of aerators can be switched off to produce a rolling motion to assist in keeping the solids in suspension.

Diffused air aeration systems allow great flexibility in meeting the varying oxygen demands along the length of the aeration basin. While it is desirable to keep the dissolved oxygen in the aeration basin at about 2 mg/L, the profile towards the mixed liquor recycle pumps can be reduced since at that time the heterotrophic organisms are usually in endogenous respiration. By reducing the DO to about 0.5 mg/L at this point, the oxygen recycled to the anoxic basin can be reduced. At the same time, however, the DO in the mixed liquor discharged to the final clarifiers should be raised to prevent release of phosphates. For this reason it is advisable to recycle mixed liquor some distance from the end of the aeration basin, allowing a short distance for high aeration.

The placing of the diffusers over the entire floor of the aeration basin ensures that no anoxic zones will form in the aeration basin. This is also helpful in avoiding low DO sludge bulking.

Point Source Aerators

The following systems could be described as point source aeration systems:

- vertical spindle surface aerators
- horizontal brush aerators
- turbine aerators
- jet aerators

Vertical spindle aerators are used in groups in open aeration basins of

BNR plants or can be used in channel systems of the Carrousel type. In the former application they perform best with draught tubes, which ensure good mixing in the lower levels of the tanks even when the energy input is low. Oxygen is introduced into the upper part of the aeration pocket, and the mixed liquor is then forced out and down to the floor and up the draught tube. By the time a circuit is completed, the mixed liquor may be without or at very low oxygen tension, and this results in the immediate denitrification of any nitrates formed in the more highly aerated part of the aeration basin. The same applies to the application of the surface aerators in the Carrousel system, except that the mixed liquor circulates in horizontal channels. These conditions result in nitrogen removals in the aeration basin varying from 30 to 100%, as will be discussed in the section on case histories. Good nitrification and denitrification can also be observed in Pasveer channel systems, which use brush aerators.

The turbine and jet aerators use separate sources of energy for mixing and aeration and allow the control of oxygen input without interfering with mixing. Excellent control over aeration is possible even to the point where anoxic zones can be created at will. However, the efficiency in terms of oxygen transfer per unit energy is low and high SVI sludges are common.

When designing a plant with surface aeration, cognizance should be taken of the ability of the aeration device for in-tank denitrification, since ignoring this may lead to serious operating problems. The degree of nitrogen removal in the aeration basin, however, is also dependent on the characteristics of the wastewater, as discussed earlier, and while it is difficult to simulate the action of large surface aerators in a pilot plant, some deductions can be made from pilot plant data that will help with the design.

Some information about the likely distribution of oxygen demand along the length of the tank is required for design in order to correctly space the aerators according to demand. While some flexibility is possible, the system is rather inflexible and the right choice should be made at the design stage.

DISSOLVED OXYGEN (DO) CONTROL SYSTEMS

It is virtually impossible to run a BNR plant efficiently without some form of DO control. This may range from very sophisticated probe-feedback systems to using timers to switch blowers or aerators on or off. Generally the weaker the wastewater and the lower the mixed liquor temperatures, the more sophisticated the control system required. Also, fine bubble systems require more intricate controls than surface aeration systems.

It is difficult to control the oxygen supply for a large fine bubble plant from the central blower house. The only viable system is one in which the main feeder line is pressurized, and in which control valves are activated by DO meters in each zone of each section of the aeration basin of each module, thus controlling the air flow to that section. The oxygen demand profile along a tank changes with the hours of the day, and this could be accommodated by the control system mentioned above. Some backup is required for DO probe malfunctioning. This could take the form of historic air flow data that could be fed to the PLC controlling the air supply to the various aeration zones. For this purpose, reliable air flow meters should be installed in the feed pipes to each zone.

DO control for surface aerators may in some ways be easier but can be complicated. One way is to feed a signal from a DO meter to a motorized adjustable weir which changes the immersion depth of the aerators. Another is to switch aerators on and off as required, while more sophisticated means may consist of raising and lowering individual aerators. Problems may be experienced when an aerator close to a DO probe is switched out, causing the flow patterns in the aeration basin to change. This can cause greater fluctuations than the range that normally would have been used for controlling the DO in that region.

Any attempt to change the oxygen input invariably changes the amount of in-basin denitrification, leading to a learning curve for the operator of such plants. However, after some experience, the operator usually develops techniques for optimizing plant operation. One such technique involves the correlation of the DO profile during the day with the ammonia and nitrates in the effluent. If in spite of a reasonable DO level the nitrates seem to rise at periods of low loading, it may be prudent to switch off one aerator to reduce the overall aerobic zone and increase the anoxic zone in this way. In underloaded plants it may be helpful to discharge some primary sludge to the reactor during the small hours of the morning to counter overaeration. This would increase the aeration energy demand somewhat but nitrates that would otherwise be discharged also represent a form of energy, and the energy would be recovered. The backup chemical requirements for phosphorus removal also may be reduced.

SECTION III: DESIGN OF OXIDATION DITCH BNR SYSTEMS

OXIDATION DITCH DESCRIPTION

The first application of oxidation ditch systems began in the 1920s in Europe, and its widespread acceptance followed studies at the Research Institute for Public Health Engineering (TNO) in the Netherlands in the 1950s (U.S.EPA 1978). Hundreds of oxidation ditch plants are now in operation around the world ranging in size from 380 m^3/d to 1.1×10^6 m^3/d (300 mgd) (Stensel et al., 1978). The oxidation ditch system is simply a modification of the activated sludge process that is characterized primarily by its configuration.

Figure 4.18 shows a schematic of an oxidation ditch or channel flow system. The unique features of a traditional oxidation ditch system that distinguish it from conventional activated sludge treatment are (1) a long SRT and hydraulic detention time of about thirty days and twenty-four hours, respectively; (2) a simplified flowsheet lacking primary clarification and anaerobic digestion, and (3) channel flow characteristics. A further modification to reduce cost and simplify operation that has been successfully developed and applied in the 1980s is the incorporation of the final clarifier within the oxidation ditch channel. This has been referred to as an intrachannel clarifier or Boat Clarifier (Stensel et al., 1988).

The long SRT provides a well-stabilized, aerobically digested waste sludge that is often further thickened and placed on drying beds and/or applied to land. Mechanical dewatering has also been used prior to sludge disposal. The long hydraulic detention time enables the system to tolerate

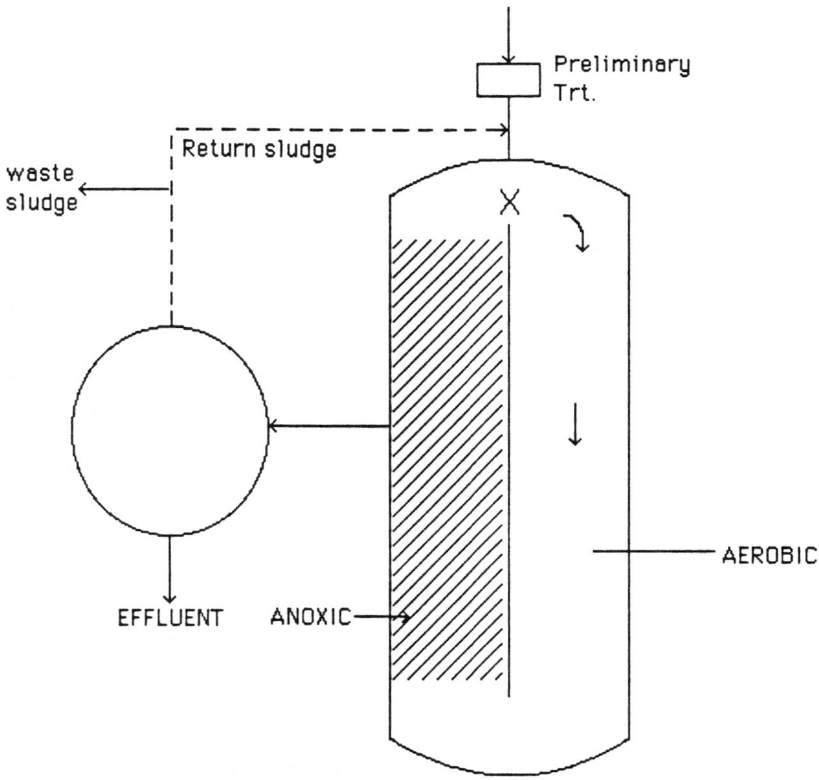

Figure 4.18 Oxidation ditch system with nitrogen removal.

variable loads with minimal operator attention. It should be recognized, however, that design with a long SRT and twenty-four hour HRT represents custom rather than process constraints. Oxidation ditches can be successfully operated with short SRTs and HRTs typical of conventional activated sludge systems, if desired.

The physical arrangement of the oxidation ditch and aerator designs leads to unidirectional mixed liquor flows with channel velocities of 0.25-0.30 m/sec (0.80-1.0 ft/sec) to keep the solids suspended. This type of mixing minimizes the energy required to maintain solids suspension for such a long detention time in the activated sludge system, and the aeration power is usually controlled by oxygen transfer needs and not mixing needs. The other unique feature of the channel flow is the ability to develop anoxic zones after the dissolved oxygen is depleted downstream of the aerator. The amount of nitrogen removal is a function of the channel configuration, DO concentration leaving the aerator, channel velocity, BOD and nitrogen loading, system SRT, and mixed liquor suspended solids concentration. The level of nitrogen removal will depend on the size of the anoxic zone, and more operator attention is needed to carry out DO control to maximize nitrogen removal. A computer-aided control system combining DO control with a daily nitrate nitrogen analysis has been demonstrated by Bahre (1987) to achieve effluent total nitrogen concentrations of less than or equal to 7 mg/L as N.

For very low loaded oxidation ditch systems, nitrification and denitrification can be accomplished without distinctive aerobic-anoxic zones (Rittman and Langeland, 1985; Sen et al., 1990). DO concentrations were maintained between 0.1 and 0.5 mg/L, and 75 to 97% nitrogen removal was measured. Measurements were made at summer temperatures when more optimal nitrification conditions existed. Nitrification rates were likely depressed due to the low DO concentration and depletion of DO within the floc. The depletion of DO within the floc provides conditions for denitrification. Due to the long SRT and low loaded operating conditions, a significant level of nitrogen removal can be accomplished at the lower than optimal nitrification and denitrification rates. Sen et al. pointed out the difficulty of depleting the DO concentration to establish an anoxic zone after an aerobic zone with a DO of 1-2 mg/L, due to low oxygen uptake rates associated with very low loaded systems. Thus, DO control is difficult and an easier option is to operate at reduced aeration with relatively low DO concentrations throughout the oxidation ditch.

Another approach for accomplishing nitrogen removal is to use intermittent aeration. The aerators are turned off and on at selected cycle intervals. The mixed liquor may be allowed to settle during the aeration off period or a propeller mixer can be added to the channel to maintain channel flow. Power requirements for such mixers may be as low as 0.001 kW/m^3.

A number of oxidation ditch configurations have been used as well as a

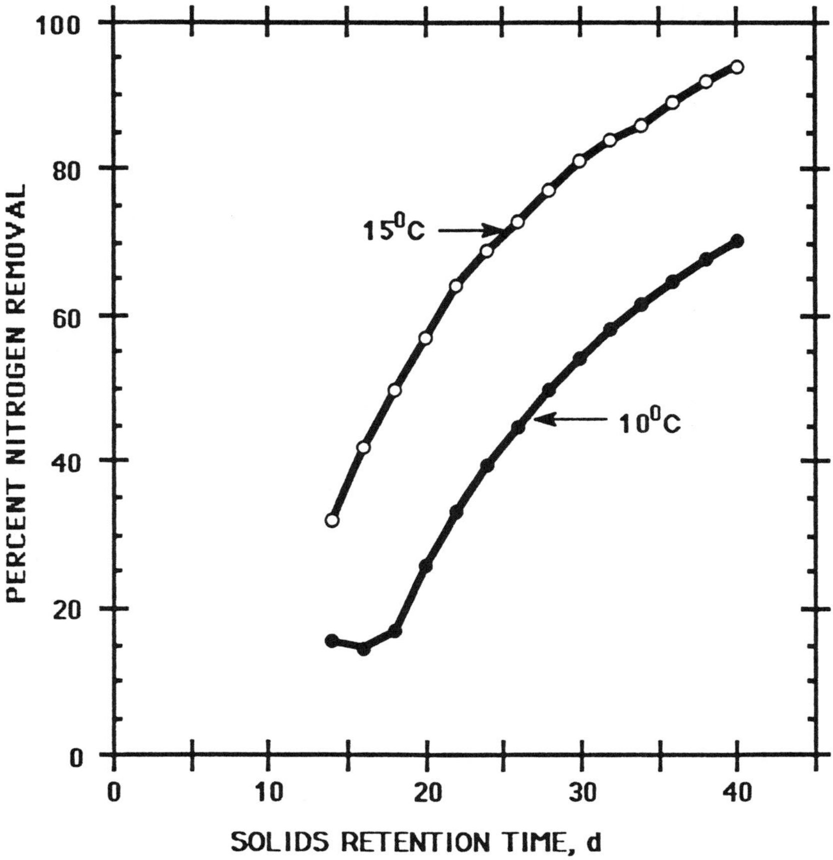

Figure 4.19 Oxidation ditch nitrogen removal design example.

variety of aeration methods. The oval shape shown in Figure 4.19 may be folded to provide two bends at one end and a wide bend at the other end. The channels may form a circular pattern around a clarifier. One proprietary process marketed in the U.S. by Envirex Inc. uses multiple concentric ditches with the influent and mixed liquor moving from outer to inner concentric channels and then to the final clarifier. Aeration discs consisting of perforated plastic are mounted in the channel to provide aeration and channel flow. Some of the concentric channels are operated under anoxic conditions to accomplish denitrification.

Common aerating systems used in the U.S. are the Carrousel-type vertical turbine surface aerator located within a channel turning zone at the end of a baffle wall and the horizontal brush or rotor aerator, located across the channel surface and submerged 10 to 30 cm below the channel surface.

Other aeration methods include jet-aeration, and diffused aeration across a section of the oxidation ditch floor with a vertically mounted internal propeller mixer to produce the channel velocity. Flexible membrane or coarse bubble diffusers have been used in this type of application. Jet aeration also requires a dual mechanical aeration design. A pump is used to pump the mixed liquor through venturi nozzles located across the channel floor to create a unidirectional flow, and a blower is used to pump air into the throat of the venturi to supply oxygen.

NUTRIENT REMOVAL PERFORMANCE

As described above, the oxidation ditch configuration and operation is conducive to nitrogen removal and can be used to obtain limited biological phosphorus removal (Sen et al., 1990). Efficient biological phosphorus removal requires external reactors in addition to the oxidation ditch. Recycle sludge or mixed liquor from the ditch itself can be directed to either an anaerobic zone or to a staged system consisting of anoxic/anaerobic zones for contacting with the influent wastewater. Sen et al. (1990) reported on the application of an anaerobic contact zone ahead of an oxidation ditch system for Bowie, Maryland. Effluent phosphorus concentrations ranged from 0.5-0.7 mg/L. The design considerations presented for biological phosphorus removal apply to the pre-anaerobic or pre-anoxic/anaerobic reactors. The remainder of this section will focus on nitrogen removal performance and design, which relates directly to the oxidation ditch.

Table 4.3 summarizes nitrogen removal performance for oxidation ditch systems treating wastewaters with relatively high influent nitrogen concentrations. Varying levels of ammonia and nitrate nitrogen exist in the effluent as a function of plant operation. When the aeration power is reduced too far to maximize denitrification, insufficient aeration volume may exist so that maximum nitrification does not occur as shown for the NTF and Winterswijk plants.

Nitrogen removal results for a multi-channel oxidation ditch system have been reported (U.S.EPA, 1978) for various operational methods using four concentric ditches. The total hydraulic detention time was fourteen to nineteen hours, the MLSS concentration was 3900 to 4800 mg/L, and the wastewater temperature ranged from 16 to 18°C. Best results were achieved with the ditches operated in an anoxic-aerobic-anoxic-aerobic sequence with return activated sludge ratios of 1.5-2.0. Ammonia oxidation ranged from 95 to 99%, and total nitrogen removal ranged from 69 to 79%. Most of the nitrate reduction was due to the use of the pre-anoxic zone with the high return sludge ratio. The above EPA report also lists a number of plants reporting nitrogen removal with percent total nitrogen ranging from 20 to

TABLE 4.3. Average Nitrogen Removal Performance in Carrousel System.

Plant	Size (mgd)	Type of Wastewater	Influent (mg/L) TKN	Effluent (mg/L) TKN	Effluent (mg/L) NH_3—N	Effluent (mg/L) NO_3—N	Total Nitrogen Removal (%)
Winterswijk	16.5	Domestic and meat packing	59.5	3.9	1.9	8.3	79.5
Lichtenvoorde	3.0	Domestic and tannery	108.3	3.1	0.8	43.5	57.0
Zutphen	6.25	Domestic and poultry	56.5	3.3	0.5	12.4	72.2
NTF, Belgium	3.8	Meat rendering	277.0	6.9	4.0	20.4	90.1
A.C. Lawrence Winchester, N.H.	0.3	Tannery	52.0	8.0	—	4.0	76.9

All plants located in the Netherlands.

85%. This variation is due to the design, the plant BOD and nitrogen loadings, and operational factors.

Table 4.4 summarizes nitrogen removal and operating conditions for two very low-loaded, brush rotor, oxidation ditch systems. The nitrogen removal was observed for low operating DO concentrations at summer temperatures. The Carrollwood Village system had six brush rotors with channel distances between them from 52 to 59 m (170 to 192 ft). At a channel velocity of about 0.3 m/sec (1.0 ft/sec) the travel time between aeration zones was about 3 min. With such a low-loaded system, the oxygen uptake rate can be estimated at about 7 mg/L-hr or only 0.10 mg/L-min.

Thus, in 3 minutes the DO depletion would only be about 0.3 mg/L, and it is necessary to operate the system at a low DO concentration to assure conditions for denitrification. The Frankfort oxidation ditch had a similar distance and detention time between four rotors in each of two ditches. With the unusually long operating SRT, the oxygen respiration rate would also result in very small DO concentration changes between rotors.

NITROGEN REMOVAL DESIGN

The actual nitrogen removal performance for an oxidation ditch system will be highly dependent on operational efforts to create anoxic zones under varying load conditions by DO control or other methods. The nitrogen removal potential can be assessed for a design by determining the necessary nitrification or aerobic volume, the possible anoxic volume, and the specific denitrification rate. Since the oxidation ditch design SRT is usually much greater than the SRT needed for nitrification, extra reactor volume is available for denitrification. The design steps are as follows:

TABLE 4.4. Nitrogen Removal in Low-Loaded Oxidation Ditch Systems at Low DO Concentrations.

	Carrollwood Village	Frankfort, KY
Plant flow rate (m^3/d)	13,250	14,200
Detention time (hrs)	17.2	41
SRT (d)	44	41
MLSS, mg/L	3060	6000
MLVSS, mg/L	2310	—
DO range, mg/L	0.05-0.85	0.0-1.1
Nitrogen removal, percent	97.6	75.7

Source: Rittman and Longeland, 1985.

(1) Establish design flow, influent BOD, TKN, and TSS concentrations and wastewater temperatures.
(2) Select oxidation ditch design SRT.
(3) Determine the amount of ammonia nitrogen oxidized by considering the amount of nitrogen used for synthesis and the ammonium concentration in the effluent. The nitrogen used for synthesis will be a function of the net yield of heterotrophic organisms and their nitrogen concentration.
(4) Use Equation (2.6) to determine the reactor volume (V_T).
(5) Determine the nitrification SRT required.
(6) From the above, determine the volume available for denitrification (V_{DN}) using the following equation:

$$V_{DN} = V_T \left(1 - \frac{\text{SRT}_n}{\text{SRT}}\right) \tag{4.9}$$

where

SRT_n = SRT required for nitrification, d
SRT = System total SRT selected, d

(7) Use Equations (2.47) and (2.50) to determine the specific denitrification rate, SDNR.
(8) Based on V_{DN}, the MLSS concentration used and the SDNR, the possible amount of nitrate reduction and total nitrogen removal can be determined.

The following design example illustrates the procedure. Table 4.5 summarizes the design inputs and assumptions.

The net heterotrophic yield is as follows:

$$Y_{nH} = \frac{Y}{1 + k_d\,\text{SRT}} = \frac{0.6}{1 + .08\,(30)} = 0.18 \text{ g/g} \tag{4.10}$$

The amount of nitrogen oxidized based on influent flow ignoring the small amount used for growth of nitrifiers is:

$$N_{ox} = \text{TKN} - \text{Effl. NH}_4\text{–N} - \text{Fn}\,(Y_{nH})\,(S_o - S)$$

$$= 35 - 0.50 - 0.10\,(0.18)\,(200 - 5)$$

$$N_{ox} = 31 \text{ mg/L} \tag{4.11}$$

$$N_{ox} = (31 \text{ mg/L})\,(5000 \text{ m}^3\text{/d}) \left(\frac{\text{kg}}{10^6 \text{ mg}}\right)\left(\frac{10^3 \text{ L}}{\text{m}^3}\right) = 155 \text{ kg/d}$$

TABLE 4.5. **Oxidation Ditch Nitrogen Removal Design Example Inputs and Assumptions.**

Parameter	Value
Flow, m^3/d	5000
BOD_5, mg/L	200
TKN, mg/L	35
Inert Solids, mg/L	80
Min. Temp. °C	15
MLSS, mg/L	3500
Y_{nH}, gr/gr	0.60
K_2, d^{-1}	0.08
Y_{nN}, gr/gr	0.15
K_o, mg/L	0.50
Nitrif. S.F.	2.0
Design SRT, d	30
Effl. BOD, mg/L	5
Effl. NH_4, mg/L	0.50

A system SRT of thirty days is assumed and from Equation (2.6) the total required volume in m^3 is

$$V_T = [0.18\,(200 - 5) + 80 + 0.15\,(31)]\,(5000\text{ m}^3)\,(30)/3500$$

$$V_T = 5132\text{ m}^3$$

$$\text{Detention Time} = \frac{5132\text{ m}^3}{5000\text{ m}^3/\text{d}} \times \frac{24\text{ hr}}{\text{d}} = 24.6\text{ hrs}$$

For nitrification the maximum specific growth rate at 15°C is calculated from the Hultman equation in Table 2.4.

$$\mu_{n,\max} = 0.34\text{ d}^{-1}$$

From Equation (2.12), μ_n is calculated assuming an average nitrification DO concentration of 1.5 mg/L.

$$\mu_n = \frac{(\mu_{n,\max})N}{K_n + N}\left(\frac{\text{DO}}{K_o + \text{DO}}\right)$$

$$\mu_n = \left(\frac{(0.34)\,(0.50)}{(0.5) + 0.50}\right)\left(\frac{1.5}{0.50 + 1.5}\right) = 0.1275\text{ d}^{-1}$$

From Equations (2.5) and (2.7) the required nitrification SRT_n is:

$$\text{SRT}_n = 1.5\left(\frac{1}{0.1275}\right) = 11.8\text{ days} \tag{4.12}$$

From Equation (4.9), the denitrification volume is then:

$$V_{DN} = 5132\text{ m}^3\left(1 - \frac{11.8}{30.0}\right) = 3113\text{ m}^3$$

From Equation (2.48), the value for A_n in Equation (2.47) is:

$$A_n = 1.5 - 1.42\,Y + \frac{1.42\,K_d\,Y\,\text{SRT}}{1 + K_d\,\text{SRT}}$$

$$= 1.5 - 1.42\,(0.6) + \frac{1.42\,(0.8)\,(0.60)\,(30)}{1 + 0.08\,(30)}$$

$$A_n = 1.25\text{ gr O}_2/\text{gr BOD removed}$$

$$\text{SDNR} = \frac{F}{2.86}\left(\frac{A_n}{Y_{nT}}\right)\left(\frac{1}{\text{SRT}}\right)$$

$$Y_{nT} = Y_{nH} + \frac{X_I}{S_o} = 0.18 + \frac{80}{200} = 0.58\,\frac{\text{gr TSS}}{\text{gr BOD}}$$

$$\text{SDNR} = \frac{0.5}{2.86}\left(\frac{1.25}{0.58}\right)\left(\frac{1}{30}\right)$$

$$\text{SDNR} = 0.012\text{ gr NO}_3\text{–N/gr MLSS-d}$$

$$\text{SDNR, }15°\text{C} = 1.02^{(15-20)}\,(0.012) = 0.011\text{ g/g-d}$$

The amount of nitrate reduced (N_R) in the anoxic zone is as follows:

$$N_R\text{, kg/d} = (\text{SDNR})\,(\text{MLSS, g/L})\,(V_{DN}\text{, m}^3)\left(\frac{10^3}{\text{m}^3}\right)\left(\frac{\text{kg}}{10^3\text{ g}}\right)$$

$$= (0.011\text{ gr/gr-d})\,(3.5\text{ g/L})\,(3.113\text{ m}^3) \tag{4.13}$$

$$N_R = 119.9\text{ kg/d or }23.9\text{ mg/L}$$

$$\text{Effluent NO}_3\text{–N} = 31 - 23.9 = 7.1\text{ mg/L}$$

$$\text{Percent N Removal} = \left(\frac{35 - 0.5 - 7.1}{35}\right)(100) = 78\%$$

Decreasing the nitrification safety factor and allowing more anoxic volume could increase the denitrification volume and percent nitrogen removal. Increasing the oxidation ditch design SRT would also increase the amount of anoxic volume and nitrogen removal. Figure 4.19 illustrates the percent nitrogen removal as a function of the system SRT for both 10°C and 15°C wastewater temperatures using the preceding design analysis with a nitrification safety factor of 1.5. Once the volume for nitrification is satisfied, additional volume can be used for nitrate reduction so that increasing the system SRT can result in more nitrogen removal. This design method determines the nitrogen removal potential and operational efforts needed to accomplish it.

SECTION IV: DESIGN OF SEQUENCING BATCH REACTOR BNR SYSTEMS

In a Sequencing Batch Reactor (SBR) system a single tank is used for activated sludge aeration and solids clarification. In addition, reaction conditions and aeration times can be varied during a cycle period. Figure 4.20 illustrates a possible SBR cycle. There is a time period required to fill the tank at the beginning of each cycle. As Step 2 shows, mixing may be used during the fill, and anoxic or anaerobic conditions may prevail during the mix period. Dennis and Irvine (1979) showed that the larger the portion of fill time employed for mixing, the lower was the resultant mixed liquor SVI. After a selected aeration period, settling occurs and supernatant is withdrawn from the tank (Figure 4.20). A modification to the above described system with its discrete feed period is a continuously fed system that is intermittently decanted. A baffle arrangement is used to prevent the feed from mixing with the settled supernatant during withdrawal (Goronsky, 1979).

Although the application of batch fed reactors for municipal treatment goes back to the early 1900s, the more general acceptance and application of SBR systems since the 1970s has been spurred on by the availability of microprocessor controllers and improved hardware. These include pneumatic, solenoid and motorized valves, level sensors, flow meters, and automatic timers (Irvine et al., 1983). A number of studies on SBR design and performance are well summarized in a technology assessment of SBRs by Irvine (1981).

Process conditions that promote biological nutrient removal in continuous flow systems can conceivably be simulated in an SBR operation. The wastewater influent can be mixed with the remaining sludge during and after the fill period to provide the anaerobic contact period shown necessary for biological phosphorus removal. Although there have been minimal

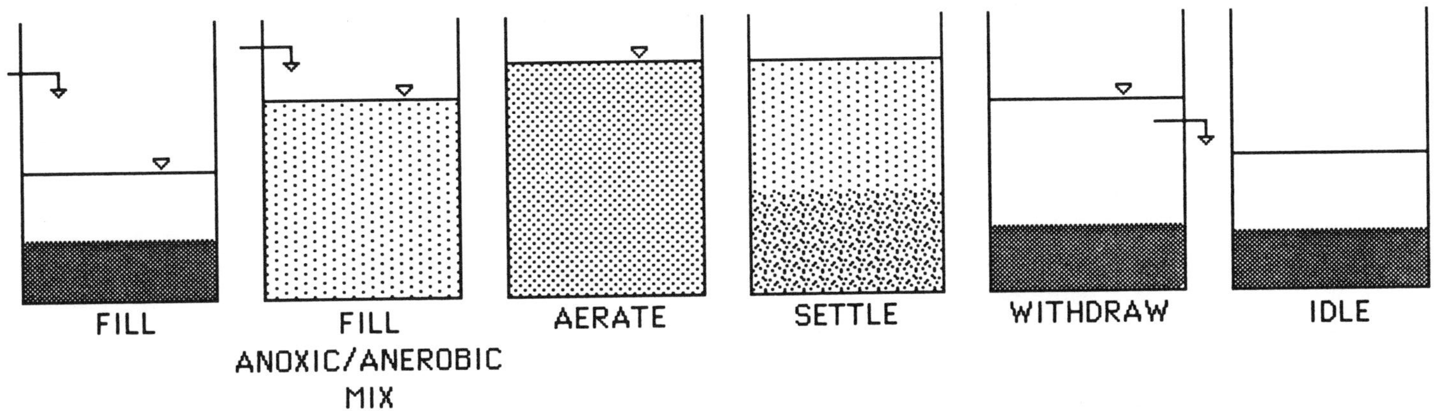

Figure 4.20 Sequencing batch reactor operating steps.

efforts to employ full-scale SBR systems for biological phosphorus removal, bench-scale studies have demonstrated the ability of SBR anaerobic/aerobic designs to accomplish biological phosphorus removal (Manning and Irvine, 1985). A two-hour mixed unaerated fill period, followed by four hours of aeration, resulted in an effluent phosphorus concentration of 0.5 mg/L when treating a synthetic feed with an influent phosphorus concentration of 13 mg/L. The absence of oxidized nitrogen was an important factor for accomplishing biological phosphorus removal.

Nitrogen removal by providing a post-anoxic period for denitrification after aeration is an obvious operating strategy. Under this condition the denitrification rate is controlled by the endogenous respiration activity of the mixed liquor, as noted by Argaman (1986) and Goronsky (1986). Argaman operated a bench-scale continuous flow intermittently decanted reactor to observe nitrification and denitrification rates. The specific denitrification rate was 0.014 g/g-d at SRTs higher than twenty days, and values ranged from 0.022 to 0.034 for much lower SRTs, typical of endogenous denitrification rates. Alleman and Irvine (1980) showed that the denitrification rate is affected by cellular storage products produced during the initial aeration phase in a laboratory SBR study. As the storage products decreased, the denitrification rate decreased. Jones et al. (1990) also showed that denitrification rates could be improved by maximizing substrate storage in laboratory studies and that rates were lower with longer SRTs and greater aeration times.

Although endogenous denitrification can be easily achieved, producing significant nitrogen removal in a pre-anoxic zone as found in conventional systems with internal recycle is much more difficult. Due to batch operation, the concurrent availability of influent organic carbon and oxidized nitrogen is limited. An attempt to approach this is the use of alternating periods of high and low DO concentration during the fill period to alternately promote ammonia oxidation and denitrification with the presence of influent carbon (Irvine, 1983). In a bench-scale study, DO concentrations were varied between 0.2 to 6.0 mg/L every hour over a ten-hour fill period. The oxidized nitrogen during the fill period varied from 0.1 to 5.0 mg/L, and the effluent oxidized nitrogen after an eight-hour post-anoxic period was less than 0.5 mg/L. This method requires longer cycle times with fewer cycles per day and results in a lower loaded system and larger SBR tank volume. The rapid DO changes required during the fill period might be difficult to attain with conventional aeration equipment.

If a low effluent nitrate nitrogen concentration is not required, then a post-anoxic phase would not be necessary. In this case a significant amount of nitrate may be removed in a pre-anoxic period during fill with the influent organic carbon driving the denitrification reaction. The ratio of fill volume to total SBR volume would determine the nitrogen removal level possible using this pre-anoxic treatment method. The smaller the ratio of the fill volume to the total volume, the greater would be the nitrogen removal, assuming all the oxidized nitrogen is reduced before aeration begins.

Full-scale plant nitrogen removal performance was reported for the Culver, Indiana SBR EPA demonstration plant for August to December 1981 (Irvine, 1983). Primary clarification was used and two existing aeration tanks, each with a volume of 380 m^3 (0.1 mgd), were converted to SBR tanks to treat an average flow of 1340 m^3/d (0.35 mgd). Although the operating conditions were not well defined, aeration during part of the fill appeared to be used with pre- and post-anoxic periods. For the study period the effluent nitrate and ammonium nitrogen concentrations averaged 1.3 and 0.9 mg/L, respectively.

NITROGEN REMOVAL DESIGN

Prior discussions on biological phosphorus removal design can be used to guide an SBR system design to create the necessary anaerobic contact time within the SBR cycle. Nitrogen removal will be required for such designs so that oxidized nitrogen will not interfere with the fermentation of necessary substrates for biological phosphorus removal. Nitrogen removal designs must consider the SBR fill volume fraction, the system SRT and the fraction of the cycle time for anoxic and aerobic reactions.

Two design approaches are presented. The first considers only a pre-anoxic time where biological phosphorus removal is not needed and where a more modest level of nitrogen removal is required. The second design approach addresses the need for a high level of nitrogen removal and requires post-anoxic denitrification.

Before describing SBR nitrogen removal design approaches, fundamental design procedures must first be reviewed. SBR system designs have varied considerably because of the large number of possible design choices including: fill volume, aeration time, settling and decant times, and SRT control. In an SBR plant survey, Arora et al. (1985) found a wide range of full-scale designs and noted the need for a comprehensive design approach. He found design SRTs ranging from fifteen to greater than eighty days, and aeration periods ranging from 0.7 to 3 hours. He found that the fraction of aeration time per cycle ranged from 0.08 to 0.25 for the plants reviewed.

Irvine (1981) suggested that only the aeration time per cycle be used to define an effective loading for an SBR system. More significantly, Argaman (1980) used the aeration time to define an effective system SRT when evaluating nitrification performance in SBR systems. The nitrification efficiency-effective SRT relationship corresponded well with that observed for conventional SRTs in continuous flow operated activated sludge systems. At 13-18°C, the critical effective SRT was found to be about 5 days. At a higher temperature range of 25-28°C, complete nitrification was obtained at effective SRTs as low as 2.5 days.

The effective SRT is defined as follows:

$$\text{SRT}_E = \text{SRT}\left(\frac{T_A}{T_C}\right) \tag{4.14}$$

where

T_A = Aeration time per cycle, hrs
T_C = Total cycle time, hrs

The SRT can be controlled by the amount of solids wasted per day during the plant operation. At steady state conditions, the solids wasted equals the solids produced and the SRT is as follows, assuming that the amount of solids produced from nitrification is very small:

$$\text{SRT} = \frac{V_T\,(X_T)}{V_F\,(Y_{NT})\,(S_o - S)N_c} \tag{4.15}$$

where

N_c = Number of cycles per day
V_T = Total aeration volume of SBR tank after fill, m^3
V_F = Fill volume per cycle, m^3
X_T = MLSS concentration at volume, V_T, mg/L

The value for Y_{NT} includes the heterotrophic net yield coefficient as shown in Equation (2.49). The net heterotrophic yield is modified here from that presented earlier to account for the effective aeration time for endogenous decay.

$$Y_{NT} = \frac{Y}{1 + K_d\,\text{SRT}_E} + \frac{X_I}{S_o} \tag{4.16}$$

Reiber et al. (1988) and Obermeyer (1988) have presented a rational spreadsheet design model using the effective SRT concept. The model can describe the impact of different aeration times per cycle or operating SRT on SBR performance parameters including oxygen requirements, sludge production, and nitrification efficiency. This approach will be described briefly here for a nitrification system design.

Cycle and Fill Times

The total cycle time and fill time are dependent upon parameters selected as appropriate to the particular installation. Settling time, withdrawal time, and number of tanks are selected components of the design basis, while aeration time is a principal design variable. From these parameters, the total cycle time can be expressed as the summation of the individual sequence times:

$$T_C = T_F + T_A + T_S + T_W \tag{4.17}$$

where

T_C = Total cycle time (hr)
T_A = Aeration time (hr)
T_S = Settling time (hr)
T_W = Withdrawal time (hr)
T_F = Fill time (hr)

Since a tank is always available to accept fill, the total cycle time can also be expressed as the product of the fill time and the number of tanks in service (n):

$$T_C = (n)T_F \tag{4.18}$$

and thus T_F can be expressed independently of cycle time as:

$$T_F = (T_A + T_S + T_W)/(n - 1) \tag{4.19}$$

Tank and Fill Volumes

The ratio of fill to total tank volume (V_F/V_T) is a critical design variable influencing the reactor volume available for solids storage, the solids separation processes, and the decanter design. Figure 4.21 is a schematic of the relevant SBR tankage allocations.

Several decanter mechanisms are available for consideration in SBR design, including both fixed and floating devices. The operational depth of the selected decanter mechanism determines the maximum depth of fill, while the desired tank depth (design basis) will generally be established by aerator efficiency requirements, or some site-specific factor. Since depth and volume are proportional, the ratio of decanter depth to full tank liquid depth determines the V_F/V_T ratio.

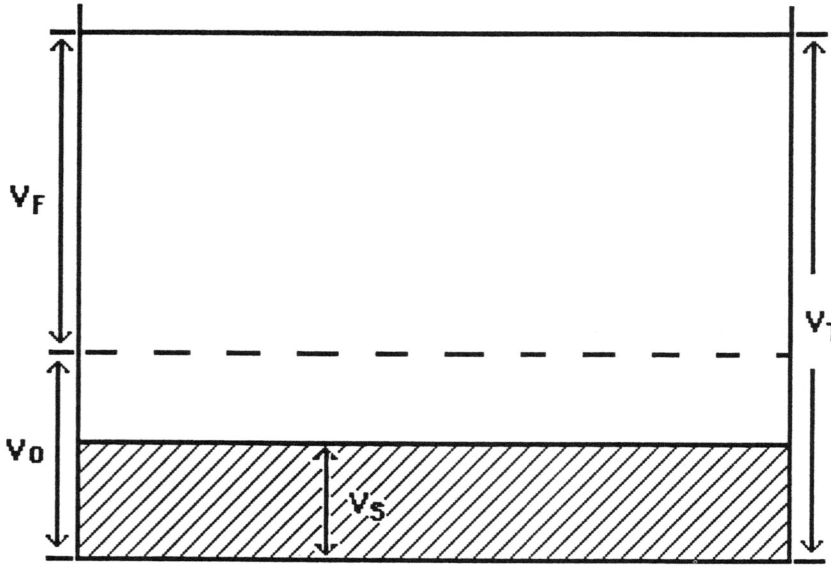

Figure 4.21 SBR tank volumes.

An internal check ensures that the fill to total volume ratio is feasible based upon sludge settling characteristics specified by the Sludge Volume Index (SVI) in the model operating assumptions. The model calculates the maximum fill to volume ratio permissible for a given SVI according to:

$$V_F/V_{T(\max)} = [1-(f)(X_T)(\mathrm{SVI}/10^6)] \qquad (4.20)$$

where

X_T = Mixed liquor biosolids concentration (mg/L)
f = Decant safety factor

Ratios larger than $V_F/V_{T(\max)}$ do not have sufficient storage capacity to hold settled biosolids, and will induce solids carryover during decantation.

Equation (4.20) shows that the Fill Volume Fraction (V_F/V_T) is related to the system SVI and desired MLSS concentration. Rearranging Equation (4.15) shows that the fill volume fraction is also related to the system SRT:

$$\frac{V_F}{V_T} = \frac{X_T}{\mathrm{SRT}(S_o - S)N_o(Y_{NT})} \qquad (4.21)$$

If a smaller fill volume is used per cycle, the system SRT will increase and the effective SRT will also increase provided that the aeration time per cycle is the same.

Equations (4.14), (4.16), and (4.21) can be combined to determine the operational SRT as a function of the selected T_C, T_A, N_C, V_F/V_T, and X values by solving the following quadratic equation for SRT:

$$[Y(1 + K_d(T_A/T_C)\text{SRT}) + (X_I/S_o)]\text{SRT} = X_T\,[(V_F/V_T)(S_o - S)(N_C)] \tag{4.22}$$

NITRIFICATION DESIGN

A nitrification design is suited for a microcomputer spreadsheet application as it requires an iterative procedure. The cycle times are selected along with the fill volume fraction and MLSS concentration. This results in a system SRT that can be used to calculate the nitrifier bacteria concentration. The use of this information and assumptions will be illustrated as follows to determine if the design selected is suitable for nitrification. The nitrifier bacteria mass in a given reactor is shown as follows:

$$X_n = Y_{nW}\,(\text{NO})\,(V_F/V_T)\,(N_C)\text{SRT} \tag{4.23}$$

where

X_n = Nitrifier biomass concentration, mg/L

As described before, the amount of nitrogen that is oxidized (NO) based on the influent concentration is a result of the influent nitrogen, the nitrogen used for synthesis, and the effluent ammonium nitrogen concentration:

$$\text{NO} = N_o - Y_{NH}\,(S_o - S)\,(F_N) - Ne \tag{4.24}$$

where

N_o = Influent TKN concentration, mg/L
F_N = Nitrogen content of biomass, g/g
Ne = Effluent ammonium nitrogen, mg/L

The oxidizable nitrogen concentration (NO_o) at the beginning of the cycle is estimated as follows:

$$\text{NO}_o = (\text{NO} + Ne)\,(V_F/V_T) \tag{4.25}$$

The aeration time necessary to achieve the desired level of nitrification is calculated by integrating the modified Monod expression:

$$\frac{-dN}{dt} = \frac{\mu_{n,\,max}}{Y_{nM}} \quad \frac{N}{K_w + N} \quad \frac{DO}{K_o + DO} \quad X_N \tag{4.26}$$

The required aeration time is then:

$$T_n = \frac{[(K_N)\ \mathrm{Ln}\ (NO_o/Ne) + (NO_o - Ne)Y_{nN}]\mathrm{S.F.}\ (24\ \mathrm{hr/d})}{(\mu_{n,\,max,DO})\ X_N} \tag{4.27}$$

where

$\mu_{n,max,DO}$ = Maximum specific growth rate corrected for DO, d^{-1}
Ne = Required effect ammonic nitrogen concentration, mg/L
S.F. = Nitrification safety factor

If T_A is greater than T_N, the design will achieve the desired level of nitrification. Otherwise, changes can be made in T_A, V_F/V_T, and X_T and the calculation is repeated until a satisfactory design is reached.

Modification and addition to the above design approach is needed to evaluate an SBR with nitrogen removal capacity for post- or pre-anoxic denitrification.

SBR Pre-Anoxic Denitrification

In this application, denitrification occurs during an anoxic period during fill, or during fill plus a mixed/non-aerated portion of the react period if needed. The nitrate reduced in this case is the nitrate remaining with the SBR volume ($V_T - V_F$) after withdrawing the effluent. Thus the effluent nitrate nitrogen concentration is determined by the fill volume fraction and the amount of nitrate reduced during the initial anoxic period. The denitrification rate would be related to the organic carbon loading and can be estimated using Equation (2.46), which relates the SDNR to the anoxic *F/M* loading. The ratio of available degradable COD added to the nitrate nitrogen to be reduced should be checked to determine if sufficient substrate is available to satisfy the desired nitrate reduction level.

The effluent nitrate nitrogen concentration and the percent nitrogen removal can be determined using Equations (4.28) and (4.29), respectively.

$$V_T NO_E = V_T\ (NO_o - Ne) + V_T\ (NO_1)$$

$$NO_E = NO_o + NO_1 - Ne \qquad (4.28)$$

where

NO_E = Effluent nitrate nitrogen concentration, mg/L
NO_1 = Nitrate nitrogen concentration remaining after pre-anoxation period, mg/L

$$\text{Percent Nitrogen Removal} = [N_o - Ne - NO_E]100/N_o \qquad (4.29)$$

The loss of nitrogen in Equation (4.29) is due to nitrogen used for synthesis and nitrate reduction. The value for NO_1 is determined by the following equations, which consider the nitrate remaining in the SBR reactor after effluent withdrawal and the amount of nitrate reduction during a pre-anoxic period. If a negative value is calculated, $NO_1 = 0$, as excess nitrate reduction capacity is available during the anoxic period.

$$NO_1 = [(V_T - V_F)NO_E - (V_T)DNO_x]/V_T$$

$$NO_1 = (1 - V_F/V_T)\ NO_E - DNO_x \qquad (4.30)$$

$$DNO_x = (SDNR)\ (X_T)\ (T_{DN})\ (1/24) \qquad (4.31)$$

$$SDNR = 0.03\ (F/M_A) + 0.029 \qquad (2.46)$$

$$F/M_A = (V_D/V_T)\ (S_o/X_T)\ (1/T_F)\ (24\ \text{hrs/d}) \qquad (4.32)$$

where

DNO_x = Nitrate nitrogen concentration reduced during anoxic period, mg/L

Substituting Equation (4.28) for NO_E in Equation (4.24) yields the following expression:

$$(V_F/V_T)NO_1 = (1 - V_F/V_T)\ (NO_o - Ne) - DNO_x \qquad (4.33)$$

If the right side of the equation is negative, excess denitrification capacity is available and NO_1 is equal to zero in Equation (4.28) used to determine NO_E.

The anoxic period time will typically be that of the fill period ($T_{DN} = T_F$).

Design Example

The assumptions used for the oxidation ditch nitrogen removal design example (Table 4.5) (SRT = 30 d) are used here with the following additional assumptions necessary for an SBR design:

Two Tanks
$T_A = 2.0$ hrs
$T_S = 0.5$ hrs
$T_W = 0.5$ hrs
SVI = 120 ml/g
$f = 1.2$

From Equations (4.18) and (4.19), the fill time (T_F) equals 3.0 hours and the total cycle time (T_C) equals 6.0 hours. The number of cycles per day, N_C, is then equal to four. The fill volume is calculated as follows:

$$V_F = Q/(N_C)\,(n) = 5000\ \mathrm{m^3/d}\ /\ (4\ \mathrm{d^1})\,(2) = 625\ \mathrm{m^3}$$

Using Equation (4.22) the fill volume fraction can be calculated:

$$\{0.6\,[1 + 0.08\,(2/6)\,30] + (80/200)\}\,30 = 3500/\,[(V_F/V_T)\,(200 - 5)\,(4)]$$

$$V_F/V_T = 0.20$$

$$Y_{NT} = 0.73\ \mathrm{grTSS/grBOD}$$

$$Y_{nH} = 0.33\ \mathrm{grTSS/grBOD}$$

From Equation (4.20), the maximum V_F/V_T allowed is:

$$(V_F/V_T)_{\max} = [1 - 1.2(3500)\,(120/10^6)] = 0.50$$

Thus, using 20% of the volume for fill and withdrawal is well within allowable fill volumes considering sludge thickening requirements.

Using Equation (4.24), the oxidized nitrogen is determined and the nitrifier mixed liquor concentration is then calculated using Equation (4.23).

$$\mathrm{NO} = 35 - 0.33(200 - 5)\,(0.10) - 0.5 = 28.0\ \mathrm{mg/L}$$

$$X_n = 0.15(28)\,(0.2)\,(4)\,(30) = 100.8\ \mathrm{mg/L}$$

Thus the nitrifier concentration is about 3% of the mixed liquor suspended solids concentration.

The available nitrogen for nitrification during the aeration period is calculated:

$$NO_o = (28.0 + 0.5)\,(0.2) = 5.7 \text{ mg/L}$$

Using the value for X_n, a $\mu_{n,\max}$ of 0.34 d^{-1} at 15°C (from Table 2.4), DO concentration of 2.0 mg/L, and K_N of 0.5 mg/L. The required aeration time for nitrification is calculated using Equation (4.27).

$$T_N = \frac{\{[(0.5)\text{Ln}(5.7/0.5) + (5.7 - 0.5)]0.15\}1.5(24)}{(0.272)100.8}$$

$$T_N = 1.3 \text{ hrs.}$$

The aeration time selected was 2.0 hours per cycle so the selected aeration time and SRT are more than adequate for full nitrification, assuming proper aeration design. Now the pre-anoxic nitrogen removal calculation can be made. The first step is to determine the SDNR and DNO_x possible during the anoxic fill period of 3.0 hrs., using Equations (4.32), (2.46), and (4.31).

$$F/M_A = (0.2)\,(200/3500)\,(1/3)24 = 0.09 \text{ d}^{-1}$$

$$\text{SNDR} = 0.03\,(0.09) + 0.029 = 0.03 \text{ g/g-d}$$

Correcting for temperature, the SNDR at 15°C equals:

$$\text{SNDR}_{15} = 0.03(1.06) = 0.022 \text{ g/g-d}$$

$$DNO_x = (0.022)\,(3500)\,(3)\,(1/24) = 9.6 \text{ mg/L}$$

From Equation (4.33), NO_1 is determined to be equal to zero, since a negative value is calculated indicating excess denitrification capacity.

$$V_F/V_T(NO_1) = (1 - 0.2)\,(5.7\ 0.5) - 9.6 = -5.4$$

The effluent nitrate nitrogen concentration from Equation (4.28) is 5.7 − 0.5 mg/L or 5.2 mg/L, and the percent nitrogen removal from Equation (4.29) is:

$$(3.5 - 0.5 - 5.2)100/35 = 84\%$$

A small *VF/VT* ratio may result in significant nitrogen removal if the denitrification capacity is available. This design shows that an iterative process is needed to develop optional designs and to evaluate different loading conditions. A spreadsheet model is useful for such evaluations.

Post-Anoxic Denitrification

The nitrogen removal level in the above example problem can be further increased by providing anoxic periods during the react period in addition to aeration. The denitrification rate will be controlled by the endogenous respiration demand which will be a function of the system effective SRT. The effective SRT will be due to the aeration and anoxic times per cycle. The following are modifications of Equations (4.14) and (4.17) to account for the anoxic period.

$$\mathrm{SRT}_E = \frac{(T_A + T_{DN})\,(\mathrm{SRT})}{T_C} \tag{4.34}$$

$$T_C = T_F + T_A + T_S + T_W + T_{DN}$$

where

T_{DN} = Anoxic reaction time, hrs.

In the design example the nitrate nitrogen concentration after nitrification was 5.2 mg/L and the ammonium nitrogen concentration was 0. 5 mg/L. The aeration time selected was 2.0 hours, but only 1.3 hours was needed for nitrification including the design safety factor of 1.5. Thus to continue the same design example, we will assume that the extra 0.7 hours is used for denitrification (T_{DN} = 0.7 hr.).

Equation (2.50) relates the SDNR to SRT for endogenous denitrification but can not be used to calculate the SDNR in this case because that equation was applicable to a complete mix activated sludge system or one that approached completely mixed conditions. In this case, the endogenous respiration rate is used with the biomass portion of the mixed liquor suspended solids to estimate a denitrification rate.

$$\mathrm{DNR} = F(K_d)\,(Y_{nH}/Y_{nT})\,(X_T)\,(1/24) \tag{4.35}$$

where

F = Fraction of biological respiration rate using nitrate instead of oxygen, g/g

DNR = Denitrification rate, mg/L-hr

DNR = 0.5(0.08)(0.33/0.73)(3500/24) = 2.64 mg/L-hr

The additional nitrate removed is then:

$$(\text{DNR})\,(T_{DN}) = (2.64)\,(0.7) = 1.8 \text{ mg/L}$$

With this post-anoxic denitrification added, the effluent nitrate nitrogen concentration is estimated to be 3.4 mg/L and 89% nitrogen removal is calculated.

If biological phosphorus removal was required in this example, the post-anoxic denitrification time would be depended on to further reduce a large portion of nitrate to minimize the negative impact of oxidized nitrogen on biological phosphorus removal during a fill and contact period. Thus, more time would be needed for the post-anoxic operation.

In summary, the design approach presented applied fundamental biological kinetic considerations for pre-anoxic and post-anoxic denitrification along with consideration to the system SRT and nitrification capabilities. The SBR has many design variables that must be evaluated when developing nutrient removal designs.

SECTION V: DESIGNER PLANT CASE HISTORIES

EXPERIENCE OVERVIEW

While an awareness grew in the U.S.A. that biological phosphorus removal was taking place at various plants in the country, none of these plants were designed for that purpose. The patented Phostrip process was introduced and a number of successful applications were reported. In other applications, plants were designed for biological nitrogen removal using methanol as the carbon source for denitrification after nitrification. None of these plants attempted simultaneous biological nitrogen and phosphorus removal.

The first full-scale plant in the world specifically designed for high-rate biological nitrogen removal without the addition of chemicals was the Goudkoppies plant in Johannesburg, South Africa, designed in 1972 to treat about 150,000 m^3/d or about 750,000 population equivalents, and began operating that year. It is illustrated by Figure 4.22.

The first plant designed for simultaneous nitrogen and phosphorus removal was the Meyerton plant in South Africa in 1974, followed by the addition of anaerobic basins to the Goudkoppies plant in 1975. Two more plants of similar size were constructed in Johannesburg shortly afterwards, those being the Bushkoppie plant and the Johannesburg Northern Works.

During this period it became clear that future restrictions on the discharge of plant nutrients would become obligatory and most local authorities opted

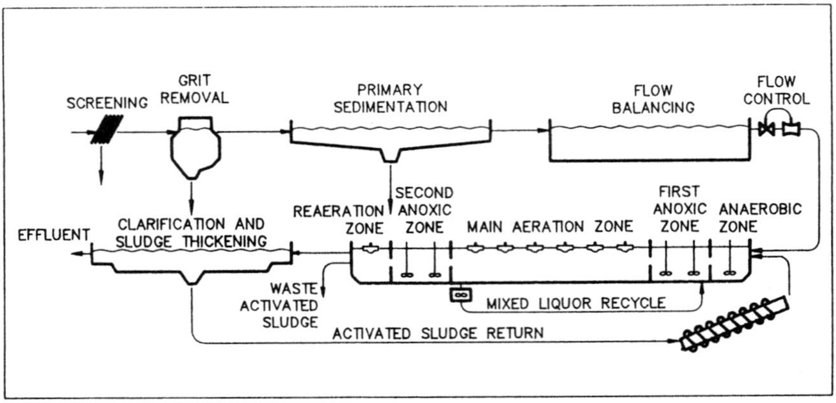

Figure 4.22 Goudkoppies flow diagram.

for BNR plants even though no full-scale experience was available, while awaiting future legislation restricting nutrients in the effluents.

These first generation plants had a number of shortcomings that resulted from a lack of understanding of the processes. This resulted in numerous papers that gave, at times, a slanted impression of the success of the BNR principles. Rather than looking at individual case histories, problems experienced with these plants will be discussed under the following headings.

UNDERLOADING AND OVERDESIGN

Most of these first-generation plants were underloaded and overdesigned due to excessive caution on the part of the designers. The idea of simultaneous nitrification and denitrification was so new at the time that the Goudkoppies plant had sufficient aerators for full nitrification not accounting for the energy derived from the recycled nitrates. The large 110 kW aerators used could not be cut back sufficiently during low loading conditions, leading to overaeration, problems of nitrates in the RAS, and resultant loss of phosphate removal, especially after weekends when the organic load was quite low. This resulted from the fact that the plants served mainly the downtown area, which was virtually vacated on weekends, and were initially underloaded.

Due to misconceptions fueled by research findings that a COD:TKN ratio of at least 14 was required for good denitrification, most earlier plants were designed as extended aeration plants without primary sedimentation. This resulted in an average energy input of less than 20 W/m^3. This added to the problems of underdesigning and overloading, and in some plants it was not possible to reduce the oxygen input without sludge settling in the aeration basin. Underloading also led to severe sludge bulking problems, which will be discussed below.

In later designs such as at Olifantsfontein, primary sedimentation with acid fermentation and a sludge age just sufficient to nitrify led to energy levels in the aeration basin of up to 100 W/m^3, which left ample room for aerator turndown to control denitrification.

EXCESSIVE AERATION OF THE INFLUENT WASTEWATER AND THE RAS

Most plants in South Africa are built on sloping sites, necessitating drops of up to 15 m from the primary sedimentation tank to the lowest aeration basin. This led to excessive aeration of the influents. At the Goudkoppies plant the influent passed through a number of cascades with screw pumps for recycling the RAS. Later, screw pumps were abandoned in favor of directly coupled centrifugal pumps, and control gates were used to restrict the turbulence and aeration of the influent. At Olifantsfontein, a level activated control valve destroys the energy of the 15 m drop under the liquid level in such a way that little air is added. In Kelowna, a control gate as shown in Figure 4.23 was introduced to almost submerge the launders of the PST, which eliminated aeration but most importantly also eliminated the odors.

POOR FINAL CLARIFIER DESIGN

A number of the earlier plants had poorly designed clarifiers, resulting from accepting "proprietary designs." It was thought that the phosphate-

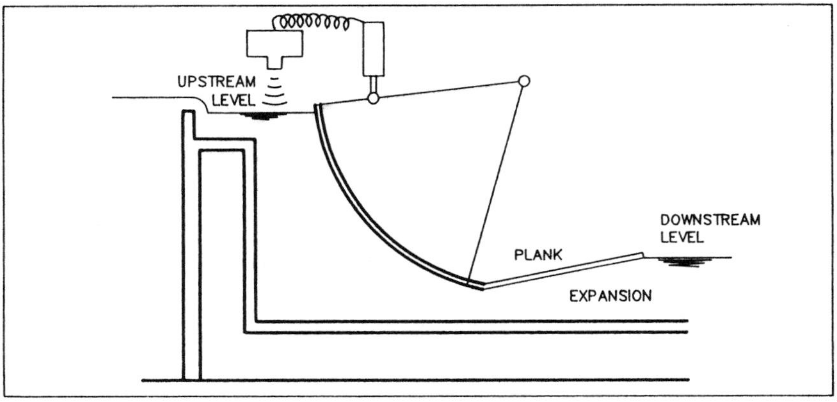

Figure 4.23 Control gate Kelowna.

containing sludge should be removed as fast as possible, and suction lift clarifiers were selected. Clarifiers were mostly shallow, had small central stilling wells, and only one suction arm per clarifier. The worst results were observed at the Northern Works of the city of Johannesburg, which was almost identical in design to the Goudkoppies plant shown in Figure 4.22, where the inability to thicken the sludge in the clarifiers led to high rates of return activated sludge recycle. This had a detrimental effect on the behavior of the anaerobic zone, much of which is recorded in numerous papers. High suspended solids in the effluent contributed to even higher effluent phosphate.

Later designs returned to deep, central feed, scraped clarifiers with much improved results. At the Vereeniging plant, a five-stage Bardenpho plant with fine bubble aeration that had to comply with a nitrogen standard of less than 5 mg/L and a Total P standard of less than 0.15 mg/L, it was possible to reduce the sludge recycle and have a sludge blanket in the final clarifiers that allowed all nitrates in the RAS to be reduced. No phosphate release to the effluent was evident. This assisted in complying with the phosphate standard of 0.15 mg/L TP for about eight months of the year with only filtration and no chemical backup. During the remainder of the time, which coincided with the rainy season, groundwater containing nitrates resulted in effluent phosphates from the BNR plants of between 1 and 2 mg/L, which required chemical backup.

Reduced RAS recycle rates ensure that some nitrates are reduced in the sludge blanket in the clarifiers while the low rate of recycle will return less nitrates and dissolved oxygen when they are present. A BNR plant can be severely compromised by not using state-of-the-art clarifiers.

SLUDGE BULKING

Sludge bulking appears to be common to nutrient removal plants, yet there are some that produce excellent SVIs. One of the main causes of sludge bulking in the first-generation plants could be attributed to over-design and underloading, as well as the initial tendency to use extended aeration. Without primary sedimentation, the mixed liquor contains a large portion of inert material, requiring larger basins, longer MCRTs, and resulting in a much lower energy input into the aeration zone. When cutting back oxygen supply under these conditions to regulate denitrification, unmixed and underaerated pockets were formed in the aeration basin which seemed to stimulate the growth of low DO filaments such as *Microthrix parvicella*. The experiments by Osborn et al. (1979), shown in Figure 4.24, confirm this. By switching out aerators to cut back on oxygen input into the Goudkoppies plants, the SVI was observed to gradually increase. This was reversed when the aerators were switched back on. When the plant is

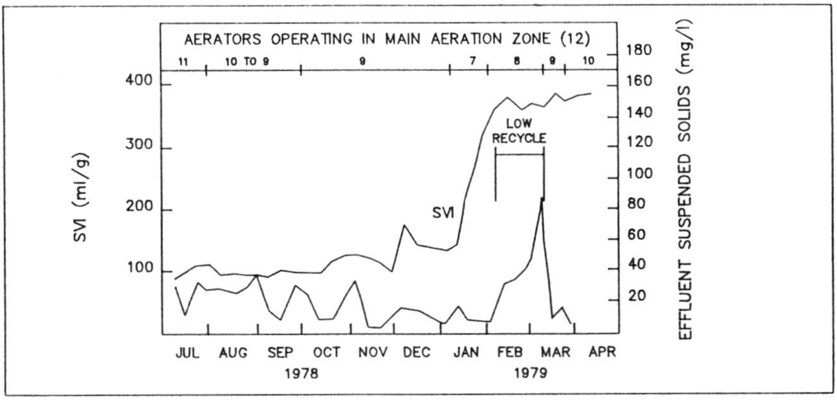

Figure 4.24 Effect of switching off aerators on SVI.

underloaded, this may cause a severe problem in that the need to cut back on oxygen may result in the clarifiers not being able to remove all the solids even at low loads.

At the Heidelberg, South Africa plant, diluted SVIs of up to 600 mg/L were measured before the aerators were lowered in order to provide improved mixing.

In later plants with primary sedimentation and lower SRTs, the energy input is sufficient to ensure that low DO filaments are not encouraged. This does not mean that denitrification can be completely prevented in the aeration basin, as will be discussed below.

The Bushkoppie plant, which consists of an anaerobic/anoxic/aerobic sequence and is fitted with a fine bubble aeration system, has SVI values of between 40 and 60 mg/L. Due to the presence of a high percentage of easily degradable material in the influent, good P removal is possible in spite of a high DO in the aeration basin and more than 10 mg/L of nitrates in the effluent.

New plants with surface aerators supplied with draught tubes to prevent zones of permanently low DO do not perform as well as the fine bubble aeration plants in terms of sludge bulking but the SVI is seldom above 150 mg/L. In comparing the results of about fifty plants, it appears that while low SVIs are not common to BNR plants, excessively high SVIs are caused by bad mixing in the aeration zone.

DENITRIFICATION IN THE AERATION BASIN

Observations and measurements indicate that some degree of denitrification takes place in the aeration zones of BNR plants and that these could vary from about 20% of the total nitrogen in the influent to all the nitrogen not captured in the sludge. A mass balance around the aeration basins of the five-stage Vereeniging plant with fine bubble aeration showed a loss of 20% of the nitrogen in the aeration basin. This may have resulted from fine tuning of the DO for optimal nitrate reduction in the anoxic zones.

The Randfontein plant was originally designed as a five-stage Bardenpho plant. Only about 40% of the flow was passed directly to the plant while the remaining 60% was treated in the existing trickling filters and returned to the aeration basin of the BNR plant. Over a three-year monitoring period no nitrates or ammonia were present in filtrate of the mixed liquor being recycled to the first anoxic zone. Thus, no nitrates were passed to the second anoxic zone. The mixed liquor in the second anoxic zone became anaerobic and in the absence of a source of VFA a release of phosphates took place that could not be taken up again upon reaeration. This secondary release was discussed earlier.

The total nitrification and denitrification of all the available nitrogen could be explained by the characteristics of the wastewater as discussed earlier. A quarter of the load to the plant consisted of wastewater from an edible oil plant. The edible oil had a slower rate of breakdown than domestic wastewater, as demonstrated in Figure 4.25. For domestic wastewater, most of the adsorbed carbon is used up in the first section of the aeration basin before nitrification has progressed, thus leaving little carbon for denitrification other than by endogenous respiration, a process that has been demonstrated to be quite slow.

When sufficient carbon is available, as in this case, a reasonably high rate

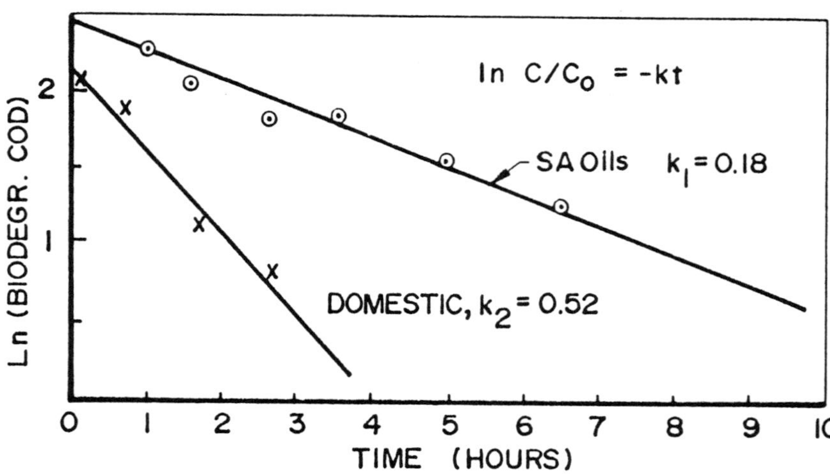

Figure 4.25 Slower breakdown of wastes.

of denitrification can be maintained through the entire aeration basin, resulting in denitrification in all the underaerated parts of the basin. Phrasing that statement differently, one can say that in a plant having surface aerators and draught tubes, such as in this case, the constant oxygen demand through the aeration basin could not be completely fulfilled by the aeration system, resulting in good denitrification in the aeration basin.

This might be considered a bonus, but if not provided for in the design it could lead to other problems, such as the secondary release of phosphates. In this particular instance it took a long time to identify the problem and then purchase new equipment to solve the problem.

About 50% of the nitrogen removed was lost in the aeration basin of the Olifantsfontein plant with an anaerobic/anoxic aerobic arrangement (Van Huyssteen et al., 1989). Because this loss was expected, no provision was made for a second anoxic zone, and excellent performance of the plant was achieved. Because of the high loss of nitrogen in the aeration basin, only 17% of the nitrogen was removed in the anoxic zone. When applying the IAWPRC model for BNR plants to this configuration and assuming that the aeration basin is fully aerobic, 13 mg/L effluent nitrates are predicted and no phosphate removal should be possible. The COD:TKN ratio of the influent averaged about 9.

In the Kelowna plant in British Columbia, the use of turbine aerators allowed the downturn of the DO without reducing mixing. The tanks are 6 m deep and with the down/up mixing pattern and point source aeration a similar loss of nitrogen took place. While the total nitrogen in the effluent remained below 3 mg/L and the nitrate nitrogen averaged less than 2 mg/L over a year, the IAWPRC model predicted that at best the nitrates in the effluent should be 7 mg/L, which would have made phosphate removal impossible.

The success of the Bardenpho plants in the U.S.A. in achieving good nitrogen removal consistently, as shown in Figure 4.26, can be attributed to the use of point source aerators in the Carrousel units for nitrification.

Where a high removal of nitrates is not required, one of the many flow sheets for ridding the RAS of nitrates could be used to protect the anaerobic zone against nitrates. However, from full-scale experience where good nitrogen removal is required, problems with excessive nitrate return are eliminated and good phosphate removal is possible.

SCUM GROWTH

Scum in the form of *Nocardia* or *Microthrix parvicella* presents problems at nutrient removal plants because of the requirement for unaerated zones. The floating organisms are trapped by the vigorous surface action of the aerated zones and cause nuisance conditions. The more organisms that are

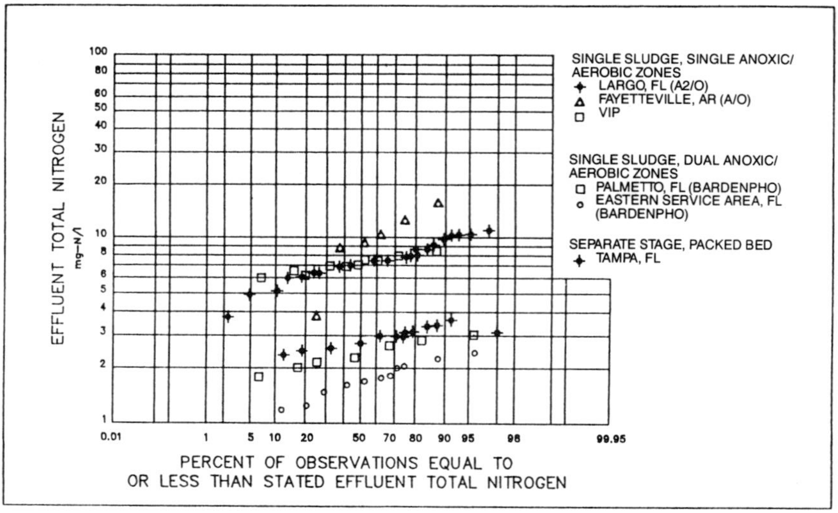

Figure 4.26 Removal of nitrogen in BNR plants in the U.S.A.

trapped, the greater the potential for growth unless they can be removed from the system. While sprays and chlorine application can assist in suppressing the growth, the proven way to avoid problems is to design the system in such a way that the scum is continually removed.

Some of the earlier plants did not have this feature, and thick layers of scum accumulated on the anoxic zones. In some cases it was possible to create a small drop in the liquid level between the anoxic and aeration zone. In another plant, a scum removal trough with pump was installed to alleviate the problem.

Scum baffles and scum removal mechanisms were provided for in clarifiers of BNR plants from the start. Scum hoppers above the liquid level were not successful mainly because of the difficulty of scraping the scum up the beach. Submerged hoppers with outlet control were more successful in removing the scum. The scum should be removed and never returned to the aeration basin since return would result in prolific growth.

CONTROL OF OXYGEN INPUT

Full-scale experience indicated the need for control of oxygen input for reliable BNR operation.

Effective control of the oxygen input was the single most troublesome problem with first generation nutrient removal plants. Because of the relatively small size of the South Africa market, imported instruments were sold without proper backup service, and DO control had a dismal record. Fortunately, most of the plants used surface aeration and it was possible to fall back on timer switches for control.

Lightning can be severe in the northern part of the country and, despite all safety precautions, the control system of the Vereeniging plant was destroyed twice and that of the Rooiwal plant in Pretoria, once.

The availability of new membraneless-type DO meters made reliance upon instrumentation for DO control possible. Before that, most plants relied upon timer switches for control. A reliable method to follow is to set the timers according to a diurnal pattern and take twenty-four-hour samples of nitrate and ammonia in the effluent on a regular basis. This has led to consistently good phosphate removal in a number of plants. At the Heidelberg plant near Johannesburg, the operator achieved an annual average effluent phosphate concentration of less than 0.5 mg/L by monitoring ammonia and nitrates in the effluent and adjusting aerator timers accordingly.

With more reliable DO meters, later plants using surface aerators relied more on the use of a mechanized adjustable overflow weir, controlled by a PLC using a feedback signal from a DO meter for controlling the immersion of the aerators. This has been very successful. Problems may arise where one of the aerators near the probe is accidentally switched off or trips, which distorts the normal flow pattern and control range. Provision was also made in a number of plants for a long static overflow weir that would prevent drowning and tripping of the aerators in the event of malfunctioning of the control gate. Tripping of aerators can lead to a massive release of phosphates to the liquid state.

In locations with weaker, colder, wastewaters such as Kelowna, a more sophisticated reliable DO control system was required for accurate control of DO throughout the plant. This entailed regular maintenance and inspection of the probes and the control system. A similar system at Rotorus in New Zealand for a fine bubble aeration system not only records the DO levels, but also the air flows to the various zones throughout the day. In cases of malfunctioning of the metering systems, it is possible to use the historic data on air flows to control the system by controlling the feed valves to the zones.

In multiple module units using fine bubble diffused air with a single blower house at the Rooiwal plant near Pretoria, a pressure sensor ensures that the blowers keep a constant pressure on the central supply system, while local DO meters control butterfly valves on the supply lines to each zone of each module. This has the advantage that there has been less interference by lightning since the distances that the control cables span were reduced. Optical cables are now used for transmitting information in high risk lightning areas.

VFA GENERATION AND ADDITION

Activated primary sedimentation tanks or a variation of this process are now used as standard design practice. In smaller plants where the owner may not want primary sludge production, the raw sewage is fed to a two-stage fermenter with stirrers from which all the sludge is removed to the BNR reactor after acidification. This has led to very satisfactory operation at a number of small plants. The waste activated sludge is then aerobically digested and irrigated on farm land. In another situation, primary sedimentation tanks were provided with sludge recycle with the idea that for future expansion digesters may be added. In the meantime the sludge accumulates in the primary tanks and is then pumped to the BNR plant during the low load periods. A plant recently completed at Benoni near Johannesburg has this arrangement. The waste activated sludge is temporarily irrigated on farm lands. During the past two months the phosphates in the effluent were undetectable.

At the Olifantsfontein plant near Pretoria, settled sludge is recycled to the inlet of the primary sedimentation tanks, thus providing excellent phosphate removal in a plant in which almost half of the flow passes through a set of trickling filters and is then discharged to the aeration basin. Without the recycle around the primary sedimentation tank, there would be no phosphate removal, as discussed in Chapter 3.

At the Secunda plant in South Africa, neither the original Bardenpho plant nor two UCT extensions added later removed phosphates satisfactorily. There was a waste acetate stream from a coal plant nearby and this was added to the anaerobic basin with the results shown in Figure 4.27. Note the short interruption when no acetates were added. This plant was a classic case in which no primary sedimentation tanks were provided and the wastewater arrived at the plant in a fresh state. There was no difference in the behavior of the Bardenpho and UCT designs both before and after the addition of acetates.

There are a few existing plants for the removal of phosphates which do not rely on some form of prefermentation for reliable phosphate removal. Many of these are retrofits, but all the later plants have designed prefermentation units. Exceptions are plants where prefermentation takes place in the sewer system, mostly in regions with a number of pumping mains or where the main outfall sewers are long and almost septic. One of the earliest plants in this category is the Klerksdorp plant in South Africa, a three-stage Bardenpho plant designed in 1974. This plant was constructed in the belief that a phosphate standard was to be promulgated but, as yet, no standard applies. No special effort is made to achieve phosphate removal, yet the rather sparse information reveals continued excellent P removal. The wastewater is collected in a large sump, stored, balanced, and pumped

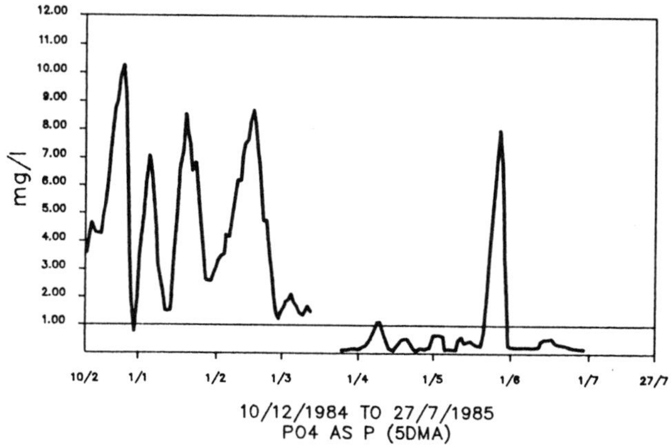

Figure 4.27 Acetate addition – Secunda plant.

through a 3 km pipeline to the plant. Acid generation in the sump and force main is the only explanation for the reliable P removal.

CHEMICAL BACKUP

While the state of the art has reached the point where reliable BNR plants can be designed and operated to perform as expected, it is still necessary to provide chemical backup systems for phosphate removal. The removal of phosphates is very much dependent on the reliable performance of the mechanical and control equipment. The failure of an aeration unit or a clarifier mechanism can suddenly increase the effluent phosphates. The failure of a pump could have a similar affect.

At first, plants were designed with chemical addition to the activated sludge plant as standby. In many ways this is counterproductive, and especially so where very low effluent P is required. Chemical posttreatment is now preferred. Chemicals added to the activated sludge seemed to interfere with the biological mechanisms more than was previously believed, causing problems with phasing out the chemical addition. At the Baviaanspoort plant of the city of Pretoria, the effluents from a new BNR plant, an older first-generation BNR plant, and a retrofitted plant, were combined and post-chemical treatment provided. Next the effluent passed through a flotation unit for the removal of suspended solids. This has proven to be a workable solution and, in spite of a number of mishaps at the different plants, the annual average effluent phosphorus concentration was

0.95 mg/L. The chemical addition is activated by on-line monitors. Most of the time no chemicals are added but excellent suspended solids removal in the flotation units ensures that the stringent effluent standards are complied with.

At Vereeniging, the BNR plant is followed by chemical treatment, chemical clarifiers, and sand filters. No chemicals are added for about eight months of the year, but the removal of suspended solids is sufficient to comply with the effluent standard of 0.15 mg P/L. During the rainy season some chemicals are needed because of nitrates in the influent.

These are a few of a large number of examples of the behavior of BNR plants at full scale. It is now possible to design such plants for reliable nutrient removal, provided all the pitfalls can be avoided. An attempt was made in this section to point to a few pitfalls in the hope that they can be avoided by the reader.

REFERENCES

SECTION I—TWO-STAGE BIOLOGICAL PHOSPHORUS REMOVAL SYSTEMS

Brannan, K. P. 1986. "Substrate Stabilization in the Anaerobic Stage of a Biological Phosphorus Removal System," Ph.D. Dissertation, Virginia Polytechnic Institute and State University, Blacksburg, VA, USA.

Daigger, G. T., C. W. Randall, G. D. Waltrip, E. D. Roman and L. M. Morales. 1987. "Factors Affecting Biological Phosphorus Removal for the VIP Process, a High-Rate University of Cape Town Process," in *Biological Phosphate Removal Wastewaters*. R. Ramadori, Editor. Oxford: Pergamon Press. pp 185-200.

Dold, P. L. 1991. "Incorporation of Biological Excess Phosphorus Removal in a General Activated Sludge Model," Dept. of Civil Engineering and Engineering Mechanics, McMaster University, Hamilton, Ontario, Canada. 32 pages.

Ekama, G. A., G. v. R. Marais and I. P. Siebritz. 1984. "Biological Excess Phosphorus Removal," Chapter Seven, *Theory, Design and Operation of Nutrient Removal Activated Sludge Processes*. Pretoria, South Africa: Water Research Commission.

Ekama, G. A. and G. v. R. Marais. 1984. "The Influence of Wastewater Characteristics on Process Design," Chapter Three, *Theory, Design and Operation of Nutrient Removal Activated Sludge Processes*. Pretoria, South Africa: Water Research Commission.

Ekama, G. A., I. P. Siebritz and G. v. R. Marais. 1982. "Considerations in the Process Design of Nutrient Removal Activated Sludge Processes," *Proceedings, IAWPRC Post-Conference Seminar on Phosphorus Removal, Pretoria, South Africa*.

Gibson, J. and P. L. Dold. 1991. "Wastewater Characterization for Activated Sludge Modelling." Department of Civil Engineering and Engineering Mechanics, McMaster University, Hamilton, Ontario. 26 pages.

Kavanaugh, R. G. 1991. "Investigation of Bacterial Populations in a Biological Nutrient Removal System," Ph.D. Dissertation, Virginia Polytechnic Institute and State University, Blacksburg, Virginia, USA. 170 pages.

McClintock, S. A., C. W. Randall and V. M. Pattarkine. 1991. "The Effects of Temperature and Mean Cell Residence Time on Enhanced Biological Phosphorus Removal," *Environmental Engineering, Proceedings of the 1991 Specialty Conference on Environmental Engineering*. ASCE. pp. 319-324.

Pattarkine, V. M. 1991. "The Role of Metals in Enhanced Biological Phosphorus Removal from Wastewater," Ph.D. Dissertation, Virginia Polytechnic Institute and State University, Blacksburg, Virginia, USA. 328 pages.

Pully, T. 1991. "Investigation of Anaerobic-Aerobic Activated Sludge Treatment of a High-Strength Cellulose Acetate Manufacturing Wastewater." Unpublished research, Department of Civil Engineering, Virginia Polytechnic Institute and State University, Blacksburg, VA, USA.

Randall, C. W., K. P. Brannan and L. D. Benefield. 1985. "The Oxygen Requirements of Biological Nutrient Removal Processes," *New Directions and Research in Waste Treatment and Residuals Management, Volume 1*, pp. 487-501. The University of British Columbia, Vancouver, B.C., Canada.

Randall, C. W., K. P. Brannan and L. D. Benefield. 1987. "Factors Affecting Anaerobic Stabilization During Biological Phosphorus Removal," *Biological Phosphage Removal from Wastewaters*. R. Ramadori, Editor. Oxford: Pergamon Press.

Randall, C. W., K. P. Brannan, S. A. McClintock and V. M. Pattarkine. 1991. "The Case for Anaerobic Reduction of Oxygen Requirements in Biological Phosphorus Removal." Submitted to the *Water Research Journal*, Water Pollution Control Federation. 35 pages.

Reddy, M. 1991. "The Concept of Phosphorus Storage Capability and Its implications for Design of Systems for Enhanced Biological Uptake of Phosphate," *Water Science and Technology*, 23:577-584. Oxford: Pergamon Press.

Reynolds, T. D. 1982. *Unit Operations and Processes in Environmental Engineering*. Brooks/Cole Engineering Division, Wadsworth, Inc., Monterey, California, USA.

Sell, R. L. 1981. "Low Temperature Biological Phosphorus Removal," Presented at the *54th Annual Conference of the Water Pollution Control Federation, Detroit, Michigan*. Air Products and Chemicals, Inc., Allentown, PA, USA.

Sen, D., C. W. Randall and T. J. Grizzard. 1990a. *Biological Nitrogen and Phosphorus Removal in Oxidation Ditch and High Nitrate Recycle Systems*. CBP/TRS 47/90, Chesapeake Bay Program, U.S.E.P.A. 120 pages.

Sen, D., C. W. Randall and T. J. Grizzard. 1990b. "Alkalinity and Turbidity Measurements in Feedback Controls to Improve Performance of a Biological Nitrogen and Phosphorus Removal System," *Instrumentation, Control and Automation of Water and Wastewater Treatment and Transport Systems*. R. Briggs, Editor. Oxford: Pergamon Press. pp. 291-298.

Siebritz, I. P. 1983. "Biological Excess Phosphorus Removal in the Activated Sludge Process," Ph.D. Dissertation, University of Cape Town, South Africa. 650 pages.

Tracy, K. D. and A. Flammino. 1985. "Kinetics of Biological Phosphorus Removal," Presented at the *58th Annual Conference Water Pollution Control Federation, Kansas City, Missouri*. Air Products and Chemicals, Inc., Allentown, PA, USA. 19 pages.

Waltrip, G. D. 1991. "Full-Scale Nutrient Removal Demonstration Project at the York River Treatment Plant with the VIP Process," Presented at the *45th Annual Conference, Virginia Water Pollution Control Association, Williamsburg, Virginia*. Hampton Roads Sanitation District, Virginia Beach, VA., USA.

Wentzel, M. C. , R. E. Lowenthal, G. A. Ekama and G. v. R. Marais. 1988. "Enhanced

Polyphosphate Organism Cultures in Activated Sludge Systems – Part I: Enhanced Culture Development,'' *Water SA*, 14:81-92.

SECTION II—DESIGN OF THREE-STAGE, FOUR-STAGE, AND FIVE-STAGE BNR SYSTEMS

Barnard, J. L. 1984. ''Activated Primary Tanks for Phosphate Removal,'' *Water S.A.*, 10:121-126.

Van Huyssteen, J. A., J. L. Barnard and J. Hendriksz. 1989. ''The Olifantsfontein Nutrient Removal Plant,'' *Proceedings, International Specialized Conference on Upgrading of Wastewater Treatment Plants, Munich, Sept. 3-7.* IAWPRC, 1 Queen Anne's Gate, London SW1H 9BT, England. pp. 1-8.

SECTION III—DESIGN OF OXIDATION DITCH BNR SYSTEMS

Bahre, G. 1987. ''Computer Aided Control of Nitrogen Removal in Small Activated Sludge Wastewater Treatment Plants,'' *Wat. Sci. Tech.*, 19:657.

Rittman, B. E. and W. E. Langeland. 1985. ''Simultaneous Denitrification with Nitrification in Single-Channel Oxidation Ditches,'' *J. Water Pollut. Control Fed.*, 57:300.

Sen, D., C. W. Randall and T. J. Grizzard. 1990. ''Developing a Multiple Parameter Approach to Improve Process Control for Operating High and Low Rate Biological Nutrient Removal Systems,'' *Proceedings, 1990 Amer. Soc. of Civil Eng. Environ. Eng. Spec. Conf., Washington D. C.*

Stensel, H. D., D. R. Refling and J. H. Scott. 1978. ''Carrousel Activated Sludge for Biological Nitrogen Removal,'' *Advances in Water and Wastewater Treatment; Biological Nutrient Removal.* M. P. Wanielista and W. W. Eckenfelder, Jr., editors. Ann Arbor Science.

Stensel, H. D. 1981. ''Biological Nitrogen Removal System Design,'' *Water Amer. Inst. of Chem. Engrs.*, p. 237.

Stensel, H. D. and M. Arcemont. 1988. ''Evaluation and Design of an Intrachannel Clarifier,'' *Proceedings, 1988 Joint Can. Soc. for Civil Eng. and Amer. Soc. of Civil Eng. Conf. on Env. Eng., Vancouver B. C.*

U. S. Environmental Protection Agency. 1978. ''A Comparison of Oxidation Ditch Plants to Competing Processes for Secondary and Advanced Treatment of Municipal Wastes,'' EPA-600/2-78-051, March 1978.

SECTION IV—DESIGN OF SEQUENCING BATCH REACTOR BNR SYSTEMS

Alleman, J. E. and R. L. Irvine. 1980. ''Nitrification in the Sequencing Batch Biological Reactor,'' *J. Water Pollut. Control Fed.*, 52:2747.

Argaman, Y. 1986. ''Nitrogen Removal in a Semi-Continuous Process,'' *Water Res.* 20:173.

Arora, M. L., E. F. Barth and M. B. Umphres. 1985. ''Technology Evaluation of Sequencing Batch Reactors,'' *J. Water Pollut. Control Fed.*, 57:867.

Dennis, R. W. and R. L. Irvine. 1979. "Effect of Fill: React Ratio on Sequencing Batch Biological Reactors," *J. Water Pollut. Control Fed.*, 51:255.

Goronszy, M. C. 1979. "Intermittent Operation of the Extended Aeration Process for Small Systems," *J. Water Pollut. Control Fed.*, 51:274.

Goronszy, M. C. 1986. "Design Considerations for Cyclically Operated Activated Sludge Systems Treating Domestic Wastewaters," *Proceedings, 41st Purdue Ind. Waste Conf., Lafayette, Ind.*

Irvine, R. L. 1981. EPA Technology Assessment of SBR Systems., Cincinnati, Ohio EPA Municipal Environmental Research Laboratory.

Irvine, R. L. et al. 1983. "Municipal Application of Sequencing Batch Treatment," *J. Water Pollut. Control Fed.*, 51:484.

Jones, W. L., E. D. Schroeder and P. A. Wilderer. 1990. "Denitrification in a Batch Wastewater Treatment System Using Sequestered Organic Substances Res.," *J. Water Pollut. Control Fed.*, 62:259.

Manning, J. F. and R. L. Irvine. 1985. "The Biological Removal of Phosphorus in a Sequencing Batch Reactor," *J. Water Pollut. Control Fed.*, 57:87.

Obermeyer, L. 1988. "A Model for the Design of Sequencing Batch Reactors." M. S. Thesis, University of Washington, Seattle, WA.

Reiber, S. H., L. Obermeyer and H. D. Stensel. 1988. *Proceedings, 1988 Joint Can. Soc. for Civil Eng. and Amer. Soc. of Civil Eng. Conf. on Env. Eng., Vancouver B. C.*

SECTION V—DESIGNER PLANT CASE HISTORIES

Osborn, D. W., A. R. Pitman and G. L. V. Venter. 1979. "Design and Operating Experience with Nutrient Removing Activated Sludge Plant in Johannesburg." *Nutrient Removal from Municipal Effluents*, a Technology Transfer Seminar held at the Conference Centre of the CSIR on 17 May 1979.

Van Huyssteen, J. A. and J. L. Barnard. 1989. "The Olifantsfontein Nutrient Removal Plant," *Proceedings, Inter. Specialized Conf. on Upgrading of Wastewater Treatment Plants, Munich.* pp 1-8. IAWPRC, 1 Queen Anne's Gate, London SW1H98T, England.

CHAPTER 5

Clarifier Design

THE important role of efficient clarification in advanced wastewater treatment cannot be overstated. This seemingly simple process has proven to be the weak link in many wastewater treatment plants, and the technology is intermixed with arbitrary and incorrect "rules of thumb." Well-designed final clarifiers can and are producing effluent qualities once reserved for sand-filtered effluents. This chapter will provide guidelines for the design of hydraulically balanced clarifiers of all types.

INTRODUCTION

While the role of the clarifier in wastewater treatment is well understood, the interactions of the clarifier in the design of the biological system and the interaction components within the clarifier itself are less definitive. That is, due to conflicting opinions and data, the interrelationships of the tank configuration, equipment for flow dispersion and removal, and sludge collection are still controversial topics.

The final effluent of an advanced wastewater treatment plant will generally have a $CBOD_5$ associated with the effluent TSS ($BOD_5 \cdot TSS$ or $f \cdot TSS$) which is much greater than the soluble BOD_5 ($SBOD_5$). The effluent $CBOD_5$, like the influent and primary effluent, can be expressed as follows:

$$CBOD_5 = SBOD_5 + f \cdot TSS \tag{5.1}$$

where in the case of the final effluent, f is the endogenous respiration of the effluent TSS and is modified by the SRT of the system and the operating temperature. Figure 5.1 is a representation of the impact of the effluent TSS on effluent $CBOD_5$.

Progress in the understanding of the activated sludge population dynamics (bulking and foaming organisms) and improvements in clarifier

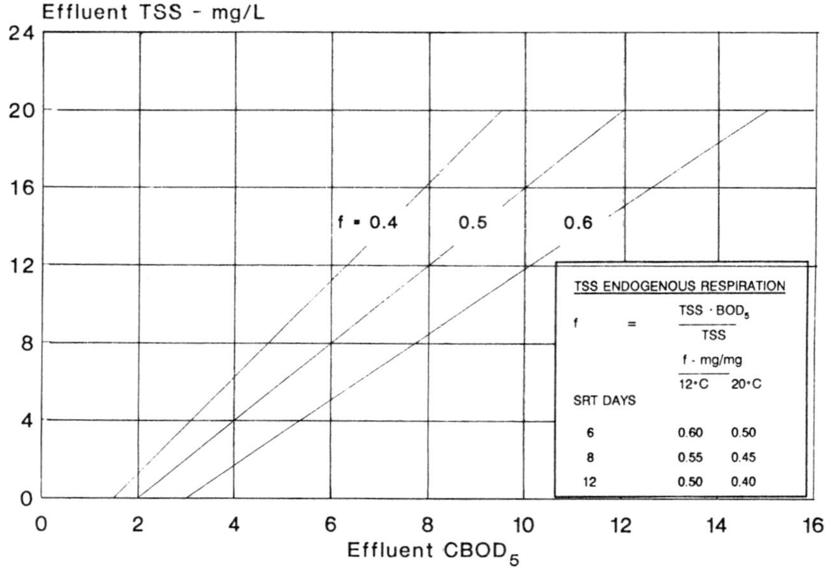

Figure 5.1 Effect of secondary effluent TSS or total $CBOD_5$.

design have resulted in the ability to consistently produce an effluent TSS of less than 10 mg/L for AWT plants, which will result in an effluent BOD_5 of $\leq$5-8 mg/L.

The emphasis of this section will be on secondary clarification. However, those principles that produce good efficiency in the secondary clarifiers work equally well for primary clarifiers. The importance of effective final clarification can never be overstated. Good clarification can often alleviate the need for a costly step of effluent sand filtration. The final clarifiers, rectangular and circular, at the Jackson Pike and Southerly WWTP in Columbus, OH each average 6 mg/L TSS (Albertson et al., 1991) and eliminate the need for sand filters that would have cost $50,000,000.

THEORETICAL AND PRACTICAL CONSIDERATIONS

The basic theory of settling has been well defined by Hazen (1904), Camp (1936, 1946, 1953), Fischerstrom (1955, 1958), Knop (1951), Fitch (1956), Kalbskopf (1970), and many other researchers and practitioners. There is no need to resummarize their important inputs in this publication.

TYPES OF SETTLING REGIMES

Four types, or classifications, of settling are recognized. They are: Type I, free settling; Type II, flocculent setting; Type III, hindered settling; and

Type IV, compaction. These four regimes occur in wastewater clarification processes to a greater or lesser degree depending on the design employed.

- Type I: Free or discrete particle settling typically occurs in grit chambers and primary clarifiers that are not chemically aided. In this type, the particles are assumed to settle independently and at a constant velocity as a function of Newton's and Stoke's Laws. The design of the Type I basin should minimize turbulence in the clarification.
- Type II: Flocculent settling is defined as a condition where particles initially settle independently, but flocculate in the depth of the clarification unit. As a result, the velocity of the settling particle can increase as aggregates of particles occur. There are no mathematical relationships to predict the settling velocity or changes in velocity. Laboratory long column tests are required to define the effects. This type of settling can occur in primary clarification, with and without chemical aids and in fixed film effluents.
- Type III: Hindered or zone settling is typical of the settling characteristics in an activated sludge clarifier or a gravity thickener, wherein the initial suspended solids concentrations are much higher than in flocculent settling. As a result, the mass of solids form and settle as a zone or blanket and the particles remain in relatively the same position to each other. As a result of this zone settling, a relative clear water zone (CWZ) or free settling zone (Type I-II) is formed above the sludge blanket.

  As the sludge blanket settles, a graduation in the sludge density occurs and the settling velocity decreases as the bottom layer approaches the compression zone. The upper layer in the flocculent settling regime will be settling at a faster rate. The settling velocity of the upper zone is decreasing with time as the concentration increases.
- Type IV: Compression settling occurs when there is a structure formed by the particles and further consolidation occurs by compression as a function of time. This type of settling occurs in primary clarifiers, thickeners, and also secondary clarifiers to a limited degree, depending on the design, operation, and type of mechanism employed; i.e., center drawoff or rapid sludge removal.

DETENTION EFFICIENCY

The detention time in clarifiers is significantly less than the displacement time for both rectangular and circular units. The design of the basin and its appurtenances will have a significant effect on the detention efficiency, which is the ratio of the actual to the theoretical displacement time. The actual retention time may refer to either the modal peak of tracer recovery or the time of recovery of a specific amount of the tracer. The loss in detention efficiency is often referred to as short-circuiting.

The detention efficiency in clarifiers can range from as low as 25% to more than 70%. However, most clarifiers operate in the range of 30-50% efficiency, with the best efficiencies produced by the longer (50 m$^+$) and deeper (4 m$^+$) rectangular basins and the smaller-diameter, deep circular basins.

An unpublished study by Erdman (c. 1958) of primary pulp and paper clarifiers for Dorr Oliver, Inc. provided insight into the effect of the ratio of the clarifier diameter to center or side water depth. The conclusions of that study are presented as Figure 5.2. In order to maintain the hydraulic retention efficiency in the range of 40-50%, the clarifier depth must be increased with an increase in the basin diameter. For example, a 2.5 mϕ × 2.0 m deep pilot clarifier will have a detention efficiency of 70-80% and thus can be operated at a higher rate than a full-scale clarifier. Similarly, the quiescent settling rate of dilute activated sludge flocs is 7-8 m/hr, but clarifiers typically operate between 0.5 and 2.0 m/hr due to hydraulic inefficiency and lower settling rate of higher MLSS concentrations.

Rectangular basins have the opposite hydraulic characteristic, short basins short-circuit almost uncontrollably while longer basins have good

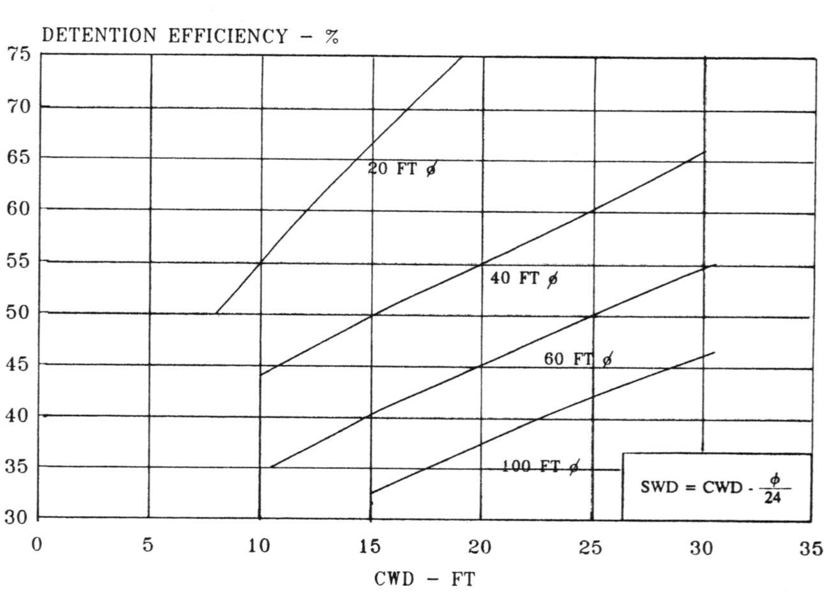

Figure 5.2 Clarifier detention efficiency vs. center water depth.

retention characteristics. While designs have been developed on the basis of maintaining a length/width ratio of ≥ 10:1, more properly, the efficiency is a function of basin length and best performance is produced by basins of 50-80 m long with depths appropriate to the design loadings for solids and flow.

The detention efficiency does not govern the clarification efficiency, but it will limit the clarification capacity. Shallow basins, such as those employed in Holland due to the high groundwater, can and do provide excellent clarification. However, they are operated at relatively low overflow rates to compensate for the inherently reduced detention efficiency.

To maximize the detention efficiency, the clarifier design must minimize energy gradients at the influent and effluent, control density currents, maximize the cross-sectional utilization of the basin, and prevent the sludge blanket from encroaching on the clarification volume.

SHORT-CIRCUITING OF FEED

Short-circuiting occurs in two modes—that is, the feed prematurely reaches the underflow and the effluent. The ideal clarifier would have both feed liquid and feed solids leave the clarifier on a first in-first out basis. While this is not achievable, the goal remains.

Lively et al. (1968) studied the sludge and liquid retention characteristics of a center drawoff 34.1 mϕ × 3.66 m deep clarifier. The results of this study are shown in Figure 5.3, where the dye appeared in the underflow within 10 minutes and peaked in the overflow at 40% of the theoretical displacement time. The clarifier was operated at a 1.55 m/hr rise rate and sludge loading rate (SLR) of 3.9 kg/m^2·hr.

Pfeffer et al. (1968) conducted dye tests of the rectangular secondary clarifiers at Pontiac, MI. The clarifiers were 33.1 m long × 10.1 m wide × 2.90 m deep. As shown in Figure 5.4, there was short-circuiting to both underflow and overflow. Peak (modal) dye concentration was recorded in the overflow in 22% of the theoretical (nominal) displacement time.

Lumley and Horkeby (1988) studied the liquid and sludge retention time in a 60 m long rectangular clarifier. They found that the sludge retention time was 76-91% of the nominal retention time. The modal peak of tracer in the effluent occurred at 32-38% of the nominal retention time (available above the sludge blanket) and the average/nominal retention ratio was 54 to 76%.

Short-circuiting can only be controlled in clarifiers, not eliminated. The later discussion on clarifier configuration and sludge transport is directed to control those conditions that cause short-circuiting. Only the batch reactor in the clarifying mode and small, deep clarifiers in mostly vertical flow will have a very high actual/nominal efficiency.

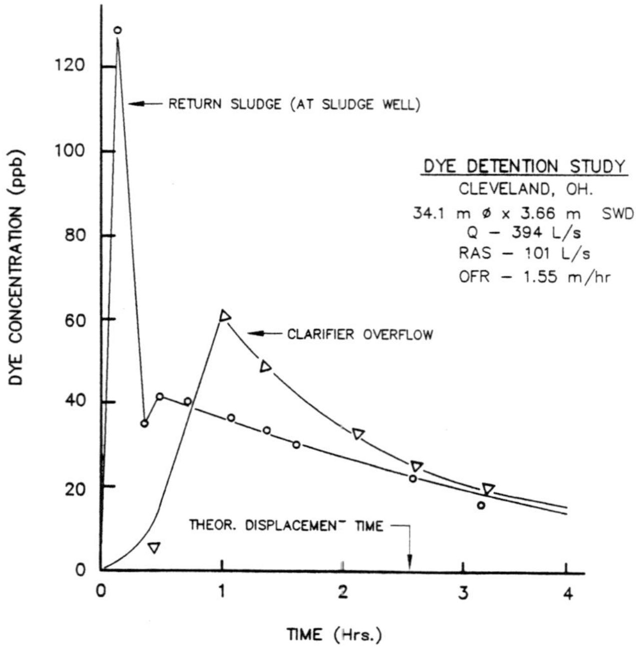

Figure 5.3 Short-circuiting in a circular clarifier (Lively et al., 1968).

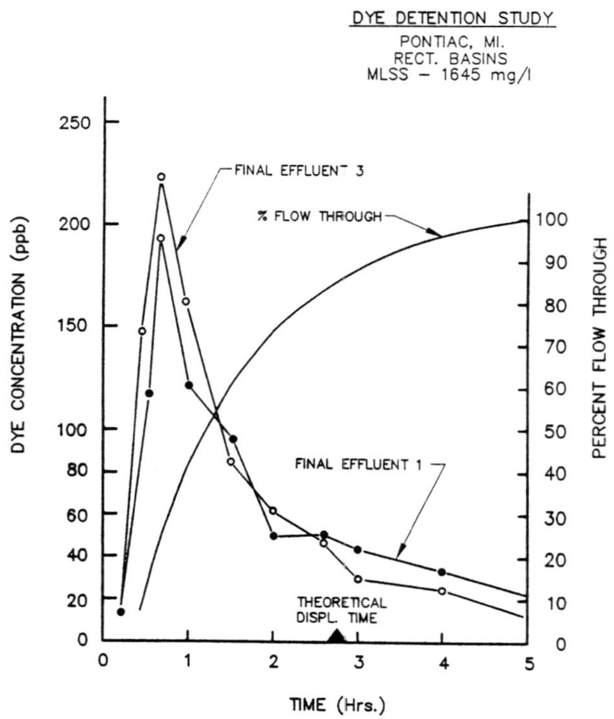

Figure 5.4 Short-circuiting in a rectangular clarifier (Pfeffer et al., 1968).

FLUX THEORY AND APPLICATION

THEORY

The final clarifier must serve a dual role—first, to remove essentially all of the solids from suspension, and second, to concentrate these solids for recycling to the aeration basin. While effluent quality cannot be sacrificed for sludge concentration, it is necessary to provide sufficient concentration to control the recycle rate to a reasonable level. Either the area for clarification, the area for sludge concentration, or specific site conditions will govern the design. The clarification area will be that area required to remove the slowest settling particle consistent with effluent criteria.

Dick (1976) points out that traditional rules of thumb in designing clarifiers should be thrown out, but one cannot ignore factors that detract from the fundamental approach to design and operation of final clarifiers. The flux approach first described by Dick and Ewing (1967) is the basic concept employed in determining the clarifier area required for concentrating activated sludge in final clarifiers, and this practice continues today.

The flow-solids balance in an aerator-clarifier system is described by the following steady-state relationship:

$$(Q + R)\,\text{MLSS} = \text{RAS(RSSS)} + Q(\text{TSS}_e) \qquad (5.2)$$

where:

$Q(\text{TSS}_e) \rightarrow 0$

In order to minimize the flux loading on the clarifier, RAS/Q must be managed; preferably RAS is $\leq 0.7Q$. Higher RAS/Q ratios can result in excessive blanket depths without benefit to the process, as shown by West (1978). The return sludge suspended solids concentration (RSSS) will be a function of the flux loading, kg/m^2·hr, the SVI, SSVI, or DSVI of the sludge, mL/g, and the type and operation of the clarifier. Thus, the RSSS, as a function of the maximum RAS/Q, will limit the MLSS concentration.

The flux loading or sludge loading rate (SLR) is defined as follows:

$$\text{SLR} = \frac{(Q + \text{RAS})\,(\text{MLSS})}{A} \qquad (5.3)$$

or

$$= \frac{\text{RAS (RSSS)}}{A}$$

Here A is the clarifier area, m^2, Q and RAS are in m^3/d, and MLSS and RSSS are in kg/m^3. The flux, or SLR, at any point (p) in the concentration gradient of the return sludge ($RSSS_p$) is also defined by the following equation:

$$SLR = (RSSS_p)(V_p) \tag{5.4}$$

where:

V_p is the settling rate, m/hr at concentration RSSS.

The well known flux curve, as shown in Figure 5.5, is derived from laboratory settling of the mixed liquor as a function of time and then replotted as hindered settling velocity vs. concentration and ultimately as SLR (flux) vs. concentration plots. These procedures were originally defined by Coe and Clevenger (1916), Kynch (1952), Yoshioka (1957), and refined by Dick (1967, 1970), Dick and Young (1972), and Keinath et al. (1977) for secondary clarifiers. A summary of the development of the procedures is provided in the *WPCF Clarifier Design Manual of Practice, FD-8* (1985).

It is important to understand that the flux test procedures do not involve any sludge transport, but the inherent assumption is that the concentrated solids are continuously removed as they reach the design underflow concentration. This factor is most often ignored or not understood in the design of sludge collectors.

APPLICATION

A primary factor in the underflow concentration is the sludge volume index (SVI). Unfortunately, there are a number of methods used to determine SVI and they produce different values, particularly at high MLSS ($\geq$3000 mg/L) and SVIs over 100-125 mL/g. There is controversy over which SVI measurement represents the approximation of the conditions within the clarifier sludge zone. It would appear that the stirred procedure (SSVI) would provide the best means of developing the thickening curve, although Koopman et al. (1983) employed a dilute SVI (DSVI) procedure with success.

Keinath and Wahlberg (1990) evaluated various methods of SVI determinations in a 1 L cylinder, 2 L settlometer and a 1.22 m settling column with and without stirring. The results indicated that the stirred SVI produced similar results in both the 1 L and 2 L vessels. The unstirred results were more consistent in the 2 L vessel. The long settling column SVIs compared

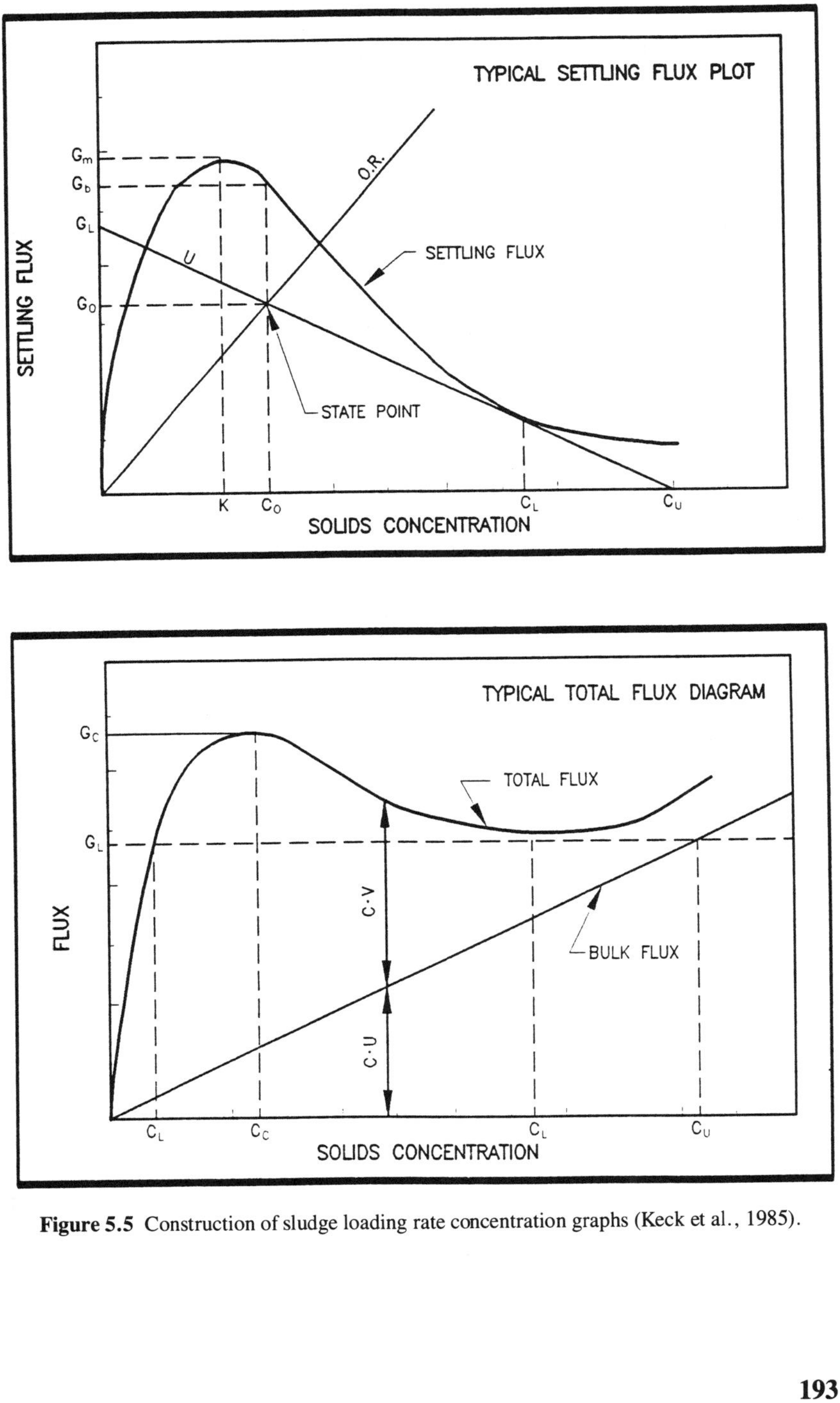

Figure 5.5 Construction of sludge loading rate concentration graphs (Keck et al., 1985).

poorly with the other test procedures. DSVI comparative tests were not conducted. Some of the SVI test results are shown in Figure 5.6.

Rachwal et al. (1982), evaluating control of bulking sludges, conducted an extensive analysis of the SVI vs. the SSVI test. The results indicated that the SVI was unreliable for MLSS analysis when the concentration was higher than 1200 mg/L at an SSVI of 80-85 mL/g. At 3500 to 4000 mg/L, the SVI was about 300% higher than the SSVI.

Rachwal et al. (1982) also demonstrated that the settling velocity of the MLSS was a function of the SSVI, and presented data to demonstrate that the maximum free-settling velocity (V_o) was a function of the $SSVI_{30}$. The correlation indicates that V_o decreases from 6.5 m/hr at $SSVI_{30}$ of 60-79 mL/g to 4.2 m/hr at $SSVI_{30}$ of 120-129 mL/g. The projected flux limits are defined by the following expression:

$$\mathrm{SLR} = V_o\, C_i\, e^{kC_i} + \frac{\mathrm{RAS}}{A} C_i \tag{5.5}$$

where:

RAS = underflow rate, m^3/hr
A = clarifier area, m^2
C_i = MLSS concentration, g/L
k = a constant, function of SSVI

The expression and coefficients were generated from several hundred tests over the SSVI and MLSS ranges set forth in Table 5.1. The r^2 values for each range of SSVI are 0.94 to 0.96, representing an excellent correlation.

Daigger and Roper (1985) described a relationship between the settling characteristics and the SVI. The objective was to correlate the test data to

TABLE 5.1. Correlation of Hindered Settling Velocity as a Function of SSVI and Concentration.

SSVI Range (mL/g)	Number of Results (n)	Solids (c) Range (g/L) Min.	Max.	$V = V_o e^{-kC}$ V_o	k	Correlation Coefficient r^2
60-79	83	0.65	9.80	6.500	0.3588	0.9607
80-99	310	1.23	11.0	5.826	0.3712	0.9591
100-119	148	1.55	9.62	5.300	0.4067	0.9374
120-139	168	1.53	8.31	4.234	0.4222	0.9618
57-146	733	0.65	11.0	5.182	0.3847	0.8732

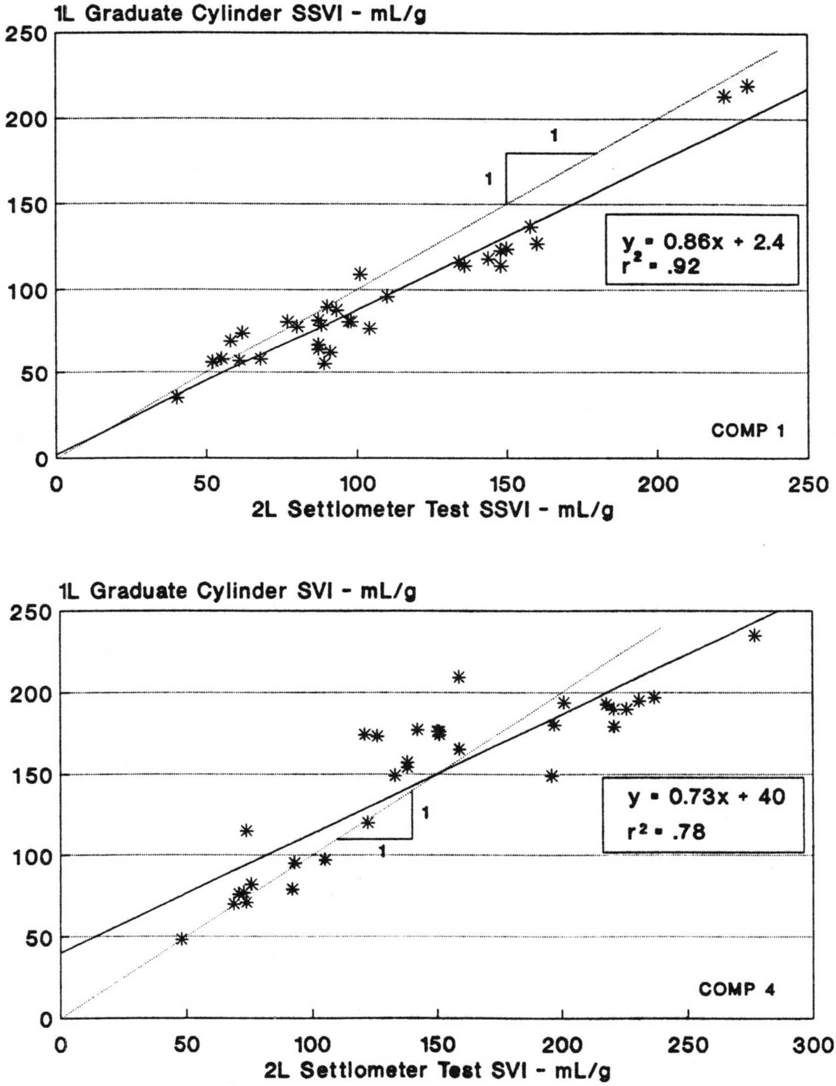

Figure 5.6 Results of SVI vs. SSVI tests in 1 L cylinders vs. 2 L settlometers.

develop a flux approach that could easily be applied to plant design and operation. The standard 1 L cylinder SVI tests were used and the SVIs were determined in 13.3 cmϕ × 1.82 m long tubes. The pilot plant clarifiers were 2.44 mϕ × 2.67 m deep and the pilot plants processed a variety of wastewater types and aeration modes.

Reduction of the data with SVIs ranging from 36-402 mL/g led to the following correlation:

$$V_i = V_o e - kC_i \tag{5.6}$$

$$V_i = 7.80e^{-[0.148 + 0.00210(\text{SVI})]C_i}$$

where:

V_i = initial settling velocity, m/hr
V_o = maximum settling velocity, m/hr
k = correlation coefficient, $r^2 = 0.997$
SVI = unstirred 1 L cylinder SVI, mL/g
C_i = initial SS concentration, g/L

This relationship was then reduced to a flux vs. underflow graph that could be employed for design and operation of secondary clarifiers. This graph is reproduced as Figure 5.7. The authors felt that adequate safety factor was inherent in the curves. However, this still requires good hydraulic design and allowance for peak sludge volumes in the clarifier.

Keinath (1990), using the SSVI data and correlation developed by Wahlberg and Keinath (1988), developed a similar design monograph for settling velocity (V_s) as a function of MLSS concentration (C_i) and SSVI. The correlation for 35 mL/g to 220 mL/g SSVI was as follows:

$$V_s = [15.3 - 0.0615(\text{SSVI})]e^x \tag{5.7}$$

where:

$$x = [-0.426 + 0.00384(\text{SSVI}) - 0.0000543\,(\text{SSVI})^2]C_i$$

This correlation was developed from a twenty-one-plant survey and produced settling fluxes substantially less than those of the Daigger-Roper formulation. The SVI was corrected to SSVI to make the comparison. Keinath opined that the Daigger-Roper curves represent special wastewater conditions as well as the imprecision of the SVI test. The Wahlberg and Keinath formulation, reduced to the same format as Daigger-Roper, is shown in Figure 5.8.

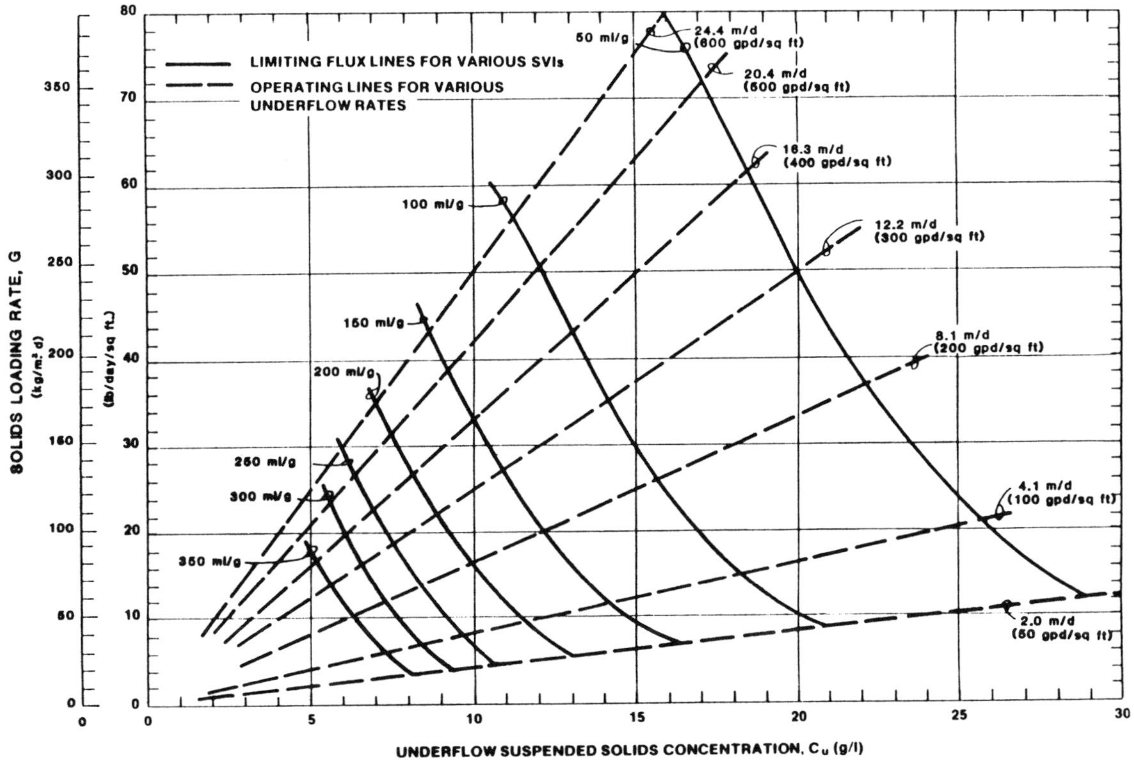

Figure 5.7 Daigger-Roper secondary clarifier solids flux curves.

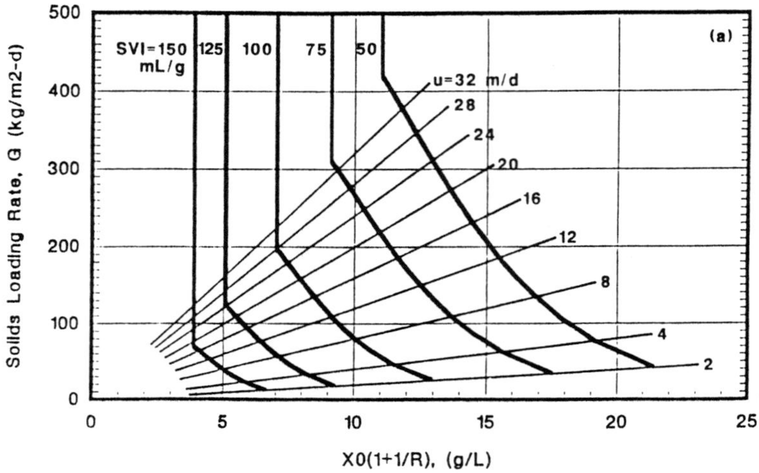

Figure 5.8 Keinath secondary clarifier solids flux curves.

Firmin (1990) evaluated the activated sludge settling characteristics of three wastewater treatment plants and represented the sludge loading rate as a function of the return activated sludge concentration in a multiple regression equation.

$$\text{SLR} = A - B(\text{RAS}) + C(\text{RAS})^2 \tag{5.8}$$

where A, B, and C are constants that vary as a function of the SVI. The SVIs in the plants studied ranged from 50-400 mL/g.

The settling velocities determined by Firmin yielded initial settling velocities (V_i) equal to or greater than those predicted by the Daigger-Roper formulation. The sludge loadings developed in the long tube tests compared favorably with those found by Daigger-Roper and are the same as those found in the *WPCF Manual of Practice* (1977) as well as those predicted by Pitman (1983). Firmin concluded that the Daigger-Roper curves were valid and provided the rearrangement shown in Figure 5.9 projecting the SLR at failure vs. SVI and RSSS.

Kalbskopf (1972) reported that the underflow concentration would be as follows:

$$\text{RSSS} = \left(\frac{100}{\text{SVI}}\right) 12{,}000 \text{ (mg/L)} \tag{5.9}$$

At 100 mL/g SVI, the Kalbskopf expression projects an operating underflow of 12,000 mg/L. The Daigger-Roper curve estimates that the failure condition for 12,000 mg/L RSSS would be at 10 kg/m²·hr or 50-70% higher than current SLR practice. In general, Daigger-Roper projects underflow concentrations about 20-25% higher than Kalbskopf.

CLARIFIER CONFIGURATIONS

The two basic types of clarifiers are (1) circular and (2) rectangular. Square clarifiers can be classified as circular if center fed and as rectangular if cross fed-takeoff. Rectangular clarifiers may also be stacked if space is limited. Stacked circular clarifiers have been employed only for industrial process applications involving fast-settling particulates.

CIRCULAR AND SQUARE

The circular clarifier is the most popular design of clarifier and is preferred for secondary clarification and for both primary and secondary

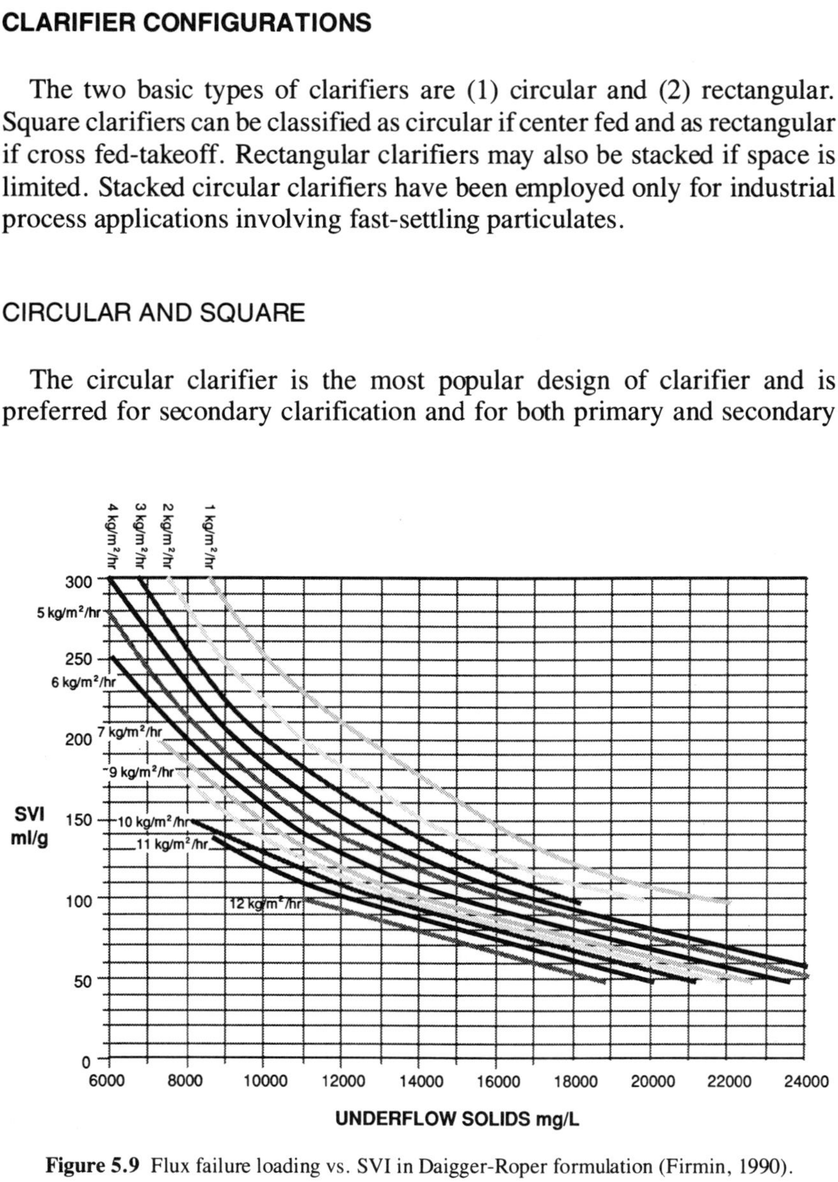

Figure 5.9 Flux failure loading vs. SVI in Daigger-Roper formulation (Firmin, 1990).

units in smaller plants. While center-fed square units have been employed, the problem of scraping the corners has never been satisfactorily resolved and few new units are being considered.

The design of the circular clarifier must consider a balance between the necessity to dissipate energy while assuring that the flow to the clarification zone is balanced. While hydraulic efficiencies are usually less than 50% in large units, current practices using energy-dissipating inlet, large flocculating feedwells, deeper basins, and Stamford baffles to eliminate wall currents and permit single weirs on even larger basins, have generally improved performance and increased capacity.

The sludge collection method for secondary clarifiers is divided among advocates for central drawoff vs. rapid sludge removal. The type of rapid sludge removal equipment design influences the floor design as well as other features. The rationale for type of sludge collection equipment will be addressed later.

Kalbskopf (1970) reported that circular clarifiers were less costly than rectangular units to construct for tank volumes of from 500 m^3 to 6000 m^3. For larger volumes, the costs were considered equal. However, fewer numbers of circular clarifiers are being constructed, which reduces the cost.

Circular clarifiers have been constructed up to 91.5 mϕ for industrial activated sludge clarification and to 70 mϕ at the Köln-Stammhein WWTP, Germany. A center feed-center drawoff circular secondary clarifier cross section is shown in Figure 5.10.

In North America, most circular clarifiers are centrally driven. Elsewhere the peripheral drive is quite popular. The decision on the drive mode has little or no influence on the performance characteristics, and is considered to be a local option.

RECTANGULAR, SQUARE, AND STACKED

The square, cross-fed unit is essentially a rectangular unit with $L = W$. These units are not efficient for wastewater treatment and they have sludge removal problems, so their use should be discouraged. Rectangular clarifiers need to be long, to provide an opportunity for energy dissipation. While $L = 10$ to $20\ W$ is often cited as a design basis, the primary factor is length with the appropriate depth as a function of the hydraulic rates. Basins that are 60 to 90 m long are more hydraulically efficient because they use distance to eliminate influent turbulence.

Inlet distribution and good energy dispersion for rectangular basins are difficult and require careful engineering designs. Literature reports that energy currents can persist for long distances and upturn at the end wall, especially in the shorter basins. Preflocculation of mixed liquor prior to the clarifier at Emscher (Knop, 1951) was beneficial to the clarification process. Design practices for primary and secondary clarifiers are similar with the exception that the quantity and volume of the sludge in the secondary clarifiers must be factored into the designs.

Secondary rectangular clarifiers with lengths exceeding 80 m are employed in several facilities. The sludge takeoff in longer basins is usually at midpoint or influent and midpoint. A few facilities remove the sludge at the effluent end, but this is not recommended for AWT plants due to denitrification. A typical rectangular clarifier cross section is shown in Figure 5.11.

Stacked clarifiers were reported by Kalbskopf (1970) to be initially constructed in Sweden and later in Hamburg, Germany as shown in Figure 5.12. This design is used extensively in Japan and recently in Westchester Co., NY (Kelly, 1989), and will be employed for Deer Island WWTP,

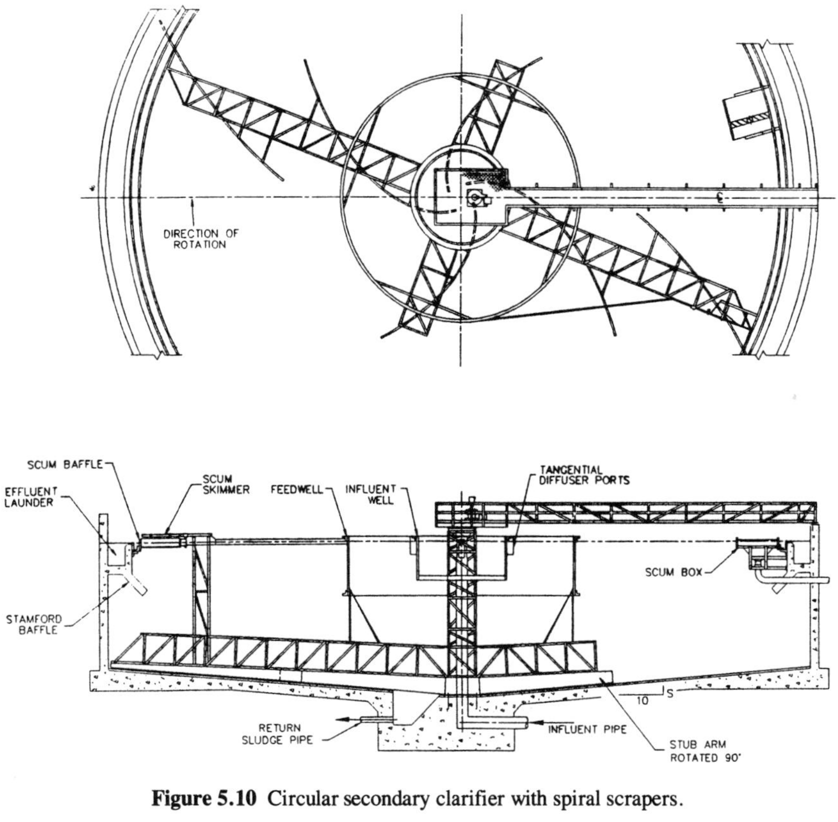

Figure 5.10 Circular secondary clarifier with spiral scrapers.

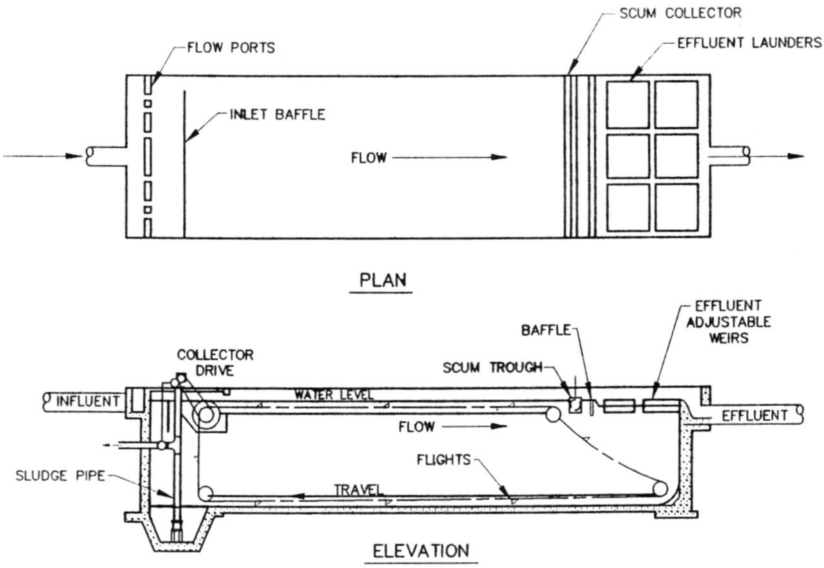

Figure 5.11 Chain and flight rectangular clarifier.

Boston, MA. These designs are used only where space limitations dictate. The inaccessibility of the units for maintenance, higher costs, and the loss of two units if there is a mechanical failure are disadvantages of the design.

These are special design considerations specific to the stacked clarifiers that necessitate careful analysis by the engineer. Due to the limited applicability, they will not be discussed further.

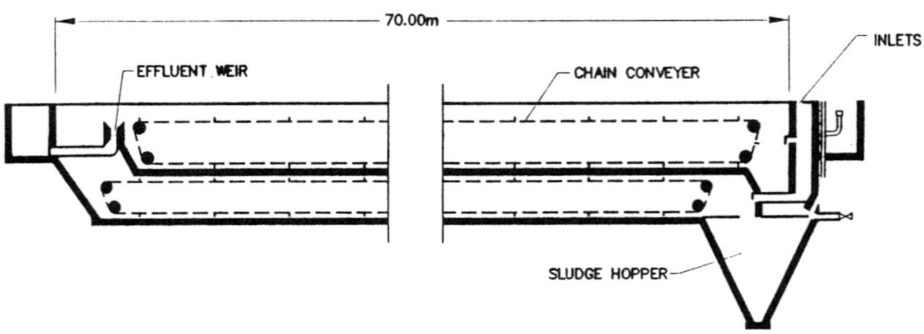

STELLINGER MOOR

DETENTION TIME : 2.5 h

OVERFLOW RATE : 0.75m/h

Figure 5.12 Stacked clarifier—city of Hamburg, Germany.

INTERCHANNEL

Clarifiers have been constructed within the orbital oxidation ditches and have been called interchannel clarifiers. Bender (1987) reported on eight different concepts of these clarifiers and notes that typical operator strategies to handle peak flows and changes in sludge settleability are not available with these designs. While Bender reported that there were ninety-two plants in operation or under construction, the limited amount of full-scale independently gathered operating results precluded a further discussion of this design. Interested engineers should carefully research the concept for data gathered independently and survey operating plants.

CLARIFIER CONFIGURATION AND FEATURES

This chapter will be more oriented to the design features of secondary clarifiers, but the importance of primary clarifier role is recognized. However, there is less controversy regarding the design of primary clarifiers, with the exception of the configuration—rectangular or circular.

PRIMARY CLARIFIERS

The use of primary clarifiers in AWT systems is a function of the BNR process requirements and influent characteristics. In order to enhance phosphorus and total nitrogen elimination, primary clarifiers have been omitted in some designs. Due to the much higher sludge yield (50-70%) without primary clarification, the biological basins are proportionally increased in size. The elimination of the primary clarifiers also produces a waste sludge that is more difficult to process. Primary clarifiers do not remove a significant amount of readily biodegradable BOD_5 or COD even when operating at optimum efficiency, and have been used to produce volatile acids by fermentation (Barnard, 1984).

Surface Area and Depth

The efficiency of primary clarifiers is more related to the detention time than the overflow rate (OFR). An extensive review (Albertson, 1963) of long-term separations (up to forty years) of many primary clarification facilities produced the following continuous empirical relationships:

$$OFR_{MX}\ (m/hr) \leq 0.182\ SWD^2\ (SWD = 1.83 - 3.05\ m) \quad (5.10)$$

$$OFR_{AVG}\ (m/hr) \leq 0.091\ SWD^2\ (SWD = 1.83 - 3.05\ m) \quad (5.11)$$

$$\mathrm{OFR_{MX}}\ (\mathrm{m/hr}) \leq 0.556\ \mathrm{SWD}\ (\mathrm{SWD} = 3.05 - 4.57\ \mathrm{m}) \quad (5.12)$$

$$\mathrm{OFR_{AVG}}\ (\mathrm{m/hr}) \leq 0.278\ \mathrm{SWD}\ (\mathrm{SWD} = 3.05 - 4.57\ \mathrm{m}) \quad (5.13)$$

The area would be determined for both average and maximum day/week flows and the larger of the two areas is used for design.

The maximum overflow rate ($\mathrm{OFR_{MX}}$) was based on plotting long-term performance data and determining the point where there was a 10% reduction in the TSS removal versus that removal achieved at 50% of the $\mathrm{OFR_{MX}}$. The data also demonstrated that there was relatively little change in raw wastewater settleability in the period 1920 to 1963.

The primary clarifiers were both rectangular and circular, but circular units dominated the base. The SWDs ranged from 1.82 to 4.73 m. Two plants, one with rectangular units 58.4 mL × 4.73 m SWD and one with 35.1 mϕ × 3.66 m SWD, were able to process flows in excess of 3.0 m/hr without significant loss of clarification efficiency. In rectangular clarifiers, the significant factors are length and depth, and the width is not significant.

Floor Slope, Diameter, and Length

The primary clarifier should be designed to maximize the underflow concentration while minimizing the inventory, if sludge concentration is the objective. Even where septicization is practiced to produce volatile acids by recirculation of underflow sludge to the clarifier influent, the following recommendations are valid. Since the sludge retention on the sloped portion of the clarifier floor is 33% of the storage volume above the cone, sludge storage/concentration should generally be within the sloped portion of the floor.

The floor slope in primary clarifiers should not be a constant value; for example, 1:12. The floor slope should always provide a minimum of 1.2 m of sludge depth over the sludge withdrawal hopper before the sludge level would extend to the wall. For example, a 1:12 floor slope in a 12 mϕ clarifier provides only 0.5 m of depth at hopper (CWD − SWD) and the floor slope should be increased to 2.4/12 to provide the 1.2 m recommended slope depth.

The *U.S.EPA Design Manual* (1987) for sludge dewatering provided recommendations for the floor configuration of small to large circular clarifiers in order to effectively concentrate the sludge and to co-thicken primary and secondary sludges. Figure 5.13 is a summary of the various primary clarifier design recommendations. The rectangular basins should have similar floor slopes ($s{\cdot}L \geq 1.2$ m) that eliminate the need for storage for sludge concentration above the effluent end SWD. Further, rectangular

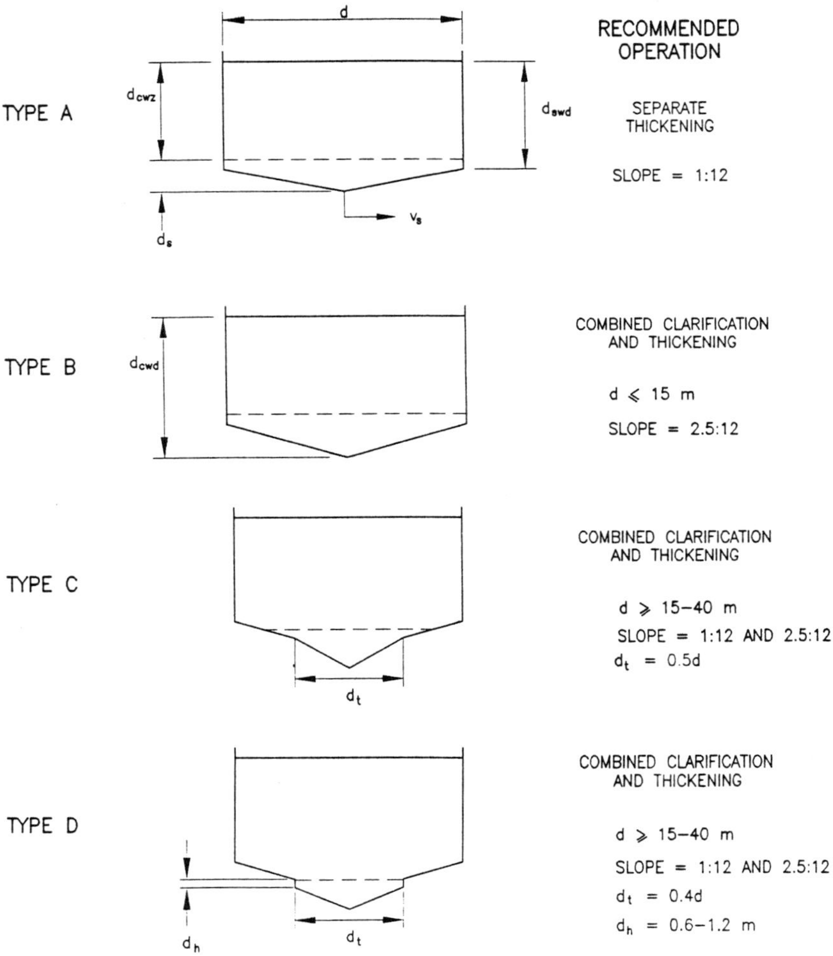

LEGEND

d = DIAMETER
d_{cwz} = DEPTH OF CLEAR WATER ZONE
d_s = DEPTH OF SLUDGE
d_{swd} = SIDE WATER DEPTH
d_{cwd} = CENTER WATER DEPTH
d_t = DIAMETER AT THICKENER ZONE
d_h = HEIGHT OF VERTICAL WALL

Figure 5.13 Recommendations for primary clarifier floor configuration.

clarifiers should have lengths ≥ 40 m, preferably 50-80 m, for best performance and depths consistent with those recommended for circular clarifiers.

Inlet and Feedwells

The method of feed introduction into clarifiers is the major cause of effluent short-circuiting. There has been inadequate attention paid to the design of influent distribution in primary clarifiers. The recommendation for circular clarifiers is a tangential energy-dissipating inlet within the feedwell and a feedwell sized on the hydraulic loading of the specific clarifier. This type of inlet structure is shown in Figure 5.14.

The energy dissipating inlet (EDI) functions to collect the influent from the center column and discharge it tangentially into the upper 0.5-0.7 m of the feedwell. The discharge ports are sized to produce a velocity of ≤0.75 m/s at maximum flow and 0.30-0.45 m/s at average flow. Where there are flow ranges in excess of Q_p/Q_A of 2.5, special designs are required, such as overflowing the inlet device into the feedwell. Movable gates are not recommended due to operator requirements.

The feedwell should be sized such that the maximum downward velocity does not exceed 0.75 m/s considering the total feedwell area, ignoring center column. The depth of the feedwell should extend about 1.0 m below EDI outlet ports. The depth of the feedwell should provide a cross-sectional area between the bottom of the feedwell and the side water depth that should be 100-120% larger than the feedwell area.

The sizing and configuration of a secondary clarifier inlet and feedwell will be provided in a later section. Calculation can be employed in a similar manner for primary clarifiers.

Baffles, Weirs, and Launders

A single weir and launder mounted on the inner side of the wall is recommended for circular clarifiers. It is necessary to provide multiple inboard launders on rectangular basins to prevent the development of velocity currents.

The circular clarifier should be equipped with Stamford baffles (Semon, 1982) either constructed as a part of the effluent launder or constructed below the effluent weir as shown in Figure 5.15. The inward projection length (SB) is not technically founded, but the current recommendations are as follows:

$$\text{SB(cm)} = 46 + 1.67\,(\phi - 9\text{ m}) \tag{5.14}$$

but not larger than 120 cm (1.2 m).

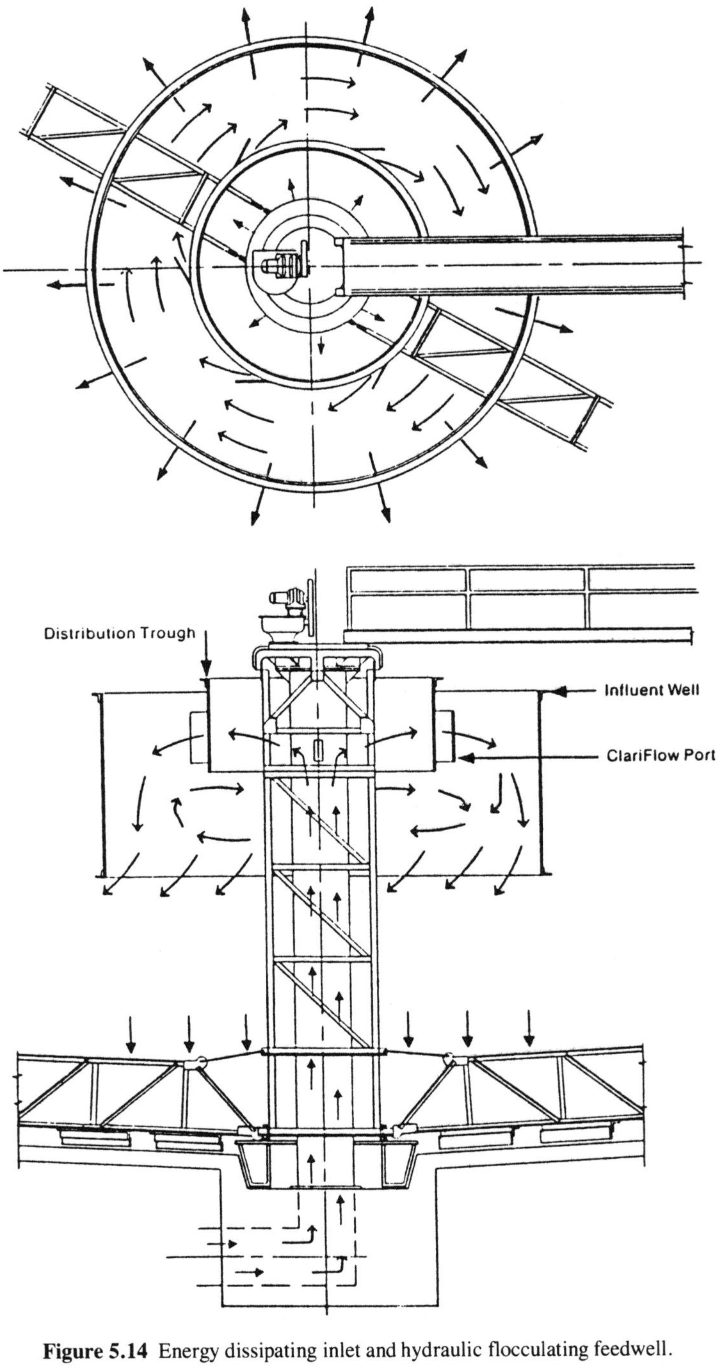

Figure 5.14 Energy dissipating inlet and hydraulic flocculating feedwell.

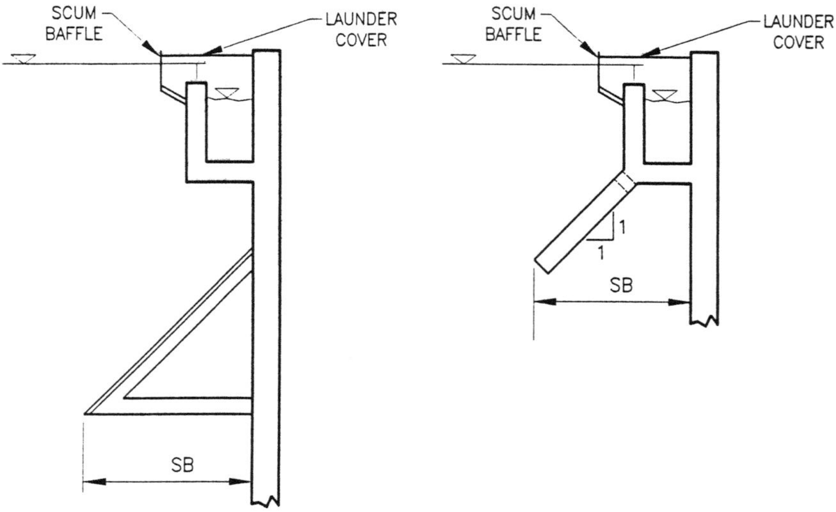

Figure 5.15 Two arrangements of the Stamford baffle for secondary clarifiers.

This same type of baffle can be placed at the effluent end of rectangular basins to prevent the upwelling currents described by Anderson (1945) from forming and carrying over suspended solids. Improvement in performance due to the Stamford baffle is shown in Figure 5.16.

SECONDARY CLARIFIERS

The same basic strategy applies to secondary clarifiers. The effluent-oriented aspects of the clarifier design must be directed to recovering the small amount of straggler floc in order to achieve low final effluent TSS of 5-15 mg/L in AWT plants. Design recommendations in this presentation are oriented to this effluent TSS level.

Surface Area and Depth

The hydraulic loading recommendations follow those set forth for primary clarifiers. However, the maximum daily flow rate should not produce an overflow rate in excess of that set forth by the equations. The overflow rate is based on the clear water zone (CWZ) or free settling zone (Type I-II) above the maximum height of the sludge blanket.

$$OFR_{MX}\ (m/hr) \leq 0.182\ CWZ^2\ (CWZ = 1.83 - 3.05\ m) \quad (5.15)$$

$$OFR_{AVG}\ (m/hr) \leq 0.091\ CWZ^2\ (CWZ = 1.83 - 3.05\ m) \quad (5.16)$$

$$OFR_{MX}\ (m/hr) \leq 0.556\ CWZ\ (CWZ = 3.05 - 4.57\ m) \quad (5.17)$$

$$OFR_{AVG}\ (m/hr) \leq 0.278\ CWZ\ (CWZ = 3.05 - 4.57\ m) \quad (5.18)$$

Where there are short-term very high peak flow conditions, an operational strategy of partial bypassing should be considered for temporarily increasing the clarification capacity. The design procedure is set forth in the operating/control strategy discussion.

The effective depth in secondary clarifiers is the clear zone above the sludge blanket. For center drawoff and rectangular clarifiers, it is recommended that the design be based on maintaining the sludge level at or before the junction of the floor wall or endwall under average conditions. At peak flow conditions, the increase in sludge blanket depth (SBD) due to higher flux and solids inventory transfer from aeration basin to clarifier should be ≤ 1.0 m. Thus, the clear water zone (CWZ) above the sludge blanket would be (SWD − SBD) m for maximum flow conditions, and this depth is denoted in the above equations. The sludge blanket depth growth at peak flows is less in shallower clarifiers due to the larger floor area and correspondingly lower SLR.

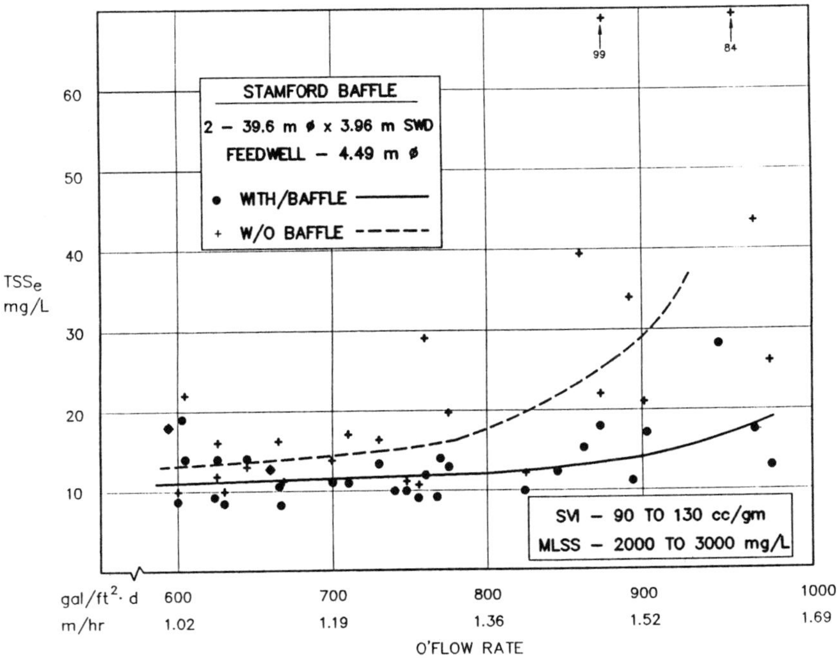

Figure 5.16 Clarification improvement with Stamford baffle.

The SWD requirements for rapid sludge removal equipment (common in North America only) can be different from those of conventional clarifiers. Two types of mechanisms are employed, one with a multiple number of uptake pipes externally controllable, and the other with a suction header and fixed openings. The first type of mechanism can and should employ the same type of floor and SWD configuration as the conventional clarifier. The latter design requires a relatively flat floor to ensure that the sludge loading is spread evenly across the floor. In this design, the sludge depth at the sidewall will be 0.6-1.5 m and the SWD recommended by Parker and Stenquist (1986) and Parker (1990) is 5.0 to 6.3 m. These tanks must be constructed about 1.0-1.5 m deeper than sloped floor clarifiers due to sludge buildup at sidewall and feedwell clearance, as discussed later.

The SVI will modify the sludge blanket depths. The above recommendations were for an SVI $\leq$ 120 mL/g. With currently available biological selector technology, this SVI criterion is generally achievable. When high-SVI sludges must be processed, the sludge zone could increase or a more dilute sludge must be removed, and this should be considered in the design.

Floor Slope, Diameter, and Length

The floor slope of a circular or rectangular clarifier should provide a minimum of 1.2 m of sludge depth over the hopper without the sludge blanket encroaching on the sidewall of circular clarifiers at average flow. Increased allowance for sludge depth for larger clarifiers is required to ensure that the sludge hopper does not rat-hole (short-circuit) feed to underflow. The recommended minimum floor slope as a function of diameter is set forth in Table 5.2.

TABLE 5.2. Floor Slope Recommendations for Secondary Clarifiers.

Diameter (m)	Minimum Floor Slope m/m	Minimum CWD – SWD[2] (m)
10	0.240	1.0
15	0.173	1.1
20	0.140	1.2
30	0.107	1.4
40	0.090	1.5
≥50	0.083	≥1.8
≥50	0.042/0.125[1]	≥1.8

[1]Dual slope, change at mid radius and evaluate whether dual sludge return hoppers required.
[2]Based on hopper @ 0.85 *R*, flat floor sludge hopper to center column.

Inlet and Feedwells

The tangential inlet structure shown in Figure 5.14 is also recommended for secondary clarifiers. It serves a twofold purpose—to distribute the flow evenly into the feedwell proper, and to produce a low-velocity circulation to encourage flocculation and aggregation of the sludge particles. The energy dissipation inlet (EDI) is designed to transfer the flow from the center column outlet ports into the upper 0.6-0.7 m of the feedwell in order to minimize the feedwell depth. The discharge velocities recommended are based on $Q_{MX} + R$ and are the same as set forth for primary clarifiers.

The sizing of the feedwell should provide for a maximum downward velocity of ≤ 0.75 m/s. The depth of the feedwell should extend about 1.0 m below the bottom of the EDI outlet ports. The CWZ below the feedwell would extend from the top of the consolidated sludge blanket to the bottom of the feedwell. Ideally, the CWZ area will be 100 to 120% of the feedwell cross section. In order to meet these requirements, it is generally necessary to minimize the feedwell depth. Deep feedwells will scour blankets as well as displace sludge away from the central area, which can cause short-circuiting to underflow and overflow.

Mechanical flocculating feedwells have been employed for activated sludge clarification. External flocculation has been employed for rectangular basins and circular basins normally use an enlarged feedwell. The mechanical flocculator designs have employed slow tip speed (0.15-0.6 m/s) centrally driven paddles or two to four individual flocculators in the feedwell. The individual feedwell mechanical flocculators operate with tip speeds of 0.6-1.8 m/s, which can break up floc and are not recommended.

Parker (1983), Parker and Stenquist (1986), and Parker (1991) reported on the performance of deep (4.9-6.1 m) secondary clarifiers with large feedwells and internal flocculators. These units provided treatment capacity up to 2.0 m/hr without significant loss in efficiency; that is, effluent TSS were on the order of 10 mg/L. The authors of these papers employ cantilevered inboard launders as part of their design philosophy to control the approach velocity, while other engineers advocate the use of the Stamford baffle. The maximum hydraulic capacity of the deep, relatively flat floor, rapid sludge removal, flocculating clarifiers is consistent with the formulation advanced earlier for both primary and secondary clarifiers.

Current practice is favoring the use of tangential energy-dissipating inlets where hydraulic energy is used to promote floc growth in the feedwell zone. Air should never be used for mixing energy as the dispersed air will cause floc breakage. Aeration of channels to the clarifiers should always be minimized for the same reason and cross-roll arrangement used, if employed.

Less significant in current practice are peripheral feed designs using a skirt around the clarifier for the feed introduction. The effluent launder has been located at the inlet area as well as the central area. Baumann et al. (1965) reported on pilot studies of center vs. peripheral feed for both discrete and flocculent solids. The peripheral-fed unit was more efficient when the center feed influent pipe was initially undersized. When the inlet feed velocity was reduced, <0.75 m/s, the center feed unit outperformed the peripherally fed clarifier over a hydraulic rise rate of 1.36 to 2.2 m/hr.

The peripherally fed and takeoff units have less stability over a wide range of flow and tend to short-circuit at low flows. Denver, Colorado modified its ten peripherally fed-takeoff units to center feed peripheral takeoff in 1990.

Baffles, Weirs, and Launders

The question of weir rate, $m^3/m{\cdot}d$, effects on effluent quality has never been satisfactorily answered. The practice of limiting the weir loading rate to 124-186 $m^3/m{\cdot}d$ in the United States dates back to work on rectangular and square basins in 1931-32 in Chicago, Illinois and Springfield, Illinois, respectively. Anderson (1945) presented results showing that the effluent TSS was a function of weir placement and overflow rate and that 122 $m^3/m{\cdot}d$ with double weirs at 0.60 R and 0.76 R were superior in performance to a single inboard launder at the wall. This author determined that the cantilevered inboard weir rates should not exceed 250 $m^3/m{\cdot}d$ and be not more than 190 $m^3/m^2{\cdot}d$ for weirs located at the wall.

This practice has continued in the U.S. for clarifiers over 30 m in diameter. The question is whether the very costly and difficult-to-service inboard launders provide a benefit. Studies at Pontiac, Michigan (Dorr Oliver, c. 1965), and Phoenix, Arizona (Hendricks, 1990) and the replacement of the cantilevered inboard launders at Memphis, Tennessee (Alfonso, 1990) by a single weir and Stamford baffle at the periphery suggest a reevaluation is in order.

The design improvements in clarifier inlets and outlets may be the reason for the contradictions. Inspection of Anderson's second-generation designs, 38.4 mϕ $\times$ 3.35 m d units, reveals a feedwell that is 6.1 mϕ $\times$ 1 m deep. The center column outlets extend perhaps 1 m below the bottom of the feedwell. The feedwell velocity would be 1.62 m/min at design conditions of 2.03 m/hr rise rate and the center column flow would jet under the feedwell. These criteria would not be considered acceptable by today's standards.

Industrial secondary clarifiers have been constructed up to 91.4 mϕ employing a single inboard launder mounted on the wall. This has been standard practice for large industrial clarifiers for forty years and continues today.

Albertson et al. (1991) reported on thirty months of data from the new 61 mϕ × 4.57 m d secondary clarifiers at Columbus Southerly, which produced an average effluent TSS of 5.5 mg/L and operated at monthly average rise rates of 1.2 m/hr, peak rates of 1.8 m/hr, and MLSS up to 5000 mg/L. There are single weirs mounted above 1.22 m wide Stamford baffles. The peak weir rates have exceeded 650 m^3/m·d and average monthly rates of 434 m^3/m·d have resulted in effluent TSS of 3-10 mg/L TSS. The average monthly data for Columbus Southerly is reported in Table 5.3.

Rachwal et al. (1982) studied a secondary clarifier with cantilevered double weirs-launder. Three studies were conducted with the double weir and four with the outer weir blanked off. These studies revealed washout only over the outer weir in the double-sided arrangement. No benefit was found, nor predicted, for a cantilevered double weir.

The Phoenix, Arizona 23rd Ave. WWTP (Albertson and Hendricks, 1991) large shallow (2.74 m) clarifiers are equipped with cantilevered launders. Observations and photos from a crane 30 m above the clarifier revealed velocity currents rising directly up to the inner or bouncing off the wall to the weir depending on scraper location. The addition of a Stamford baffle was not effective since the flow diverted from the wall to over the weir. The weir and launder must be above and behind the Stamford baffle. The new 54.9 mϕ × 4.88 m SWD units will employ this baffle arrangement.

Based on use of current clarifier design technology, there doesn't appear to be any demonstrated process benefits associated with cantilevered inboard launders in final clarifier up to 60 mϕ, perhaps larger. The inboard launders are costly and difficult to keep clean. Sludge deposits build up on the supports and periodically float. The area between the cantilevered launder and the wall cannot be skimmed by a center-driven mechanism.

In AWT plants the effluent launders should be covered to prevent the effusive growth of algae in and around the scum baffle, weirs, and launders. The covers may be constructed of concrete, fiberglass, or metal and should extend over the scum baffle and supports to eliminate areas of attachment. The clearance above the water level should only be sufficient for the skimmer. An arrangement for covers was shown in Figure 5.15 with the Stamford baffle.

BASIS OF DESIGN

Efficiency of treatment will be the same for a well-designed circular or rectangular clarifier. This doesn't imply that they are operating at the same

TABLE 5.3. Columbus Southerly Secondary Effluent Quality and Operating Parameters.

Date	Flow L/s	$CBOD_5$ mg/L	TSS mg/L	TKN mg/L	NH_4—N mg/L	TP mg/L	MLSS mg/L	RAS mg/L	SVI mL/g
Jul-88	4074	2	4	1.7	0.33	1.6	2520	8840	128
Aug	3969	2	4	1.8	0.25	0.5	1660	8150	90
Sep	4109	2	4	1.5	0.11	2.3	2000	8210	92
Oct	3505	2	5	1.7	0.12	2.4	2080	7700	96
Nov	4298	2	7	1.6	0.05	1.5	2180	8660	90
Dec	3838	3	9	1.8	0.08	2.0	2660	9570	102
Jan-89	4583	3	10	1.7	0.07	0.1	2850	9400	105
Feb	4390	3	10	1.7	0.09	2.0	3190	9140	78
Mar	4981	2	7	1.6	0.11	1.0	3210	12,500	65
Apr	5218	2	7	1.3	0.07	0.9	3580	14,500	70
May	5240	2	5	1.2	0.08	1.2	3410	14,300	80
Jun	5073	1	4	1.2	0.08	0.9	3370	14,000	79
Jul	4320	1	3	1.2	0.09	1.5	4090	15,300	69
Aug	4254	1	4	1.4	0.15	2.0	4700	17,700	58
Sep	4048	1	5	1.4	0.11	2.1	3920	14,200	60
Oct	3741	1	5	1.5	0.11	1.5	3380	12,500	63
Nov	3956	1	5	1.5	0.09	1.0	3470	12,200	70
Dec	4109	2	6	1.6	0.12	0.7	3820	12,400	73

TABLE 5.3. (continued).

Date	Flow L/s	$CBOD_5$ mg/L	TSS mg/L	TKN mg/L	NH_4—N mg/L	TP mg/L	MLSS mg/L	RAS mg/L	SVI mL/g
Jan-90	4692	3	8	1.9	0.26	0.5	3710	13,400	74
Feb	5428	2	6	1.3	0.14	0.9	3340	11,200	90
Mar	4565	2	5	1.4	0.10	0.2	3580	12,100	88
Apr	4486	2	6	1.4	0.12	0.8	3050	10,600	77
May	5165	4	9	1.9	0.32	0.6	2250	9020	67
Jun	5047	3	5	1.4	0.10	0.4	2560	8750	80
Jul	5143	2	5	1.5	0.19	0.3	2670	9550	92
Aug	4140	1	4	1.5	0.12	0.4	3220	9520	84
Sep	4188	2	4	1.6	0.18	1.5	3260	9960	70
Oct	5196	2	6	1.7	0.11	1.1	3410	11,700	68
Nov	4123	2	5	1.9	0.19	—	4530	13,100	70
Dec	5739	2	8	1.5	0.10	—	3270	11,900	77
Avg	4521	2	6	1.5	0.13	1.1	3165	11,336	80

overflow rate nor that the cost of the facility is the same. In this area, design preferences prevail perhaps to a greater degree than the site conditions.

It is important to review the range, quality, and duration of the loading characteristics that must be processed. Normally, BNR effluent quality requirements are high (low values) and major upsets can cause effluent violations.

PRIMARY CLARIFICATION

The primary factor modifying the performance efficiency of the primary clarifiers will be the hydraulic flow characteristics. Preferably, the primary clarifiers will be able to handle 80 to 85% of the flow range with only a minor loss in the planned BOD_5 and TSS removal efficiency. Key data for the design of the primary clarifiers are as follows:

(1) Settleable solids as a function of rise rate. Often this is not evaluated, but it is assumed that 60 to 65% of the influent TSS and 30 to 35% of the $CBOD_5$ is removable domestic wastewater.

(2) The frequency characteristics of the flow profile both daily and seasonally. When there is considerable infiltration and/or inflow, the duration of the high flows should be determined. Often wet and dry seasons should be evaluated separately. Dry seasons can impose more severe loading conditions on BNR plants, particularly cold weather dry seasons, due to the higher percentage removals required.

(3) Waste sludge handling requirements may impose restraints on the clarifier. Where primary or primary plus secondary sludges are co-thickened in primary clarifiers, the tank configuration should be based on the volume of sludge to be processed. Steeper floor or dual floor slopes will be more effective as shown in Figure 5.13.

(4) Larger primary clarifiers ($>$25 mϕ) should be provided with two stub arm scrapers to assist the main scrapers and prevent rat-holing when the primary clarifier is used to thicken sludge.

The projected primary effluent quality for cold and warm weather, high and low flows, should be established for the design of the biological system. Poorer than expected primary treatment will reduce biological system SRT and could adversely impact effluent quality.

SECONDARY CLARIFICATION

The secondary clarifier performance is impacted by the influent flow, return flow, SVI, ZSV, MLSS, RSSS, and flocculation characteristics of the mixed liquor. The design of the secondary clarifiers must be based on the following data:

(1) The flow profile and frequency characteristics as set forth for primary clarifier design.
(2) The maximum level of SVI to be processed
(3) The maximum MLSS concentration to be processed as a function of flow, Q + RAS, which will establish the flux on the final clarifiers
(4) The concentration of the return sludge to maintain the MLSS
(5) The quantity of the aeration solids displaced to the final clarifiers during peak flow periods. The space for this quantity must be available in the final clarifiers without affecting needed clarification volume.

Perhaps the most difficult parameter to evaluate in the past has been the SVI. However, selector technology is available today that will maintain SVI < 120 mL/g and most probably at 60 to 100 mL/g. The poor correlation of SVI to DSVI or SSVI at SVIs > 100 mL/g and MLSS > 2000 mg/L should be considered carefully for bulking sludges.

The design must provide for a sludge consolidation zone to produce the design underflow concentration and volume for sludge storage during peak flows. This increase in blanket should not interfere with feed introduction nor reduce the planned clear zone for clarification. The relative volumes are shown in Figure 5.17.

HYDRAULIC CONFIGURATION DESIGN PROCEDURES—SECONDARY CLARIFIERS

RECTANGULAR

The design conditions for the sizing of the rectangular clarifier in this example are as follows:

- Flow = 1000 L/s (avg), 1700 L/s (max)
- RAS = 500 L/s (avg), 600 L/s (max)
- MLSS = 3500 mg/L
- RSSS = 10,500 mg/L
- SVI $\leq$ 120 mL/g
- OFR = 1.1 m/hr (avg), 1.8 m/hr (max)

The basin length selected is 40 m long. Based on this length, the area required would be as follows:

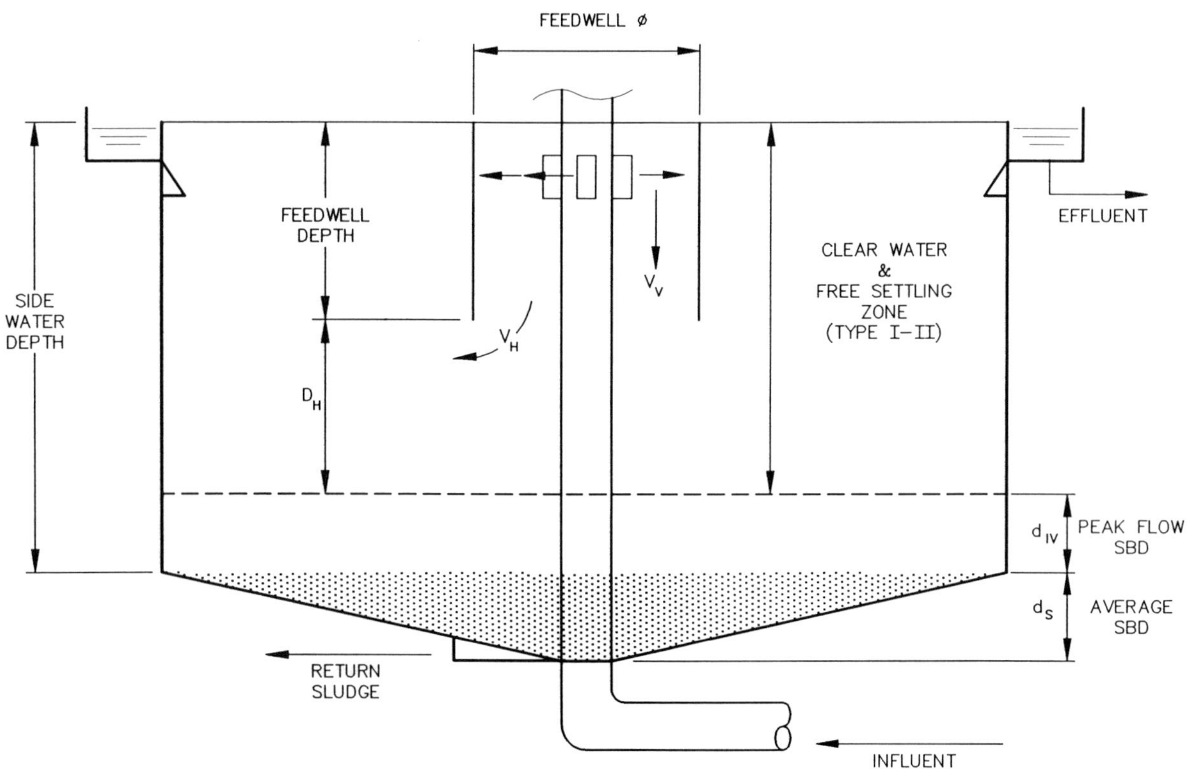

Figure 5.17 Dense sludge blanket profile – average and peak flows.

$$A_{avg} = \frac{(1.0\ m^3/s)\ (3600\ s/hr)}{1.1\ m/hr}$$

$$= 3273\ m^2$$

$$A_{max} = 3400\ m^2 \qquad \text{(governs)}$$

$$10\ \text{units} = 40\ \text{m long} \times 8.5\ \text{m wide}$$

The basin depth is a function of the normal sludge volume in the basin and the solids inventory that shifts from the aeration basin to the clarifier at peak flows. For example, if the aeration retention was six hours or 21,600 m^3 at 3500 mg/L MLSS, peak flow could reduce the MLSS at a constant RAS as follows:

$$\text{Peak Flow MLSS} = \frac{(0.5\ m^3/s)\ (10{,}500\ mg/L)}{(0.5 + 1.7\ m^3/s)}$$

$$= 2386\ mg/L$$

The quantity of solids displaced would be as follows:

$$TSS = 21{,}600\ m^3\ (3.5 - 2.39\ kg/m^3)$$

$$= 23{,}980\ kg$$

The average clarifier increased floor loading due to inventory shift would be as follows:

$$\text{Inv. S.L.} = \frac{23{,}980\ kg}{3400\ m^2}$$

$$= 7.05\ kg/m^2$$

If the additional inventory averaged 8000 mg/L, then the inventory transfer-caused increase in sludge depth (d_{iv}), as shown in Figure 5.17, would be as follows:

$$d_{iv} = \frac{7.05\ kg/m^2}{8.0\ kg/m^3}$$

$$= 0.88\ m$$

The steady state flux (SLR) at both the average and peak flow conditions would be as follows:

$$\text{SLR} = \frac{(1.7 + 0.5 \text{ m}^3/\text{s})\,(3600 \text{ sec/hr})\,(2.39 \text{ kg/m}^3)}{3400 \text{ m}^2}$$

$$= 5.57 \text{ kg/m}^2 \cdot \text{hr}$$

During this period, there will be an underlying denser sludge phase approaching the return concentration of 10,500 mg/L.

The sludge volume at this concentration, based on 75 minutes (scraper speed ≥ 0.6 m/min) retention, would be as follows:

$$V = \left(\frac{5.57 \text{ kg/m}^2 \cdot \text{hr}}{10.5 \text{ kg/m}^3}\right)(3400 \text{ m}^2)\,(1.25 \text{ hr})$$

$$= 2255 \text{ m}^3$$

Based on retention of this volume within the sloped floor section of the clarifier, the minimum slope distance (d_{sf}) of the floor would be as follows:

$$d_{sf} = \frac{(2253 \text{ m}^3)\,(2)}{3400 \text{ m}^2}$$

$$= 1.33 \text{ m}$$

During the peak flow condition, the sludge blanket will extend about 0.9 to 1.0 m above the effluent-end floor of the clarifier. The clear water zone (CWZ) above the maximum sludge blanket would be the basis for sizing the total water depth in the clarifier. A minimum clear water zone for BNR plants of 3.5 m is recommended. The basin geometry would be as follows:

	Infl End	Effl End
Water depth – m	5.8	4.4
Peak flow sludge depth – m	2.3	0.9
Min clear water zone – m	3.5	3.5

The uniform influent flow distribution to each of ten basins is difficult. Kalbskopf (1970) recommends that the influent slots ≤20 cm and notes, as do other engineers, that pre-flocculation and inlet flocculation is beneficial. If pre-flocculation is provided, then a flow control structure to regulate flow to each of ten basins is required. A flocculation period of 5 to 10 minutes is adequate. However, it would be easier and more effective to provide ten weirs discharging equal flow into ten flocculation zones rather than provide zero head flow splits from the flocculation zone into the clarifiers.

The effluent weir length would be based on a weir loading of ≤200 $m^3/m \cdot d$ at maximum flow. The total weir length would be as follows:

$$\text{Weir L} = \frac{(1.7\ m^3/sec)\ (86{,}400\ s/d)}{200\ m^3 \cdot d}$$

$$= 734\ m$$

The weirs should be placed in the last 0.3 of basin length and a 1.3 m wide Stamford baffle is recommended at the end wall to eliminate upturn currents.

CIRCULAR

The design conditions for the sizing of the circular clarifier in this example are as follows:

- Flow = 1000 L/s (avg), 1700 L/s (max)
- RAS = 500 L/s (avg), 600 L/s (max)
- MLSS = 3500 mg/L
- RSSS = 10,500 mg/L
- SVI ≤ 120 mL/g
- OFR = 1.0 m/hr (avg), 1.8 m/hr (max)
- Tank Design ≥ 4.5 m SWD or as required

Number and Basin Size

Determine the basin diameter based on three units, as follows:

$$\text{Avg Flow Dia} = \left[\frac{(1.0\ m^3/s)\ (3600\ s/hr)}{(0.785)\ (1.0\ m/hr)\ (3\ \text{units})}\right]^{0.5}$$

$$= 39.1\ m \qquad \text{(governs)}$$

$$\text{Peak Flow Dia} = 38.0\ m$$

Check the solids loading rate based on average Q + RAS of 1.5 m³/hr at 3500 mg/L MLSS:

$$\text{SLR} = \frac{(1.5\ \text{m}^3/\text{s})\ (3600\ \text{s/hr})\ 3.5\ \text{kg/m}^3)}{(3)\ (39.1\ \text{m})^2\ (0.785)}$$

$$= 5.25\ \text{kg/m}^2 \cdot \text{hr}$$

During peak flow conditions, the RAS is not changed and the shift in inventory and lower MLSS will result in the same flux at peak flows with increased sludge depth due to shift.

Recommend using 3-40 mϕ units.

$$\text{OFR} = 0.96\ \text{m/hr and}\ 1.63\ \text{m/hr (max)}$$

$$\text{SLR} = 5.0\ \text{kg/m}^2 \cdot \text{hr}$$

Energy-Dissipating Inlet

The design of the inlets to collect the flow from the center column outlets is as follows:

$$\begin{aligned} \text{Dia} &= 4\ \text{m}\ (10\%\ \text{of tank}\ \phi) \\ \text{Depth} &= 1\ \text{m (or below center column outlets)} \\ \text{Peak}\ Q + \text{RAS} &= 0.733\ \text{m}^3/\text{s per unit} \\ \text{Port Velocity} &\leq 0.5\ \text{m/s} \end{aligned}$$

$$\text{Port Area} = \frac{0.733\ \text{m}^3/\text{s}}{0.5\ \text{m/s}}$$

$$= 1.47\ \text{m}^2$$

With eight ports at 0.65 m deep, the openings are as follows:

$$\text{Width} = \frac{1.47}{(8)\ (0.65)}$$

$$= 0.28\ \text{m}$$

$$8 - 0.28\ \text{m w} \times 0.65\ \text{m d}$$

The configuration of the energy-dissipating inlet is shown in Figure 5.18.

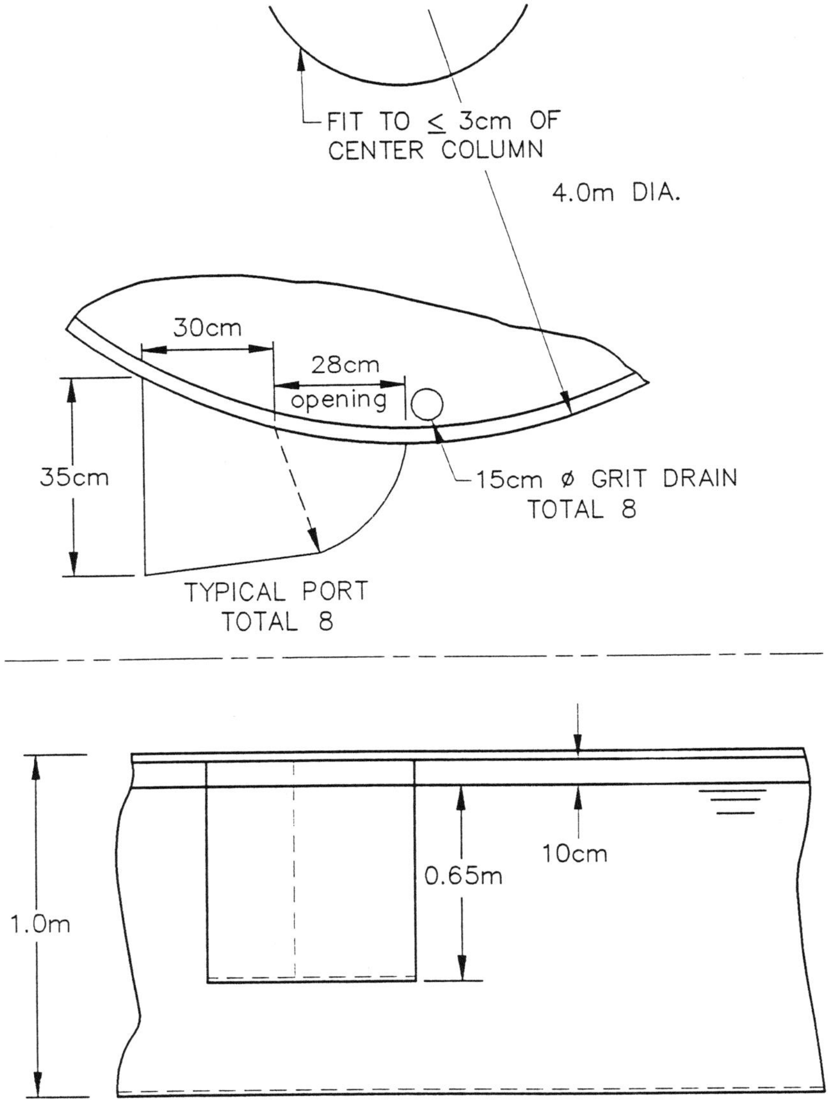

Figure 5.18 Example of energy-dissipating inlet sizing.

Feedwell

The design criteria of the feedwell are as follows:

$$\begin{aligned}\text{Peak } Q + \text{RAS} &= 0.733 \text{ m}^3\text{/s}\\ \text{Max Vert Velocity} &\leq 0.75 \text{ m/min } (V_H)\\ \text{Max Horiz Velocity} &\leq 0.80 - 0.9\, V_H \text{ m/min (CWZ)}\\ \text{Min Depth} &= \text{EDI port depth} + 1 \text{ m}\end{aligned}$$

$$\text{Feedwell Dia} = \left[\frac{(0.733 \text{ m}^3/\text{s})\,(60 \text{ s/min})}{(0.7 \text{ m/min})\,(0.785)}\right]^{0.5}$$

$$= 9.6 \text{ m}$$

$$\begin{aligned}\text{Feedwell Depth} &\geq 0.65 + 1.0\\ &\geq 1.65 \text{ m (min)}\end{aligned}$$

$$\begin{matrix}\text{Clear Water Zone Depth}\\ \text{under the feedwell}\end{matrix} \geq \frac{(0.733 \text{ m}^3/\text{s})\,(60 \text{ s/min})}{(9.6)\,(3.14)\,(0.7)\,(0.85)}$$

$$\geq 2.45 \text{ m (min)}$$

The minimum total SWD of the basin will be the minimum CWZ depth plus the peak flow sludge depth. The final review of sludge collection equipment may result in adjustment/compromises that may modify the initial sizing.

Stamford Baffle/Effluent Weir

The Stamford baffle horizontal projection from the wall of a 40 mϕ clarifier would be as follows:

$$\text{SB} = 46 + (40 - 9), \leq 120 \text{ cm}$$

$$= 97 \text{ cm}$$

Use a 0.97 m projection from the wall extending down from the bottom of the launder or in other arrangements as was shown in Figure 5.15. Provide covers over the scum baffle, weirs, and launders to prevent algal growths. Experience has demonstrated the BNR plants with the highest effluent quality are the most prolific generators of algae. Chlorine can be added to the weirs, but this produces turbidity in the effluent unless added continuously.

DESIGN OF SLUDGE TRANSPORT/COLLECTION EQUIPMENT

DISCUSSION

Center Drawoff Designs

The earlier reported studies by Pfeffer et al. (1968) and Lively et al. (1968) demonstrated that there was significant short-circuiting of feed to underflow. Myers et al. (1968), in dye studies of the 30.5 mϕ $\times$ 2.44 m d final clarifiers at Indianapolis, concluded that 50% of the feed short-circuited and significantly diluted the return sludge. Dye appeared in the underflow within 5 minutes of introduction into the feedwell. This is typical of many, if not most, larger clarifiers.

Eves (1981) reduced the sludge blankets in a secondary clarifier by modifying the two-arm, six-blade scraper operating at 3.0 m/min to a four-arm spiral collector operating at 4.1 m/min. The spiral scraper performed better than the original units except when handling dilute, high-SVI sludges. It was not reported whether agitation caused by the higher scraper velocity resulted in the lower clarifier efficiency on the bulking sludges.

Warden (1981), who analyzed the sludge transport characteristics of straight blade and spiral scrapers, gave an interesting analogy of the sludge collector. He stated,

> This situation is rather like a nightmare belt conveyor which gets narrower as it approaches the discharge end. It operates at over-capacity at all points along its length and, in fact, overflows continuously as it approaches the discharge end.

He noted that the angle of attack (α), rake speed, and blade height and length were factors the design engineer must consider. He concluded that the capacity of the scrapers must exceed the withdrawal rate to ensure satisfactory operation, i.e., eliminate short-circuiting of influent to sludge withdrawal.

Günthert (1984) reported on recent design methodology used in Germany and discussed the interrelationship of influent short-circuiting, scraper blade depth, and scraper velocity. Full-scale results at three facilities with five different clarifier configurations indicate improved sludge collection with higher blade speeds (up to 4 m/min) and deeper blades (up to 49 cm). The five clarifiers studied demonstrated that short-circuiting in circular clarifiers increased with higher return sludge flow rates and solids loading.

Later, Billmeier (1988) summarized Günthert's short-circuiting observations into the graph shown as Figure 5.19. Billmeier also provides further support for Günthert's and Warden's position that the blade angle, length, depth, and velocity of the scraper are the primary performance factors in

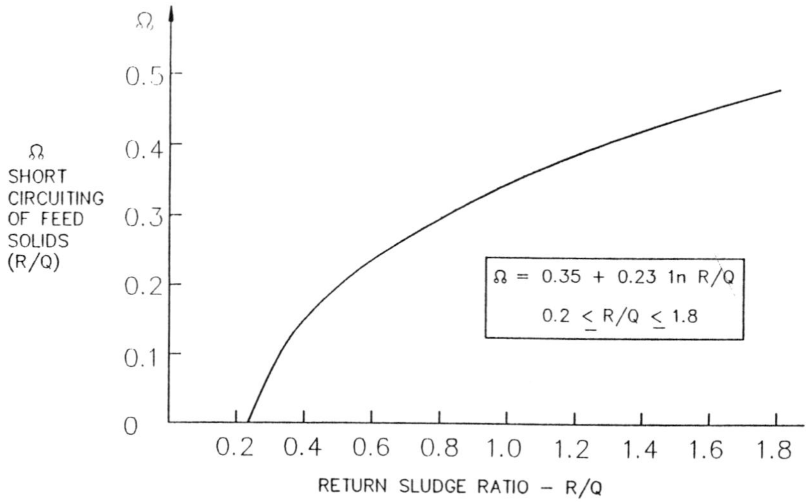

Figure 5.19 Short-circuiting feed solids as a function of *R/Q* ratio.

the transport of sludge toward the sludge hopper. He noted that the scraper blades in circular clarifiers are generally too small, but the problem can often be remedied by deepening and lengthening blades and increasing the rotational speed. The new 70 mϕ secondary clarifiers at Stüttgart, Germany were to be fitted with inner spiral scrapers that are 100 cm in depth to transport the projected deep sludge blankets.

Lumley and Horkebey (1988) data on rectangular tanks revealed a higher degree of sludge short-circuiting at higher solids loadings (>13 lb/ft^2·d) when producing a 6500 mg/L return activated sludge concentration (RASC) than at lower loadings and an RASC of 12,900 mg/L. At the lower sludge loading and lower return activated sludge/flow (RAS/*Q*) ratio, the sludge retention efficiency improved from 76 to 91%.

Albertson and Okey (1990) concluded that the widely perceived limitations of tip speed to 3-4 m/min were erroneous and cited successful operations at South Valley, Utah (45.7 mϕ $\times$ 3.96 m d) at 10 m/min while failing at 3.7 and 5.2 m/min. Also increasing the collector speed to 6.4-7.9 m/min at Phoenix, Arizona was successful in controlling the depth of the sludge blanket in large circular basins that were only 2.74 m deep. In their evaluation of the sludge transport and short-circuiting problems in the 61 mϕ Columbus Southerly clarifiers, they agreed with the earlier conclusions of the European investigators regarding inadequate scraper capacity.

The Columbus results clearly demonstrated increased short-circuiting and increased blanket depths with higher RAS/*Q* ratios. Best operation was at 35-40% *Q* even for MLSS of 3000-4500 mg/L.

From the past problems of sludge removal encountered in secondary clarifiers, several important conclusions can be made. They are as follows:

(1) The pumping effect of the collector is the primary motivating force to transport secondary sludge to the hopper.
(2) In order for efficient transport, the depth of the scraper should approximate the sludge depth to be transported.
(3) Short-circuiting to underflow dilutes the underflow and invalidates laboratory flux study conclusions which are based on first in-first out principles and are not transport limited.
(4) Floor slope is very important to control inventory and allowance must be made for peak flow storage.
(5) The sludge return rate should be minimized to limit flux and reduce feed turbulence as well as reduce short-circuiting.
(6) Sludge hoppers should be located as a function of the sludge depth approaching the hopper in order to limit the depth and speed of the scraper.
(7) Scraper velocities much higher than current accepted criteria can be employed for overloaded collectors.

Multi-bladed straight (oblique) blade mechanisms are widely used for primary and secondary clarifiers. The calculation procedure for these scrapers, shown in Figure 5.20, is similar to the spiral scraper. When conveying primary sludge, the sludge would be moved toward the hopper twice per revolution and the transport time is relatively straightforward to determine. The slippage factor, f, is considered to be in the range of 0.7-0.8. The distance transported per revolution is $2(f)(L)$.

It is more difficult to analyze the effect of the array of oblique blades on activated sludge transport. The low viscosity can result in backflow behind the scraper, thus reducing the value of f. Field studies have not addressed this issue. However, increasing the blade depth, length, and the scraper speed at Phoenix, Arizona (Stensel et al., 1991; Albertson and Hendricks, 1991) resulted in a significant improvement in sludge transport.

Due to the scraper's continuous contact with the sludge, the spiral scraper can transport sludge 200-300% faster than oblique scrapers. The total length of spiral scraper to be mounted on the rake is much less and results in less weight. For this reason, no further discussion of oblique blades is included in this manual.

Rapid Sludge Removal

The introduction of the rapid sludge removal (rsr) clarifier in 1958 by Dorr Oliver was in response to a perceived need to keep activated sludge as

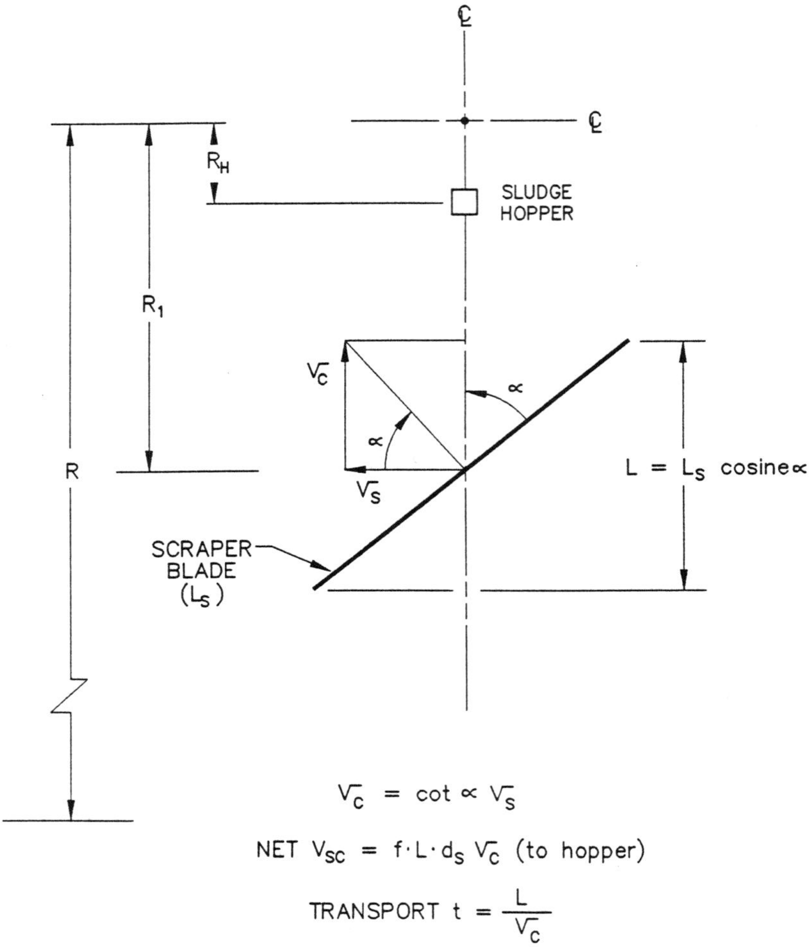

Figure 5.20 Sludge transport with an oblique blade.

fresh as possible prior to return to the aeration basin. In order to accomplish this, the RAS rate was increased 50-100% over that which would be required for a center drawoff clarifier. While these units have established wide acceptance in North America, worldwide use is quite limited. The rsr designs evolved to both circular and rectangular clarifiers and to designs with fixed opening and those with externally adjustable openings.

The original rationale for the rsr type mechanism was soon proven invalid by biological studies (Westgarth et al., 1964) demonstrating that the viability of activated sludge was not impaired by extended (12-24 hrs) periods without oxygen. In fact, current technology encourages periods of oxygen deficiency to improve the settleability of the activated sludge.

However, this did not affect the acceptability of the rsr type mechanism. This was due to the center drawoff mechanism's own deficiencies, principally that the scrapers were not matched to the requirements. Also, many clarifiers were too shallow, had poor inlet and feedwell designs, and had inadequate floor slopes. Recognition and correction of these problems was occurring at the same time as the shift to rsr mechanisms in North America.

The primary difference between the externally adjustable, multi-pickup (organ pipe) design and the fixed port suction header unit, shown in Figure 5.21, is that the suction header design must employ a flat floor. This has some disadvantages. First of all, the total basin depth must be 1-1.5 m deeper to provide the same clear water zone depth under the feedwell to prevent scouring. Secondly, the sludge inventory is 300% greater than the conical section of a conventional clarifier, a large sludge inventory resides directly under the effluent weirs where denitrification can cause solids losses, and in order to manage these adverse conditions, a high return rate is employed with a return sludge concentration generally equal to about 8000(100/SVI) mg/L. The underwater column to header seal is an added operating and maintenance problem.

The organ pipe design, schematically illustrated in Figure 5.22, and installed on a flat floor, has all of the limitations cited above in addition to the operating and maintenance of the individual uptake pipes. When located on a sloped floor, then, the varying deposition of sludge can result in the various uptake pipes producing nearly clear water to high sludge concentrations which plug the uptake pipes. Experience has shown that operators generally do not modify the flow by the adjustable valves.

The rsr type of clarifiers often have a characteristic higher solids level in an effluent zone following the collectors. This is thought to be due to the drafting of a large volume (40-50% of Q + RAS) along one or two arms to satisfy the hydraulics of sludge removal. Further, the cleaned area behind the mechanism represents a zone of least flow resistance. This was also noted at Phoenix, as cited earlier. That is, the sludge withdrawal generates the short-circuiting velocity currents that are represented in Figure 5.22. This is less pronounced in the very deep (5-6 m) SWD basins.

Albertson (1991), in analyzing the supplier designs of the organ pipe type clarifiers, found that the v-plows were generally undersized and the uptake pipes were improperly located in the v-plow. Also, the scraper designs did not transport excess sludge towards the tank center and thus caused excess retention of sludge in the outer areas. In order to improve existing units, scraper modifications to deepen the blades, and relocation of scrapers to bypass 50% of the outermost sludge toward the center and to relocate the uptake pipes were recommended. It was noted that these modifications would be much less effective on flat floors. The revised scraper and uptake pipe arrangement is shown in Figure 5.23.

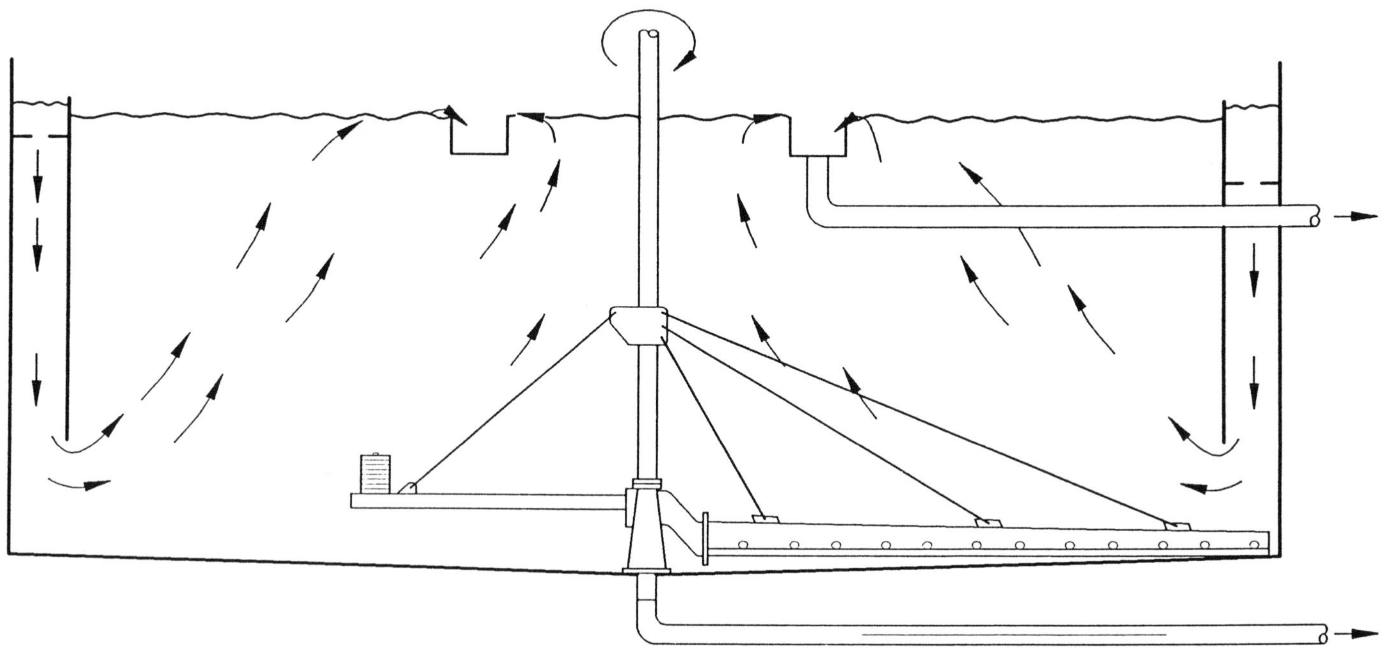

Figure 5.21 Illustration of a circular peripheral feed clarifier with hydraulic sludge removal.

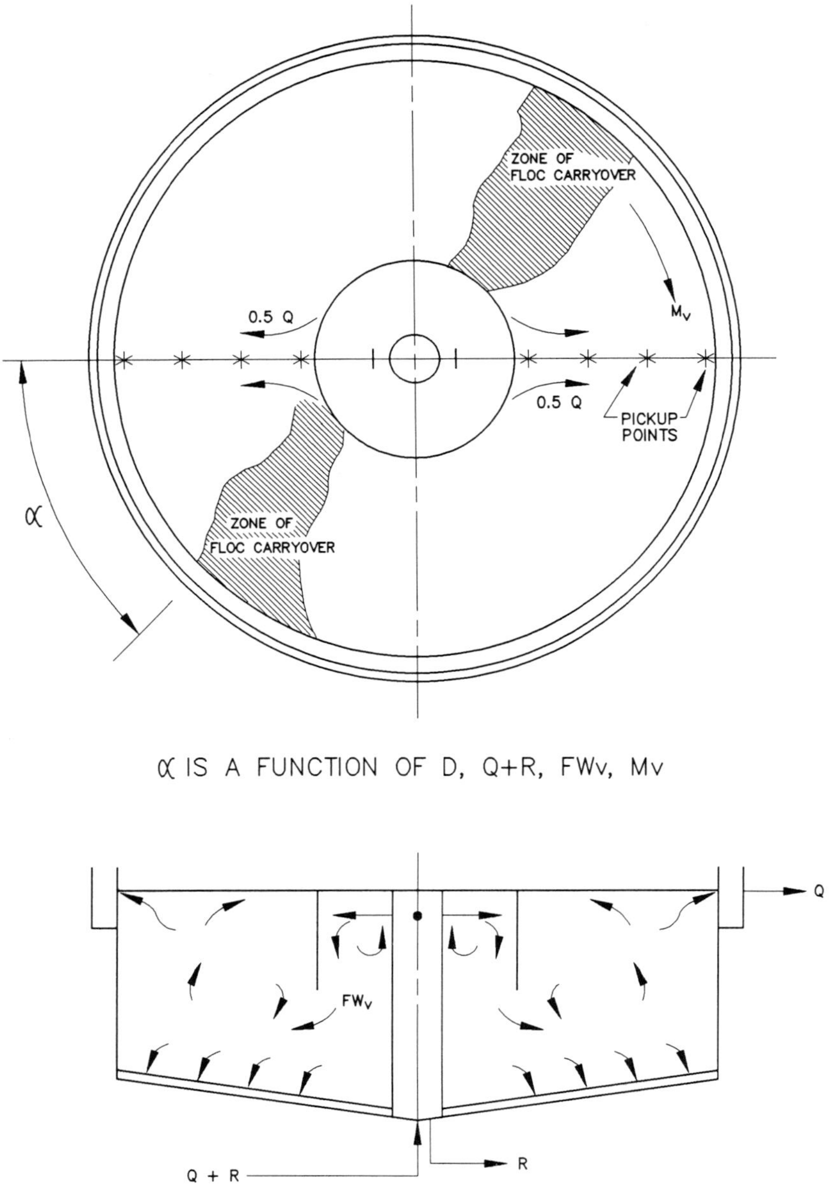

Figure 5.22 Short-circuiting in a rapid sludge removal clarifier.

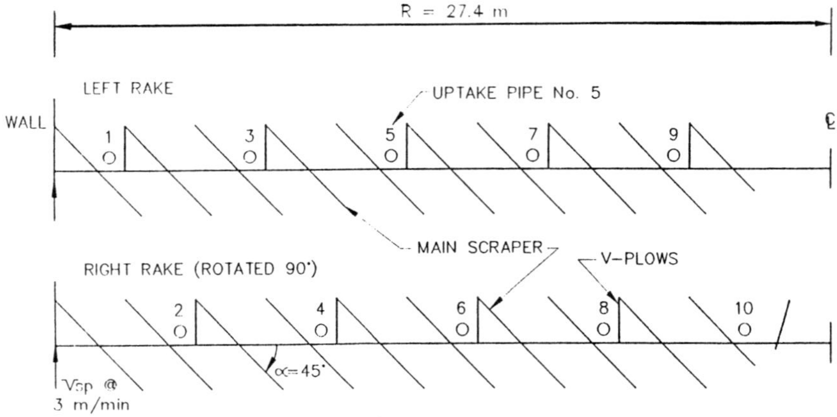

Figure 5.23 Modified scraper design with offset uptake pipes to bypass sludge to center.

The rsr type clarifiers can, and do, produce an effluent quality equal to that of a conventional clarifier. However, they are more costly, more difficult and expensive to operate, and return a more dilute, higher volume of sludge that may limit the maximum MLSS in the biological basins. Kalbskopf (1970) reported that parallel studies at Emscher Mouth WWTP demonstrated that higher return sludge ratios resulted in higher SVIs.

There are no known operating advantages that would favor use of rsr type clarifiers over conventional scraper designs where the collectors are properly sized for the loading to be transported. Thus, the subsequent design discussions will be oriented to conventional clarifiers.

PRIMARY SLUDGE COLLECTORS

The design of primary sludge scrapers is relatively straightforward and poses no significant problems. It is important that the tankage slope provide the necessary sludge depth without resulting in excessive inventory, as addressed earlier. For circular basins, spiral scrapers will transport sludge from the outer portion of the basin faster than straight blades and can represent a lower mechanism loading. Scraper calculations provided in the secondary clarifier discussion are applicable to primary clarifiers.

SECONDARY SLUDGE COLLECTORS

The key consideration is the transport capacity of the scrapers vs. the underflow rate. When the scraper capacity is less than the underflow rate, the feed short-circuits and the dense phase blanket is relatively level except in the vicinity of the sludge hoppers. When the scraper capacity is greater than the RAS rate, the dense phase level may slope towards the hopper, there will be less short-circuiting, and blankets will be lower.

Rectangular

While reciprocating (traveling bridge) single scrapers have been employed, they are not considered suitable for activated sludge clarification. The conventional chain-and-flight scraper is employed for this analysis.

In rectangular clarifiers, the scraper velocity will limit the forward flow velocity. Adjusting scraper spacing and height can fine tune the influence of the scraper, but other variables like sludge viscosity and the velocity and direction of the overriding liquid flow will also impact scraper performance. With the proper scraper arrangement, the transport efficiency (f) should be ≥ 1.0 even when handling low-viscosity, bulking sludges. The volumetric displacement capacity (V_{sc}) of the rectangular clarifier scraper shown in Figure 5.24 may be simply stated as:

$$V_{sc} = (f)(L_s)(d_s)(v_s) \tag{5.19}$$

where:

V_{sc} = transport capacity of scrapers, m^3/hr
f = transport efficiency, possibly 0.8-1.3, as a function of sludge viscosity, sludge depth and the sludge scrapers' spacing, depth, and velocity
L_s = length of scraper, m
d_s = depth of scraper, m
v_s = velocity of scraper, min

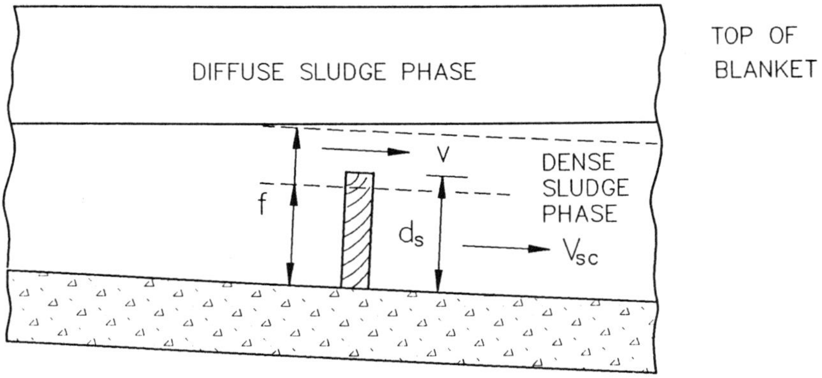

Figure 5.24 Displacement capacity of rectangular clarifier scraper.

In order to assure that the clarifier will not short-circuit influent to the underflow, the mechanism must be able to transport all of the concentrated sludge at the maximum loading to the hopper. The requirements for the rectangular scraper designed for the clarifier loadings set forth earlier for the circular clarifier can be readily determined.

$$\begin{aligned} SLR &= 5.6\ \mathrm{kg/m^2 \cdot hr}\ \text{(steady state SLR)} \\ RSSS &\geq 10{,}000\ \mathrm{mg/L}\ \text{(minimum RAS concentration)} \\ SVI &\leq 120\ \mathrm{mL/g} \\ L &= 40\ \mathrm{m}\ \text{(transport distance to hopper)} \end{aligned}$$

The steady state concentrated sludge deposition rate (d_s) as a function of loading is determined as follows:

$$d_s = \frac{(SLR)\,(10^6)\,(100\ \mathrm{cm/m})}{(RSSS)\,(100\ \mathrm{kg/m^3})}$$

$$d_s = \frac{(5.6)\,(10^6)\,(100)}{(10{,}000)\,(1000)}$$

$$= 56\ \mathrm{cm/hr}$$

If the settled sludge was evenly deposited across the tank floor (assumed prologue to flux failure), then the 56 cm of concentrated sludge must be removed and transported to the sludge hopper from the 40 m length of basin every hour. The calculated total sludge volume would be 22.4 m^3/hr per meter of tank width. Table 5.4 shows the required scraper depth as a function of scraper velocity, assuming the transport efficiency, f, is equal to 1.0.

Typically, rectangular clarifier mechanisms are supplied with 15 to 20 cm flights that move at 0.3 to 0.6 m/min. Higher scraper velocities are

TABLE 5.4. Determination of Rectangular Scraper Depth.

Scraper Velocity (m/min)	Sludge Volume m^3/hr·m	Sludge Depth at Hopper (cm)	Recommended Scraper Depth (cm)
0.3	22.4	122	—
0.6	22.4	62	—
0.9	22.4	41	42
1.2	22.4	31	32
1.5	22.4	25	26
1.8	22.4	21	22

considered unnecessary or inoperable according to U.S. equipment suppliers. Equipment suppliers have advised engineers that the bulk of flow impetus is derived from hydraulic effects even with slopes of $\leq$1:50, but have offered no proof or process calculations. In contrast, Metcalf and Eddy (1922), from their observations of several plants, concluded that if only gravity (and density) effects were relied on, the slope must be at least 2 to 1, or 60°. Other investigators reviewed earlier came to the same conclusion; hydraulic effects are minimal.

The collector mechanism in this example should have at least 32 cm flights and be capable of operating up to 1.8 m/min. A two-speed, three-speed, or variable speed drive would provide sufficient flexibility to handle a wide range of sludge loadings and characteristics.

The earlier recommendations regarding the floor slope of 1.4 m for sludge inventory of 75 minutes is considered conservative based on the higher scraper speed required to transport sludge. The peak flow inventory shift can be partially stored in the sloped zone, reducing the 0.9 m height in the SWD zone. The basin depth could be reduced to 4.0 m at the effluent end and to 5.4 m at the influent end. Reducing the floor slope to less than 1.4/40 m is not recommended.

Sludge hopper placement is important. At higher MLSS, the transport distance could not exceed 35 to 40 m. For long tanks, hoppers should be placed near midpoint and the inlet end, and both hoppers can be served by a single drive. While a hopper is often placed at the one-third point, failure tends to occur at the longer effluent end (longer sludge transport equals higher sludge depth). Such failures suggest that placement of the hoppers at midpoint (two drives) or midpoint and influent end (one drive) is best.

CIRCULAR CLARIFIERS

Center Drawoff

Circular clarifier scrapers pose a more difficult problem in defining their capacity—the sludge to be conveyed is attacked at an angle and flows circumferentially toward the hopper, the floor area shrinks, conveying velocity decreases, and sludge depth increases geometrically as the sludge progresses toward the center of the tank. While the effects of the underflow velocity in circular tanks are more dominant near the withdrawal point, it is obvious that the deeper blanket near the center necessitates circular units having a steeper floor slope, higher scraper velocity, and deeper blades.

To some extent, transporting activated sludge with a conventional circular tank scraper can be likened to "pushing a wet noodle" in that there is ample opportunity for the sludge to flow over the conveyer blade or behind and back towards the wall rather than be transported toward the hopper.

European experience indicates that the spiral scraper is more efficient for sludge transport than the multi-straight blade scraper dominant in U.S. units. Spiral scrapers, shown in Figure 5.25, represent standard practice in Europe and elsewhere, and a discussion may be found in the *Water Pollution Control Federation Manual of Practice for Clarifier Design* (1985) as well as papers by Eves (1981), Warden (1981), Günthert (1984), and Billmeier (1988).

The formula used to predict the pumping velocity (V_c) of a spiral scraper is as follows:

$$V_c = (f')\,(v_s)\,(\text{sine }\alpha) \tag{5.20}$$

where:

V_c = sludge velocity parallel to the scraper blade, ft/min
f' = efficiency of sludge transport allowing for slippage, sludge accumulation and other factors; usually 0.75
v_s = velocity of scraper at radius, r, ft/min
α = angle between tangent to the spiral scraper and the radius, degrees

Time of travel (t) of sludge along a spiral collector length (L_c) is defined by:

$$t = \frac{L_c}{V_c} \quad \text{and} \quad L_c = \frac{R - R_1}{\text{cosine }\alpha} \tag{5.21}$$

where:

L_c = length of the spiral cord, $R - R_1$, ft
t = travel time, minutes
R = tank radius, ft
R_1 = radius at center of sludge hopper, ft

The transport factor (f') of circular clarifier scrapers is more complex than that of rectangular units, since it is also a function of the slippage due to the angle of attack (α) as well as such factors as rotational speed, floor slope, sludge viscosity, the diverging spacing between rakes, the depth of the blade relative to the sludge to be conveyed, and reduced velocity as sludge is transported toward the hopper.

It is difficult to accurately project sludge depths and develop sludge profiles in circular clarifiers. There is no term or experience common to the sanitary engineering field for defining the area or distance of influence ahead of a conveyor nor any literature guidance. To determine the conveyed sludge depth profile, it is necessary to define the zone of transport and the sludge volume moving ahead of the scraper. Logically, the transport zone

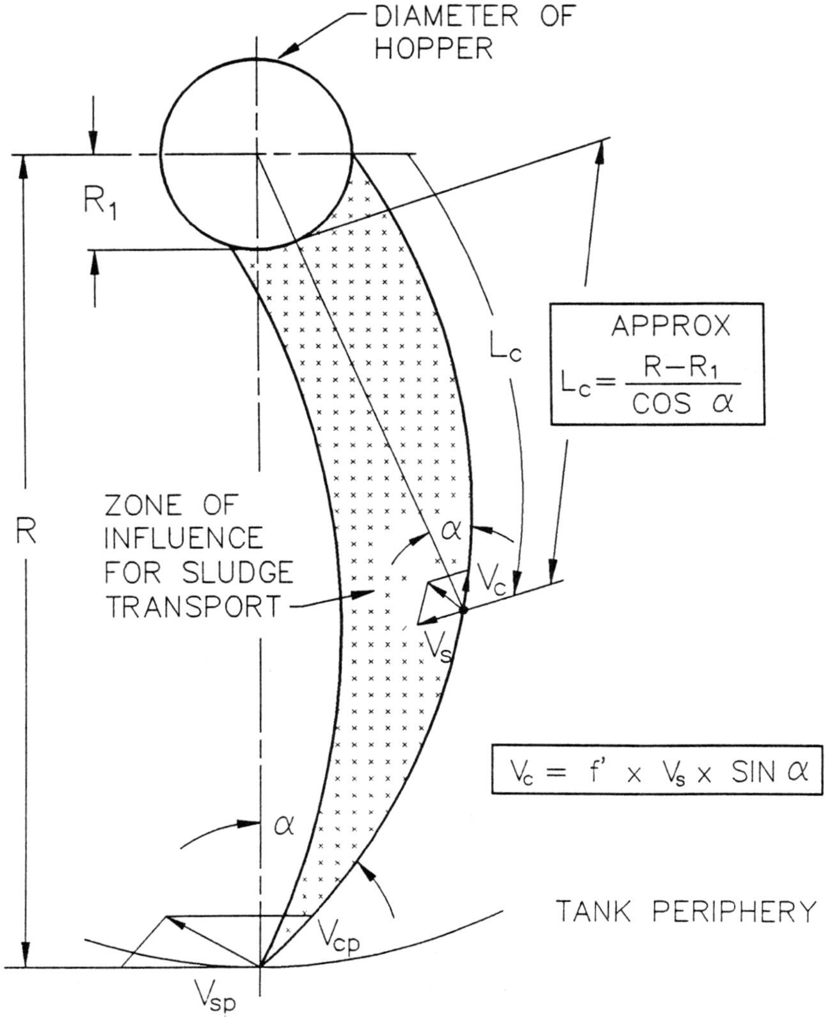

Figure 5.25 Spiral sludge conveyor design and operation.

is smaller with high-SVI sludge and larger with dense, low-SVI sludge. Conventional practice of rectangular scraper placement indicates a ratio of 10 to 20 times the blade depth to ensure an effective transport distance (ETD) zone.

One method (Albertson and Okey, 1990) of analyzing the clarifier sludge profile is to estimate the depth of the sludge transported through a series of decreasing circles until the sludge reaches the hopper. It cannot be assumed that the sludge flows evenly across the entire floor at a velocity dictated by the scraper depth. In the outer area, the sludge may move through only 15-20% of the circumference (effective transport perimeter) while in the inner reach, sludge may flow through the entire circumference and should be encouraged to do so to prevent rat-holing.

While the first six Columbus clarifiers were 61 mϕ, the scraper depth provided was only 28 cm deep. Subsequent calculations projected that the sludge depth would exceed 1.5 m approaching the hopper at a maximum loading of 6.5 kg/m^2·hr and an SVI of 150 mL/g. Initial startup of these conditions resulted in severe short-circuiting, poor transportability, and sludge blankets of 3-4 m. Tracer testings proved that the estimated value of $f' = 0.2$ (80% slippage) was correct, and outside to inside transport time was 4.5-5.5 hours. The sludge would move in the excessive circumferential path predicted by Warden and shown in Figure 5.26. The two new clarifiers will have dual speed drives and deep blades. The projected performance characteristics are also shown in Figure 5.26.

When the selectors reduced the Columbus SVI to 70-90 mL/g, SLR reduced to ≤ 4 kg/m^2·d, the RAS/Q was reduced to 0.45-0.50, and the sludge blanket was eliminated. But it returned at an SVI of 105-110 mL/g as the sludge volume exceeded the transport capacity. The existing units will have the scrapers deepened and provided with new two-speed drives.

This same procedure was used to develop a blanket depth profile analysis that can be used to develop the scraper depths required as a function of the solids loading and the size of the clarifier. This procedure was used to define the scraper sizing for the 3-40 mϕ clarifiers sized earlier. The input for the design is set forth in Table 5.5.

Since the clarifier was less than 50 mϕ, a single sloped floor was employed. At 0.090 m/m floor slope (Table 5.2), at least 1.5 m of sludge depth over the hopper results when the level sludge depth reaches the side water depth. The approximate sludge inventory in this truncated cone zone would be as follows:

$$V = \frac{(3 \text{ units})\,(40\text{ m})^2\,(0.785)\,(1.53\text{ m})\,(0.85\text{ m})}{3}$$

$$= 2260\text{ m}^3$$

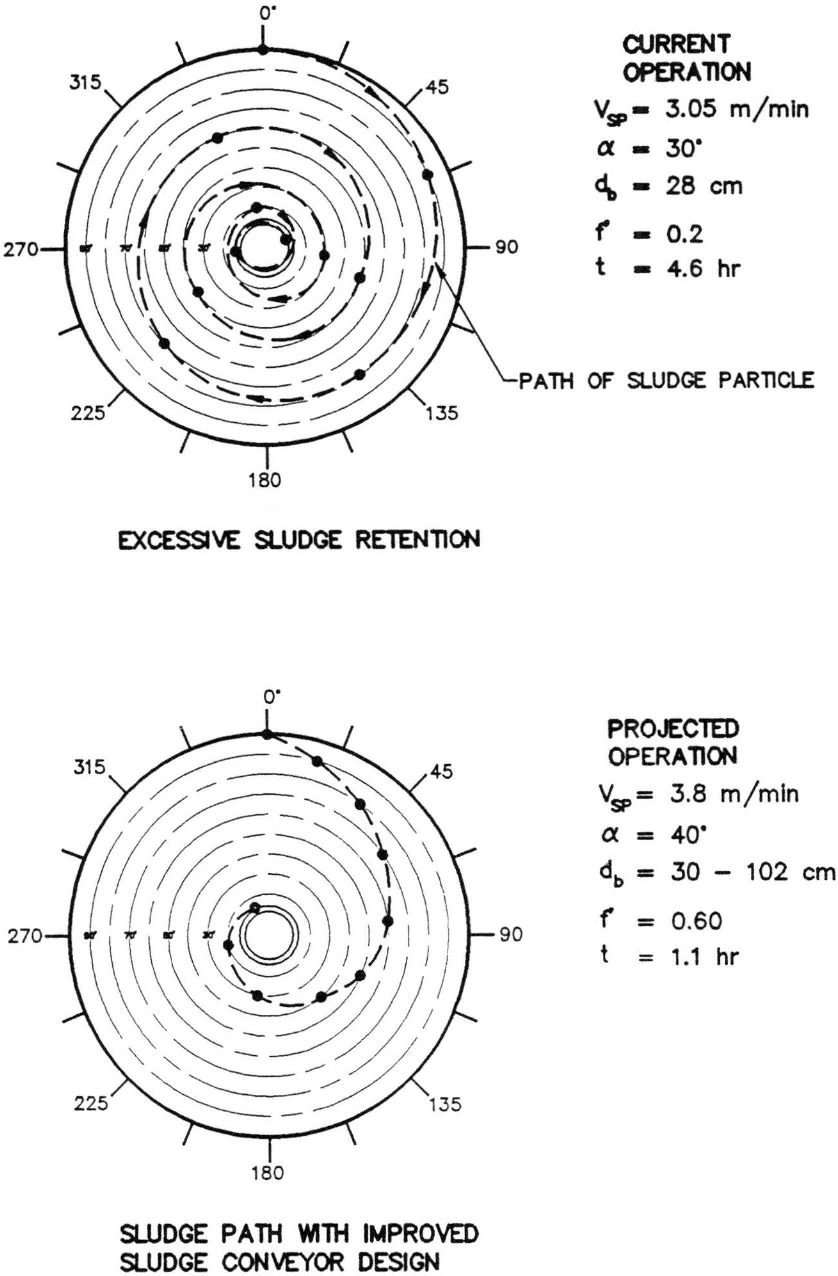

Figure 5.26 Projected sludge path of original and new 61 mϕ Columbus clarifiers.

TABLE 5.5. Input Data for Sludge Transport Analysis.

Clarifier Data	
No. clarifiers	3
Diameter	40 m
Side depth	≥4.5 m
Slope from	20 m radius
	0.090 m/m
	2 scrapers
Slope from	10 m radius
	0.090 m/m
	4 scrapers
Slope from	4 m radius
	0.000 m/m
	4 scrapers
Hopper at	3 m radius
Center column	0.7 m rad
Center depth	4.92 m
Process Data	
Sewage flow	1000 L/s
MLSS conc	3500 mg/L
RAS flow	500 L/s
DSVI = or <	120 mL/g
Spiral Collector Data	
Inner attack angle	35 degrees
Outer attack angle	30 degrees
f, effectiveness	0.75
Tip speed	3.2 m/min
Scraper depth	
at wall	25 cm
at hopper	75 cm
Scraper influence	15 d_{sr}
Scraper taper	0.324 m/m
Feedwell	9.6 m dia × ≥1.65 m d
ED inlet	4.0 m dia × 1.00 m d

The underflow rate is 500 L/s or 1800 m^3/hr. Thus, this conic volume would provide for about 1.3 hour of sludge residence time. The transport time, wall to hopper, is less than one hour. Table 5.6 is the summary of the transport calculations for a spiral rake tip speed of 3.2 and 4.8 m/min. The clarifier was provided with two 50% arms to assist transportation in the inner area.

The calculated and projected sludge depth profiles for tip speeds of 3.2 and 4.8 m/min are shown in Figure 5.27. The projected sludge level considers the leveling effect of the hydraulic forces in the zone where there are four scrapers as shown by the sharp dip of the calculated sludge blanket due to doubling of the transport capacity of the scrapers. Since the sludge depth approaches the height of the scraper at 3.2 m/min (a), the scraper capacity at 4.8 m/min (b) will reduce the sludge blanket.

Earlier, the minimum clear water zone (Type I and II settling) under the 1.65 m d feedwell was 2.45 m for a total SWD − SBD_p of 2.45 + 1.65 or 4.1 m. The SBD_p at peak flow would result from a transfer of 23,980 kg of MLSS from the aeration basin. Based on an average concentration of 8000 mg/L, temporary storage of 3000 m^3 in the clarifiers is required. The projected sludge height in the 3-40 mϕ clarifiers with a 1.5 m truncated cone depth is as follows:

$$\begin{aligned} \text{RAS Rate} &= 1800 \text{ m}^3/\text{hr} \\ \text{Inventory Shift} &= \underline{3000 \text{ m}^3} \\ \text{Total Storage} &= 4800 \text{ m}^3 \end{aligned}$$

$$\begin{aligned} \text{Cone Volume} &= 2260 \text{ m}^3 \\ \text{Full Dia Volume} &= \underline{2540 \text{ m}^3} \\ \text{Total Volume} &= 4800 \text{ m}^3 \end{aligned}$$

The 2540 m^3 of sludge would require 0.67 m of sludge depth above the SWD. Therefore, the minimum SWD of the clarifier is as follows:

$$\begin{aligned} \text{Feedwell Depth} &= 1.65 \text{ m} \\ \text{Feedwell Outlet} &= 2.45 \text{ m} \\ \text{Peak SBD} &= \underline{0.67 \text{ m}} \\ \text{Min. SWD} &= 4.57 \text{ mm} \end{aligned}$$

Check maximum allowable overflow rate:

$$\begin{aligned} \text{OFR}_{\text{MX}} &= 0.556 \text{ CWZ} \\ &= 0.556\,(1.65 + 2.45) \\ &= 2.27 \text{ m/hr} \end{aligned}$$

TABLE 5.6. Determination of Sludge Scraper Depth.

Radius (m)	Scrapers (No.)	Rake Velocity (m/min)	Scraper Area (m^2)	Sludge Deposit (kg/d)	Effective Sludge Volume (m^3/min)	Transport Perimeter (m)	Sludge[1] Depth (cm)
20.0	2	3.2/4.8	295	35,503	—	—	8/5
17.5	2	2.8/3.6	255	30,769	2.35	9.6	23/15
15.0	2	2.4/3.6	216	26,035	4.39	11.8	41/27
12.5	2	2.0/3.0	177	21,302	6.09	13.9	57/38
10.0	4	1.6/2.4	138	16,568	7.51	32.1	40/27
7.5	4	1.2/1.8	98	11,834	8.61	36.3	52/35
5.0	4	0.8/1.2	50	6045	9.38	31.4	92/61
3.0	4	0.48/0.72	27	3222	9.77	18.8	[2]
0.7		—			—	—	

[1]Calculated rake velocity at lo speed/high speed.
[2]Location of hopper.

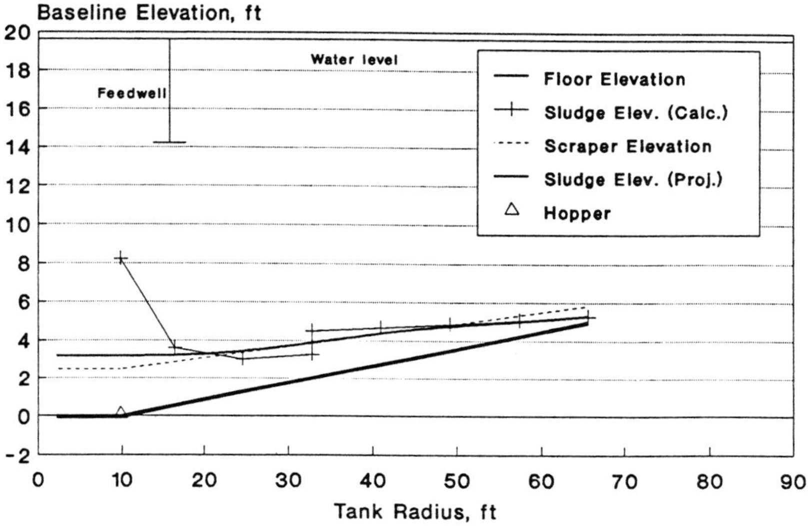

(a)

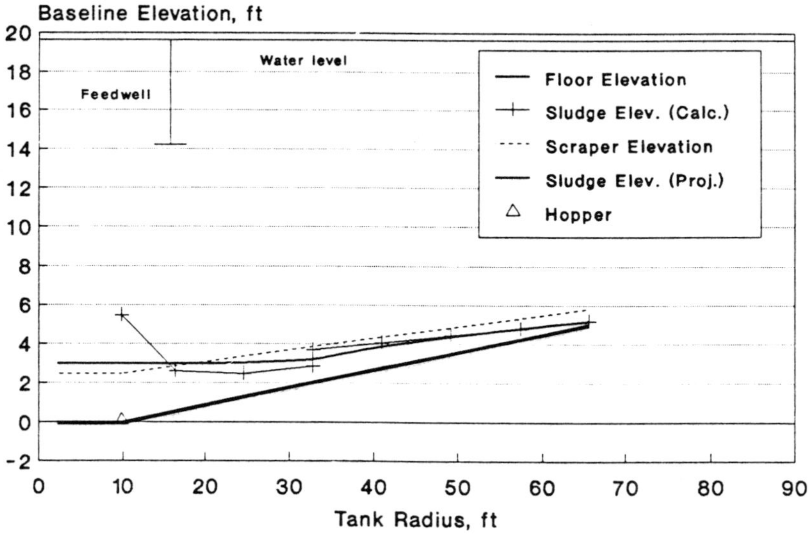

(b)

Figure 5.27 Sludge profile for two conveyor speeds. (a) Tip speed at 3.2 m/min and (b) tip speed at 4.8 m/min.

$$\text{Design} = \frac{(1.7\ \text{m}^3/\text{s})\ (3600\ \text{s/hr})}{(3)\ (1256\ \text{m}^2)}$$

$$= 1.62\ \text{m/hr}$$

The clarifier diameter could be reduced to 37 mϕ, which would result in an SLR of 5.8 kg/m^2·d and OFR_{MX} of 1.95 m/hr. The final decision in the range of 37-40 mϕ would rest with site specific conditions that cannot be evaluated in this discussion.

RETROFITTING THE COLLECTORS

Existing collectors can be retrofitted with new scrapers to enhance their performance. These modifications may also be required in order to operate at higher MLSS. In some cases, increasing the speed to 5-8 m/min will provide the added capacity. Generally, straight scrapers have been found to be too shallow and too short. Improvements to existing scrapers include one or more of the following modifications:

(1) Replace the oblique scrapers with tapered and deeper spiral scrapers.
(2) Increase the mechanism speed to 5-8 m/min.
(3) Add on to the existing straight blade depths to match the projected sludge depth.
(4) Lengthen the scrapers to 3.5-5 m and reduce numbers; larger clarifiers would employ the longer scrapers.
(5) Add two stub arms on large units ($\geq$30 mϕ) to improve sludge transport to the hopper.

The decision to modify the mechanism should be accompanied by transport calculations to indicate the specific problem and the effects of various remedial steps.

Secondary sludge scrapers can be deepened as shown in Figure 5.28 (Albertson, 1991). Generally it is necessary to lengthen the scrapers at the same time. On clarifiers larger than 40 mϕ, the addition of two intermediate stub arms can be beneficial.

It is not recommended that primary sludge scrapers be constructed within the framework of the mechanism, as the load on the mechanism would increase significantly. Most often, it will only be necessary to increase the scraper length and operate at a higher speed. Short stub arms extending beyond the sludge hopper can reduce rat-holing.

The fundamental problem of most rsr type clarifiers is that the scrapers are too shallow to contain the sludge, the scrapers aren't staggered or

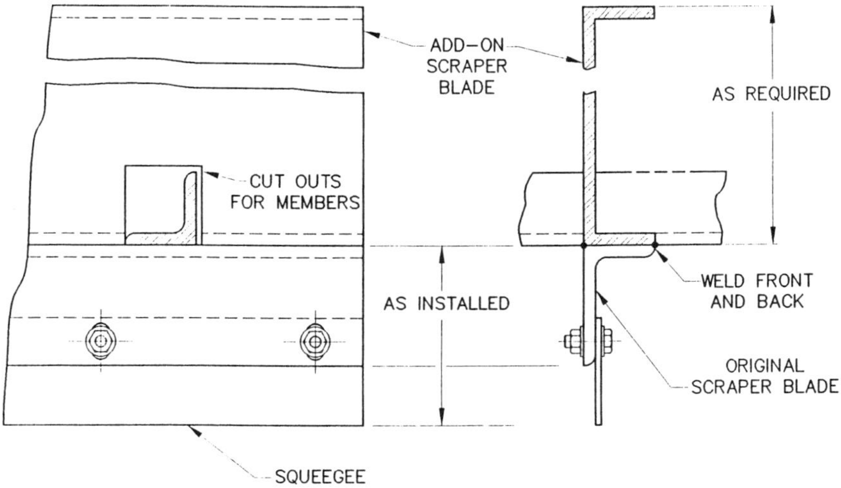

Figure 5.28 Deepening of the secondary scraper blade.

arranged to transport excess sludge to the center, and the uptake pipes are not located at the proper point in the v-plow, as is shown in Figure 5.23 and Figure 5.29. The specific design needs are to be determined by transport calculations.

A number of rsr clarifiers have been converted to center-drawoff. This is an expensive procedure and has been proven not feasible for clarifiers with less than 1:12 floor slope. The cost to increase floor slope using a repoured floor is equally costly and loss of hydraulic capacity resulting from a shallower SWD may result.

PROCESS CONTROL STRATEGIES

PRIMARY CLARIFIERS

The control strategies for primary clarifiers are relative to the sludge handling practices. Where the sludge is thickened separately by gravity, the practice is to remove sludge continuously at the final dilution concentration of 0.5-1.0% TSS for the thickener.

When the primary clarifier is used to thicken the primary or mixed sludges, the underflow rate should not be more than 50% of the scraper transport capacity in the vicinity of the sludge hopper. Cross-collectors in rectangular tanks and four scrapers in the area of the hopper are helpful. Dual slopes and depressed floors (Figure 5.13) are used to enhance the thickening and enhance control of underflow concentration.

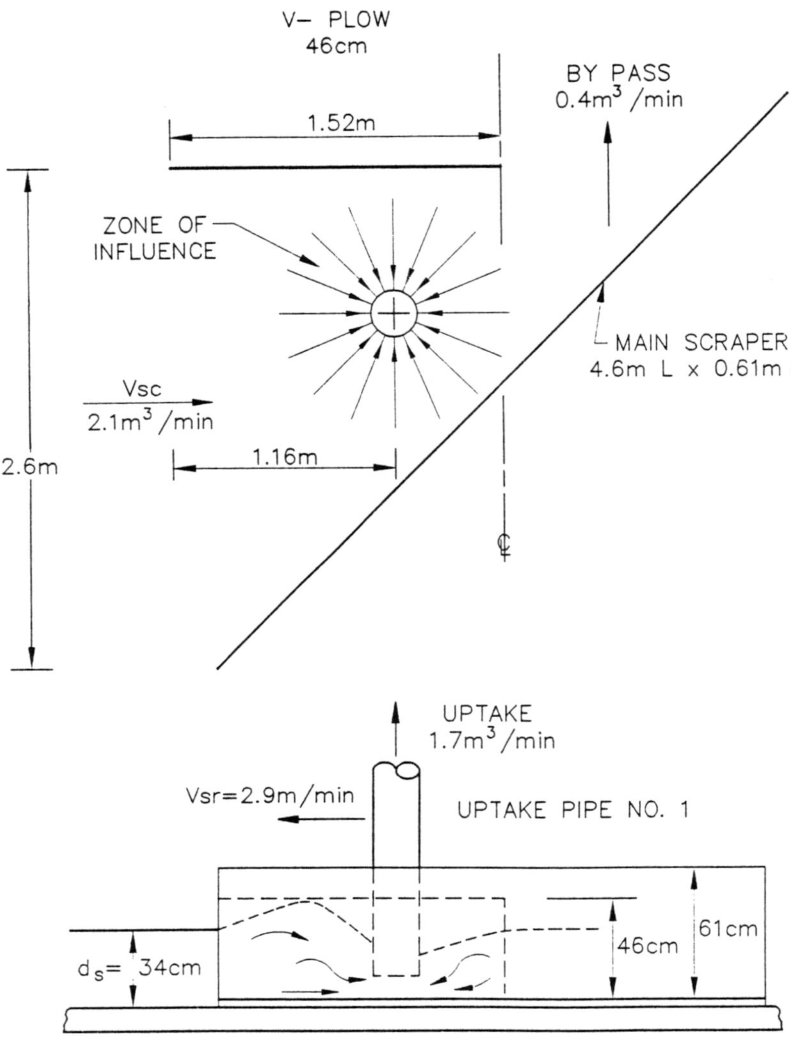

Figure 5.29 Proper location of uptake pipe.

SECONDARY CLARIFIERS

Once constructed, there are limited means of adjusting final clarifier operation. For this reason the design engineer must fully evaluate the range of operating conditions to be encountered.

The low effluent values consistent with BNR requirements make it difficult to average out an upset for weekly and monthly limits. Further, some permits call for not-to-exceed values. Specific steps can be taken in the design and operation to control effluent quality during peak flows without constructing oversized clarifiers for 0.5-5 percentile flows. Thus, all design and operating strategies are directed at controlling peak flow effects.

RAS

The RAS rate should never be controlled as a constant ratio of the influent flow. This places a severe burden on the final clarifier and has been the direct cause of clarifier flux failures. There should be no attempt to prevent the shift of aeration MLSS to the clarifier during the peak flow periods. This shift must be part of the clarifier design volume. The flux comparison of the two modes of operation is shown in Table 5.7.

The blanket will expand considerably as the SLR increases (West, 1978) and this leads to an increase in effluent TSS. Failure in secondary clarifiers is usually a mixture of flux and discrete particle/floc carryover. This occurs as the blanket height increases, limiting the clarification zone, which then causes scour of the blanket surface and increasing levels of solids carryover. If the situation worsens, the upper level of the blanket carries over.

The RAS rate should not be increased with increasing flows. If the clarifier is being overpumped, reduction of the RAS rate could result in a subsidence of the sludge blanket. The RAS should be minimized to control the RSSS within the range of 11,000-14,000(100/SVI), mg/L.

If the scraper capacity is limited, this limitation cannot be corrected by increasing the RAS rate. The basic problem must be addressed with modifications to increase the scraper depth and length and/or increase the scraper speed.

Sludge Blanket

The basic strategy of operation is to maintain the average sludge blanket level at or below the SWD or to maintain a free and flocculent settling zone that is consistent with the overflow rate. If the blanket is maintained within these limits, there are no other remedies for poor clarification within the clarifier itself. Lower rise rate may improve results where TSS with very

TABLE 5.7. Flux Loadings vs. RAS Mode.

RAS Mode	Q_P + RAS (m^3/s)	MLSS (mg/L)	Flux (kg/hr)	SLR ($kg/m^2 \cdot hr$)
Constant ratio	2.55	3500	32,130	8.5
Constant flow	2.20	2380	18,840	5.0

low subsidence rates are encountered. Dispersed hydrous flocs are the first lost to the effluent.

The preferred operation would be with the scraper and RAS at the lowest rate consistent with maintaining the blanket at or below the SWD at average flow conditions. As the sludge blanket increases, the first response would be to increase the mechanism speed, hence scraper capacity. This will ensure that the clarifier is not transport-limited and that short-circuiting is minimized.

Continuous monitoring of the sludge blanket in the area of the effluent launders is recommended. With sonic types, dual units can be used to sense the elevation of both the diffuse and dense blanket phases.

Back River WWTP, Baltimore, Maryland uses continuous sludge blanket monitoring of 12-36.5 mϕ final clarifiers to make the operating judgment. As the blanket depth increases in the rsr type clarifiers, the operating options are to shift the flow from the overloaded clarifier and/or to modify the RAS rate. The option of increased scraper speed is not amenable to rsr type mechanisms.

Settleability of MLSS

Sekine et al. (1989) described a continuous method and apparatus for testing the settleability of the sludge prior to clarification. Using a suspended solids sensor and optical means of detecting the subsiding sludge blanket, the device is capable of determining the ZSV and SVI of the mixed liquor. This device could be used to trend the SVI and provide warning of deleterious changes.

Aeration

In the last resort, blanket or solids carryover can always be prevented by shutting off the aeration, internal recycle, and mixing equipment. This will immediately prevent additional solids from reaching the clarifier and permit draw-down of the blanket and return of the biomass to the aeration basin. However, the equipment must be amenable to shutoff of aeration without becoming fouled. This is a very successful preventative step where aeration is provided by surface aeration, submerged turbines, coarse bubble diffusers, and nonfouling membrane diffusers.

This mode of operation can continue for a time equal to 50 to 60% of the plug-flow reaction retention time (Q + RAS) of the aeration basin, which generally results in removal of the total sludge blanket from the clarifier,

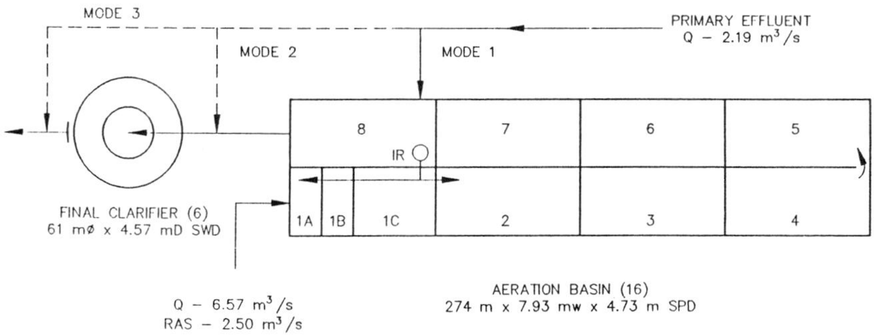

Figure 5.30 Flowsheet arrangement for bypass-contact.

allowing a restart of equipment. Very little effluent quality degradation will occur and the cycle can be repeated if the flow peak persists.

BYPASS TREATMENT

In many cases it is necessary to treat short-term peak flows much greater than 1.8-2.0 average flow. In those cases, the biological treatment-clarifier design can be modified to process more wastewater flow with little or no disruption of the process efficiency. This is an alternative to shutting down the aeration and mixing equipment. This concept is most applicable to plug-flow basins or staged aeration, but can be employed with complete mix plants although with lower effectiveness. Essentially, the approach is to convert a plug-flow plant into a combined plug-flow contact-stabilization plant in order to store solids in the aerator and have a much lower MLSS to clarify. The process flowsheet is shown in Figure 5.30.

Laboratory studies were conducted at Columbus (Albertson, 1987) to determine the efficiency level achieved by this mode of operation whereby 6570 L/s was treated in the staged aeration basins, and 2190 L/s primary effluent was either processed by (1) feeding to the last 12.5% of the aeration volume, (2) mixing with the aeration effluent prior to the clarifier, or (3) mixing with the final effluent. The results of eight tests are shown in Table 5.8.

As shown, there was only a negligible loss in process efficiency of treatment by using the last 12.5% of the aeration basin volume in a contact mode. Nitrification rates were also good in this zone. Mixing the primary effluent with the mixed liquor after aeration provided much better treatment than combining the two flows after clarification. As expected, there was no nitrification and some stored phosphorus release in the latter two modes.

As the mixed liquor concentration is reduced, the settling velocity increases. Using the Daigger-Roper expression for the initial settling velocity and 95 mL/g SVI, the settling velocity for each mode is as shown in Table 5.9.

The initial effect will be a large drop in the MLSS and rise in the settling rate, which allows time for removing the excess blanket from the clarifiers. Thereafter, the MLSS will come to steady state conditions at the new flow conditions with a concentration of 3100 mg/L in the aeration basin or 89% of the average flow mass.

WAS

During peak flows, temporary or permanent wasting of MLSS can alleviate solids overloading on the clarifiers. Generally, it is important to waste as much as possible, especially during extended peak flow conditions. This reduces the clarifier inventory.

Experience

The city of Medford, Oregon (Hammond, 1987) reported on a successful storm flow strategy that results in MLSS storage in the aeration basin rather than the clarifiers. The operator can convert one or two zones to stabilization in the four-zone aeration basin for peak flow conditions.

Resnick (1987) found that by arranging to stagewise feed the four zones in series and ultimately to discharge primary effluent into the mixed liquor effluent channel as the flow increased, the treatment capacity was increased from 1140 L/s to 1840 L/s without violating discharge permits. The plant was thereby able to retain most of the solids in the aeration basins and better utilize their clarifiers due to larger clear water zones and higher settling velocities due to lower MLSS.

TABLE 5.8. Effluent Quality: Bypass-Contact Tests.

	Bay 7/8 Treated 6.57 m^3/s	P.E. Quality 2.19 m^3/s	(1) Contact Aeration 8.76 m^3/s	(2) Contact Mixing 8.76 m^3/s	(3) Combined Effluents 8.76 m^3/s
BOD_5	5.4	71	6.4	9.8	21.8
TSS	13.9	57	18.0	20.1	28.2
TKN	5.0	14.3	5.2	7.8	7.3
NH_4—N	2.6	9.5	1.8	4.1	4.3
TP	0.8	4.5	0.9	2.2	1.7
TOD[1]	22.7	186	21.1	38.5	63.4

[1]TOD = $2BOD_5 + 4.6NH_4$—N.

TABLE 5.9. Aeration-Clarification Operating Mode on Settling Velocity.

Mode	Flow (L/s)	MLSS (mg/L)	V_i (m/hr)
Avg	4990	3500	2.31
Max	6570	2890	2.86
Bypass contact (intial)	8760	1690	4.34
Bypass contact (steady state)	8760	2330	3.47

Thompson et al. (1989) reported on studies to prevent washout by using step feed and reducing the concentration of the solids to the clarifier. There was only a moderate increase in effluent TSS in the conversion to step feed.

These procedures are easily implemented in the design phase. They may or may not be readily accomplished in an existing plant without additional piping and valves.

REFERENCES

Albertson, O. E. 1963. Dorr Oliver Internal Report, Reduction of Clarifier Operating Data.

Albertson, O. E. 1987. EEI Letter Report to URS, Columbus, OH Laboratory Studies to Evaluate Treatment Efficiency of Partial Bypass of Primary Effluent.

Albertson, O. E. 1991. "Improving the Rapid Sludge Removal Collector." To be presented at the *64th Annual Conference of the WPCF, Toronto, Ontario, Canada*.

Albertson, O. E. and P. Hendricks. 1991. "Bulking and Foaming Organism Control at Phoenix, AZ WWTP." Submitted to the IAWPRC for 1992 Biannual Meeting, Washington, D.C.

Albertson, O. E., R. F. Scott, H. D. Stensel and R. W. Okey. 1991. "Expansion and Upgrading of Columbus, OH WWTPs to Advanced Wastewater Treatment." presented at the *6th IAWPRC Conf. on Large Wastewater Treatment Plants, Prague, Czech*.

Albertson, O. E. and R. W. Okey. 1990. "Rationale Design of Secondary Clarifier Collectors." Presented at the *63rd Annual Conference of the Water Pollution Control Federation, Wash., D.C.*

Alfonso, P. 1990. Maxson WWTP, Memphis, TN. Personal communication with O. Albertson.

Anderson, N. E. 1945. "Design of Final Settling Tanks for Activated Sludge," *Sewage Works J.*, 17(1):50.

Barnard, J. L. 1984. "Activated Primary Tanks for Phosphorus Removal," *Water SA*, 10(3):121-126.

Baumann, E. R., J. L. Cleasby, P. E. Morgan. 1965. "Performance of Sedimentation Tanks Evaluated in Model Tests." To be presented at the *National Symposium on Sanitary Engineering Research, Development and Design at Penn State*.

Bender, J. H. 1987. "Assessment of Design Tradeoffs When Using Interchannel Clarifiers," *J. Wat. Poll. Con. Fed.*, 59(10):871-876.

Billmeier, E. 1988. "The Influence of Blade Height on the Removal of Sludge from Activated Sludge Settling Tanks," *Wat. Sci. Tech.*, 20(4/5):165-175.

Camp, T. R. 1936. "A Study of the Rational Design of Settling Tanks," *Sewage Works J.*, 8(5):742.

Camp, T. R. 1946. "Sedimentation and the Design of Settling Tanks," *Transactions-ASCE, Vol III*, p. 895.

Camp, T. R. 1953. "Studies of Sedimentation Basin Design," *Sewage and Industrial Wastes*, 25(1) pg. 1.

Coe, H. S. and G. H. Clevenger. 1916. "Methods for Determining the Capacities of Slime Settling Tanks," *Transactions of the Amer. Inst. of Mining Engrs.*, 55:356.

Daigger, G. T. and R. E. Roper. 1985. "The Relationship between SVI and Activated Sludge Settling Characteristics," *J. Wat. Poll. Con. Fed.*, 57:859.

Dick, R. I. 1967. "Fundamental Aspects of Sedimentation, Part 2," *Water & Waste Eng.*, 6:44.

Dick, R. I. 1976. "Folklore in the Design of Final Settling Tanks," *J. Wat. Poll. Con. Fed.*, 48(4):633-644.

Dick, R. I. and B. B. Ewing. 1967. "Evaluation of Activated Sludge Thickening Theories," *Jour. Sanit. Engr. Div. Proc. Amer. Soc. Civil Engrs.*, 93(EE4):9.

Dick, R. I. and K. E. Young. 1972. "Analysis of Thickening Performance of Final Settling Tanks," *Proc. 27th Ind. Waste Conf.*, Purdue Univ, p. 33.

Dorr Oliver. c.1965. Field Tests of Launder and Feedwell Arrangements to Prevent Short-Circuiting at Pontiac, MI. Dorr Oliver, Inc.

Erdman, A. c.1958. Internal Dorr Oliver report on effect of CDW/Dia ratio on clarification efficiency. Dorr Oliver Inc., Stamford, CT.

Eves, E. G. 1981. "Hogsmill Valley Works: Twenty Years On," *Water Pollution Control*, pp. 453-459.

Firmin, A. C. 1990. "Activated Sludge: An Integrated Approach to Design and Operation." Presented at the *63rd Ann. Conf. WPCF, Wash., D.C.*

Fischerstrom, C. 1955. "Sedimentation in Rectangular Basins," *Proceedings-ASCE*, Vol. 81, pp. 1-29.

Fischerstrom, C. H. 1958. "The Significance of Detention in Sedimentation: Discussion," *Sewage and Ind. Wastes*, 30(11):1357-1363.

Fitch, E. B. 1956. "Flow Path Effect on Sedimentation," *Sewage and Ind. Wastes*, 28(1):1.

Fitch, B. 1966. "A Mechanism of Sedimentation," *Ind. & Eng. Chem. Fundamentals*, 5:129.

Günthert, F. W. 1984. "Thickening Zone and Sludge Removal in Circular Final Settling Tanks," *Wat. Sci. Tech.*, 16:303.

Hammond, G. 1987. "Preventing Solids Loss during High Flows," *WPCF Operations Forum*, pp. 24-27 (May).

Hazen, A. 1904. "On Sedimentation," *Transactions-ASCE, Vol. LIII*, p. 45.

Hendricks, P. 1990. Personal communication, Superintendent of Phoenix WWT plants.

Kalbskopf, K. H. 1970. *European Practices in Sedimentation, Water Quality Improvement by Physical and Chemical Processes*, Austin, TX: U. of Texas Press.

Kalbskopf, K. H. 1972. ''Discussion of the Design Parameter for Secondary Sedimentation Tanks,'' *Water Res.*, 6:429-431.

Keinath, T. M. 1990. ''Diagram for Designing and Operating Secondary Clarifiers According to the Thickening Criterion,'' *J. Wat. Poll. Con. Fed.*, 60(3):254-258.

Keinath, T. M., M. D. Ryckman, C. H. Dana and D. A. Hofer. 1977. ''Activated Sludge-Unified System Design and Operation,'' *J. Enviro. Eng. Div., Proc. Am. Soc. Civ. Eng.*, 103:829.

Keinath, T. M. and Wahlberg, E. J. 1990. Private communication providing test results to O. Albertson for plots.

Kelly, K. 1989. "Space Saving Techniques in Wastewater Treatment: Innovative Clarifier Helps." Presented at *62nd Ann. Conf. of WPCF, San Francisco, CA.*

Knop, E. 1951. ''Measurement of Treatment-Tank Flow-through Periods by Means of Radioactive Matter,'' *Die Wasserwirtschaft*, 41:117.

Koopman, B. and K. CaDee. 1983. ''Prediction of Thickening Capacity Using Diluted Sludge Volume Index,'' *Water Res.*, 17:1427.

Kynch, G. J. 1952. ''A Theory of Sedimentation,'' *Transactions of Faraday Soc.*, 48:166.

Lively, L. D. et al. 1968. ''Phosphate Removal Studies by Activated Sludge, Amenability Studies at Cleveland, OH.'' NTIS No. PB-226 382. Robert S. Kerr Water Research Center, Ada, OK.

Lumley, D. J. and Horkeby. 1988 ''Detention Time Distribution of Sludge in Rectangular Secondary Settlers.'' Handout summary of poster presentation, *IAWPRC 14th Biennial Conference, Brighton, England.*

Metcalf, L. and H. Eddy. 1922. *Sewerage and Sewage Disposal, 1st Edition*, 7th Printing. New York: McGraw-Hill Book Co. pp. 434-435.

Myers, L. H. et al. 1968. ''Phosphate Removal by Activated Sludge, Amenability Studies at Indianapolis, Indiana,'' NTIS No. PB-226 381, Robert S. Kerr Water Research Center, Ada, OK.

Parker, D. S. 1983. ''Assessment of Secondary Clarification Design Concepts,'' *J. Wat. Poll. Con. Fed.*, 55(4):349.

Parker, D. S. and R. N. Matasci. 1989. ''The TF/SC Process at Ten Years Old: Past, Present and Future.'' Presented at *62nd Ann. Conf. of WPCF, San Francisco, CA.*

Parker, D. S. 1991. ''The Case for Circular Clarifiers,'' *Water/Engrg and Mgmt*, pp. 22-25 (April).

Parker, D. S. and R. J. Stenquist. 1986. ''Flocculator-Clarifier Performance,'' *J. Wat. Poll. Con. Fed.*, 58(3):214-219.

Pfeffer, F. M. et al. 1968. ''Phosphate Removal by Activated Sludge, Amenability Studies at Pontiac, Michigan,'' NTIS No. PB-226-379, Robert S. Kerr Water Research Center, Ada, OK.

Pitman, A. R. et al. 1983. ''Practical Experience with Biological Phosphorus Removal Plants in Johannesburg,'' *Water Sci. Tech.*, 15:233.

Rachwal, A. J., D. W. M. Johnstone, M. J. Hanbury and D. J. Critchard. 1982. ''The Application of Settleability Tests for the Control of Activated Sludge Plants,'' from *Bulking of Activated Sludge*, B. Chambers and E.J. Tomlinson, editors. Chichester, England: E. Horwood Ltd.

Resnick, J. 1987. Supt. of Davenport, IA WWTP. Private communication regarding storm flow handling.

Sekine, T. et al. 1989. "Evaluation of Settleability of Activated Sludge Using a Sludge Settling Analyzer," *Wat. Res.*, 23(3):361-367.

Semon, J. 1982. Private communication on installation of first Stamford baffle, Stamford, CT.

Stensel, H. D., O. E. Albertson, P. Hendricks, D. Wendell and P. Petryk. 1991. "Anoxic-Aerobic Activated Sludge Treatment." To be presented at *Nat. Conf. Envir. Engrg, ASCE, Reno, NV.*

Thompson, D., D. T. Chapman and K. L. Murphy. 1989. "Step Feed Control to Minimize Solids Loss during Storm Flows," *Res. J. WPCF*, 61(11/12):1658-1665.

U.S.EPA. 1987. *Design Manual: Dewatering Municipal Wastewater Sludges*, EPA/625/1-87/014, Off. of R and D, Cincinnati, OH.

Wahlberg, E. J. and T. M. Keinath. 1988. "Development of Settling Flux Curves Using SVI," *J. Wat. Poll. Con. Fed.*, 60:2095.

Warden, J. H. 1981. "The Design of Rakes for Continuous Thickeners," *Filtration and Separation*, 18(2):113-114, 116.

Water Pollution Control Federation. 1977. *Wastewater Treatment Plant Design Manual of Practice No. 8.* Washington, D.C.

Water Pollution Control Federation. 1985. *Clarifier Design Manual of Practice FD-8, Facilities Development.*

West, A. W. 1978. "Clarifier Sludge Blanket Responses to Clarifier Sludge Flow Percentage Control." From Updated Summary of the Operational Control Procedures for the Activated Sludge Process, U.S.EPA, also presented at the *51st WPCF Annual Conference, October 1978.*

Westgarth, W. C., F. T. Sulzer and D. A. Okun. 1964. "Anaerobiosis in Activated Sludge Process." *Proceedings of the Second International Conference, Advances in Wastewater Research, Vol. 2.* Pergamon Press, August.

Yoshioka, N. et al. 1957. "Continuous Thickening of Homogeneous Flocculated Slurries," *Chem. Eng.* (Jap.), 21:66.

CHAPTER 6

Retrofit of Existing Plants

INTRODUCTION

THE necessary considerations for the retrofit of existing conventional activated sludge plants to some form of BNR, and some actual case histories of such retrofits, are presented in this chapter. Existing plants can be modified (1) to remove phosphorus biologically, without the removal of nitrogen, (2) to remove nitrogen biologically without the biological removal of phosphorus, or (3) to biologically remove phosphorus and nitrogen simultaneously. For some plants, the nutrient removal objectives can be achieved simply by changing the way the plant is being operated. For other plants, extensive process modifications may be necessary to meet the treatment requirements, or the plant capacity may have to be substantially reduced. Examples of the different circumstances are presented and discussed in this chapter.

PLANT MODIFICATION FOR BIOLOGICAL PHOSPHORUS REMOVAL

It is very simple to modify conventional, plug-flow activated sludge plants for BPR, and this can nearly always be accomplished without decreasing the treatment capacity of the plant. In fact, retrofit may increase the capacity of the plant, particularly if capacity is limited by the aeration equipment or the clarifier capacity. Superior treatment and performance are nearly always achieved after BPR conversion. For example, sludge settleability improves, the effluent concentrations of BOD_5 and TSS are lower, the aeration requirements decrease, and process stability is better. In particular, filamentous growth upsets are much less likely to occur. There is little reason not to change to BPR, and it can be seriously stated that owners and operators of all conventional, municipal activated sludge plants should consider modifying to BPR regardless of whether or not phosphorus removal is required. It is a superior, more economical method of treating municipal wastewaters. It is also a superior way to treat many industrial wastewaters.

RETROFIT CONSIDERATIONS

The following factors should be considered when retrofitting an activated sludge plant for BPR:

(1) Aeration basin size and configuration
(2) Clarifier capacity
(3) Aeration requirements
(4) Type of aeration system
(5) Sludge processing units
(6) Operator skills

An increase in activated sludge basin volume is not required for BPR retrofit because the removal of substrate in the anaerobic zone is more rapid than it is in an aerobic zone of equivalent size. There is no reason to increase the basin size even if the total nominal HRT is as low as 2.5 hours, if nitrification is not required. However, the anaerobic zone can be too large a fraction of the total volume and be detrimental to treatment if there is insufficient time for the polyP bacteria to metabolize the stored substrate and remove the soluble phosphorus from solution. Fortunately, these reactions occur very rapidly and the aerobic reactions are not likely to limit treatment if the anaerobic zone is 50% or less of the total volume.

While it is typically not necessary to increase the volume of an activated sludge basin size to retrofit it for BPR, the configuration of the basin can significantly affect the feasibility of the retrofit. A plug-flow basin is the easiest type to retrofit because plug-flow conditions enhance the BPR processes by ensuring good separation of the necessary anaerobic and aerobic zones. A basin designed for completely mixed conditions, particularly a square tank with surface aerators, is more difficult to retrofit because significant reconfiguration is required. Examples of how a plug-flow basin and a square basin can be retrofitted for BPR are shown in Figure 6.1.

Clarifier modification is usually not needed because implementation of BPR improves sludge settleability. However, rim-flow clarifiers are less suitable for BPR than centerfeed clarifiers because the flow is usually up through the sludge blanket. Some phosphorus release typically occurs in the clarifier sludge blanket of a BPR plant, but in a properly operated centerfeed clarifier the entire sludge blanket plus the released phosphorus is drawn off the bottom of the clarifier and recycled to the anaerobic zone. When the flow is up through the blanket, such as in a rim-flow clarifier, the

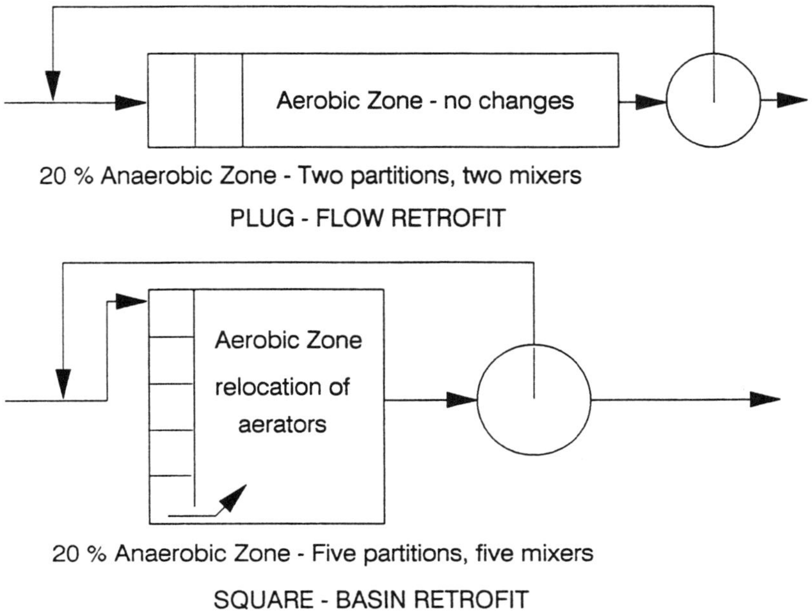

Figure 6.1 Retrofit of activated sludge basins for biological phosphorus removal.

released phosphorus is leached out of the blanket and carried over the effluent weir.

The aeration equipment is usually removed from any zone that will permanently become a part of the anaerobic zone. There is usually no need to add additional aeration equipment in the aerobic zone to compensate for the removed equipment because the processes in the anaerobic zone reduce the oxygen transfer requirements by 10 to 20% (Randall et al., 1991). In fact, the primary concern is the protection of the anaerobic zone from the recycle of too much dissolved oxygen. Therefore, DO concentrations in excess of 2 mg/L in the effluent section of the aeration basin should be avoided.

The sludge processing units are of primary concern when retrofitting an activated sludge plant for BPR. The recycle of any soluble phosphorus changes the COD:P ratio entering the activated sludge process, and may make it impossible to meet the required effluent phosphorus concentration without prefermentation or chemical polishing. The use of anaerobic digesters, gravity thickeners for waste activated sludge, and the recycle of the waste activated sludge for settling with the primary sludge in the primary clarifier are particularly detrimental if not properly managed. Typical recycles from units of this type and suggestions for appropriate management techniques are presented in Chapter 9.

Greater operator skills are generally needed for the operation of BPR plants compared to conventional, fully aerobic plants, but the necessary skills are easily learned and applied. A retraining program for the operators should be part of any retrofit project.

CASE HISTORIES

The York River, Virginia, Wastewater Treatment Plant

The York River, Virginia, activated sludge sewage treatment plant (STP), owned and operated by the Hampton Roads Sanitation District (HRSD), was modified during June and July of 1986 to evaluate biological nutrient removal (BNR) at a full-scale facility. More specifically, the emphasis of the investigation was the implementation of enhanced biological phosphorus removal (BPR) by retrofitting a conventionally designed and operated activated sludge plant. The effects of complete nitrification on BPR performance, and simultaneous nitrogen and phosphorus removal, were also investigated during the latter stages of the project, while the final stage was modification of the process for VIP/UCT operation. The results of A/O and A^2/O operation are presented in this section. The results of VIP/UCT operation have been presented by Randall et al., (1990) and Waltrip (1991).

The York River plant was originally designed to treat a flow of 56.9 million liters per day (ML/d), i.e., 15 mgd, by screening, aerated grit removal, preaeration, primary clarification, coarse bubble step-aeration activated sludge, secondary clarification, and chlorine disinfection. Sludge processing facilities provided were primary sludge gravity thickeners, secondary sludge dissolved air flotation (DAF) thickeners, two-stage anaerobic digesters, and belt filter presses. The anaerobic digesters were normally operated without supernatant recycle, but the filtrate from the belt presses and the supernatants from the thickeners were mixed and recycled to the headworks. Multiple units (two or more) were available for all processes except anaerobic digestion, but there were no separate trains. A schematic of the plant before modification is shown in Figure 6.2.

The York River activated sludge facilities consisted of six basins, each with a capacity of 2.54 ML (0.67 MG), for a total aeration basin volume of 15.14 ML (4 MG). Each basin was constructed with three equally spaced concrete partitions rising to within one foot of the water surface to facilitate four-stage step-aeration operation. The flow to the plant was averaging 29.6 ML/d (6.5 mgd) at the time of the modification, and the plant was being operated as a conventional plug-flow system, not as step-aeration. Three of the six basins were in use before modification, giving the process a hydraulic retention (HRT) of about 7.4 hours. Because the flow was low, only two of the reactors, or one-third of the units, were modified for BNR,

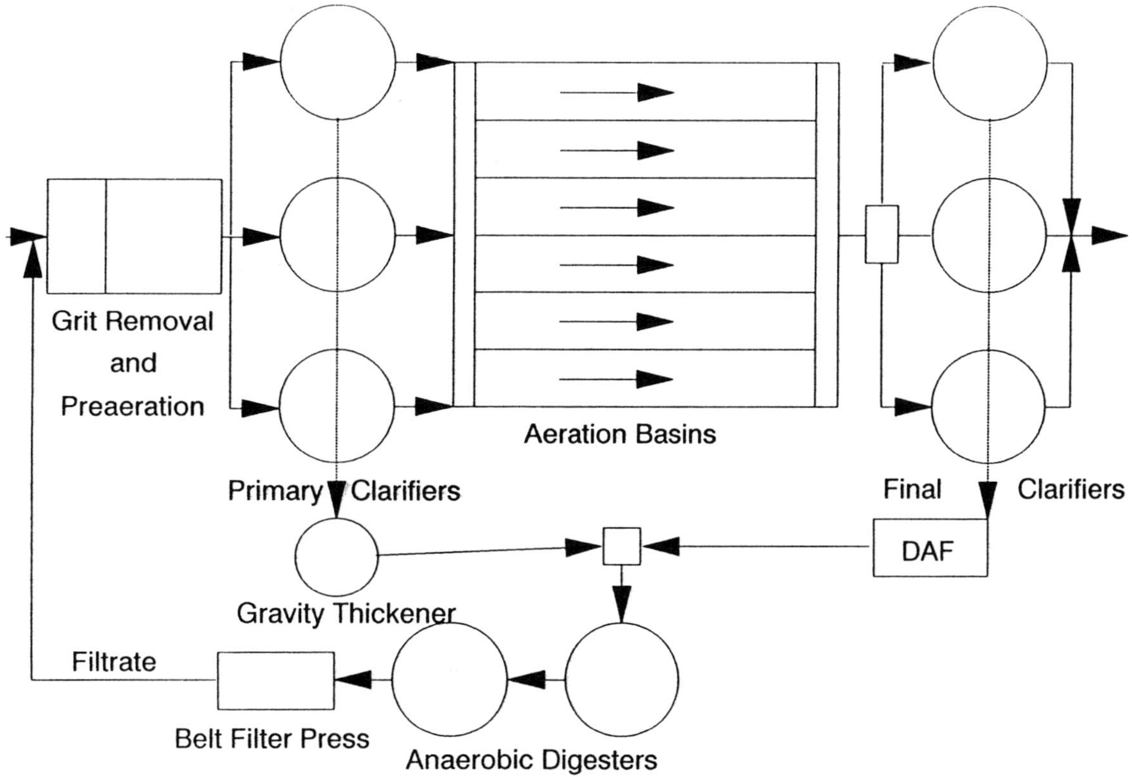

Figure 6.2 Schematic of the York River WWTP before modification.

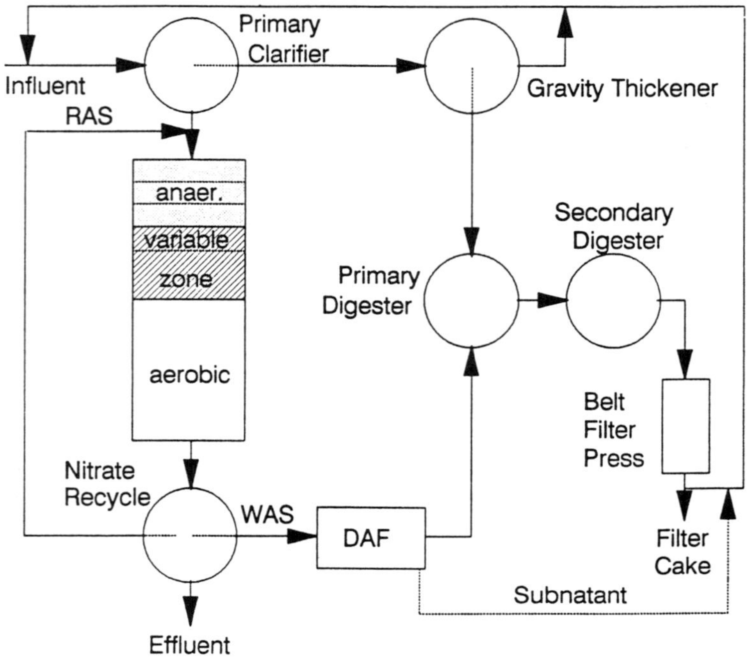

Figure 6.3 Schematic of York River, Virginia BPR WWTP.

for a total modified volume of 5.07 ML (1.44 MG). Thus, at the design flow of 24.6 ML/d (6.5 mgd) the total HRT was five hours. The costs of modification were kept at a minimum by utilizing a nominal life of two years and a maximum life of five years as a basis of design. The process schematic after modification is illustrated by Figure 6.3.

It was concluded, prior to modifying the plant, that modification for operation as an A/O or A^2/O process would require the fewest changes and implementation would be the most economical. As Figure 6.3 shows, two baffles were added to the first section of each of the two retrofitted reactors, dividing that section into three cells of equal size. A baffle was also constructed at the one-third point of the second section to divide it into two cells with the second cell twice the size of the first. An under-and-over flow pattern was established, beginning with the first baffle. The baffles were constructed of cement block and were installed without removing the aeration equipment. The baffles were placed at the one-third points so they would fit between the horizontal aeration headers. The intent was to always operate the three cells in the first section anaerobically, but retain the option of operating the two cells in the second section anaerobically, anoxically, or aerobically, as desired. Two recycle pumps and appropriate pipes were installed to enable recycle of nitrate-rich mixed liquor from the aeration basin effluent back to the anoxic cells to enhance denitrification. A propeller mixer was installed at the center of reach of the five modified cells using I-beam mountings. No modifications were made to the effluent half of the reactors and those two sections were always operated aerobically.

Table 6.1 lists the actual costs of the materials used for the modifications, and the approximate cost of the labor associated with construction. The actual labor was performed by HRSD personnel. It is projected that the entire 57 ML/d (15 mgd) plant could be permanently retrofitted for biological nutrient removal at a cost of about $2.0 million, representing a cost of approximately $350,000 per 10 ML/d of plant capacity ($133,000 per mgd).

Limitations of BNR at York River Plant

The York River STP was selected for this demonstration recognizing that it had several disadvantages for effective BPR, particularly with respect to achieving low concentrations of phosphorus in the effluent. Those disadvantages were:

(1) There were low concentrations of organics in the raw wastewater compared to TKN and phosphorus.

TABLE 6.1. Cost of BNR Modifications, York River, Virginia WWTP 24.6 ML/d Biological Nutrient Temporary Retrofit.

Item	Cost
Mixers (10)	$ 46,807.43
Mixer supports	7672.33
Baffles (6)	4468.67
Pumps (2) (nitrate recycle)	21,312.00
Pipe (nitrate recycle)	10,749.87
Electrical	11,995.19
BPR materials cost	64,946.02
Nitrate recycle materials cost	38,059.47
Total materials cost	100,005.49
Approximate labor cost	51,500.00
BPR retrofit cost	99,279.35
Total A^2/O modification cost	154,505.49
VIP(UCT) modification cost (denitrified recycle, RAS diversion)	34,000.00
Total project cost	188,503.49

(2) The organic strength of the wastewater was reduced by preaeration and primary settling before entering the biological process.

(3) Phosphorus was recycled back to the headworks from the two sludge thickening processes and from anaerobic digestion.

The average composition of the raw wastewater during the six weeks immediately preceding BPR start-up is given in Table 6.2, along with average effluent concentrations and percent removals. The BOD:TP ratio averaged 18.3, somewhat below the 20:1 ratio preferred. During the project, the raw wastewater BOD_5:TP ratio increased and averaged almost 23:1, with a range in monthly averages of 18:1 to 27:1. However, phosphorus recycled to the headworks from the sludge processing units changed the ratio to 15:1 in the combined flow, and it was further decreased to a project average of 12.5:1 by primary treatment. Thus, while the raw wastewater sometimes had a $TBOD_5$:TP ratio considered to be suitable for obtaining effluent phosphorus concentrations of less than 1 mg/L, primary treatment and phosphorus recycle decreased the ratio to the extent that such effluent concentrations were very unlikely, if not impossible. Table 6.3 shows the changes that occurred, based on project averages.

Because a certain amount of organic matter (BOD) is required to remove a unit of phosphorus or nitrogen, the primary effluent wastewater was "BOD limited" with respect to nutrient removal. This did not mean that biological nutrient removal processes could not be used. It simply meant that very low concentrations of phosphorus could not be achieved by biological means alone, unless prefermentation was used. An appropriate operating strategy also would be as follows. The available BOD could be used to achieve large reductions in phosphorus and nitrogen through biological mechanisms. Then, supplemental quantities of either phos-

TABLE 6.2. Conventional Activated Sludge Operation June 19-August 3, 1986: York River STP BNR Demonstration Project.

Parameter	Plant Influent (mg/L)	Plant Effluent (mg/L)	% Removal
$TBOD_5$	168	3.5	98
TCOD	411	27	93
TSS	154	9	94
TPO_4	9.2	7.7	16
SPO_4	—	6.8	26
TN	30.4	16.6	45
TKN	30.4	2.4	92
NO_X	0	14.2	—

Influent flow = 22.5 ML/d (5.94 mgd). Nominal HRT = 8.0 h.
Average MCRT = 4.7 d. Number of basins in operation = 3.

TABLE 6.3. Changes in Organics-to-Phosphorus Ratios with Recycle and Primary Treatment: York River STP BNR Demonstration Project.

Parameter	Raw Influent	Combined Influent	Change mg/L	Change %	Primary Effluent	Decrease mg/L	Decrease %
$TBOD_5$, mg/L	209	209	0	0	139	70	33
TCOD, mg/L	377	418	41	11	249	169	40
TP, mg/L	9.26	13.76	4.50	49	11.11	2.65	19
TKN, mg/L	28.4	35.7	7.3	26	29.0	6.7	19
$TBOD_5$:TP	22.6	15.2	—	−33	12.5	—	18
TCOD:TP	40.7	30.4	—	−25	22.4	—	26

phorus precipitating chemicals could be added to economically attain specific effluent standards, as desired.

The high phosphorus concentrations were actually an advantage during this project because they made it possible to evaluate the true phosphorus removal potential of the process. Phosphorus was never a limiting factor during the evaluations.

Scope of Demonstration

The demonstration project progressed through several periods characterized by operating conditions and treatment objectives. Table 6.4 lists those operating periods chronologically. The list shows that the system was initially operated for phosphorus removal only, and operated in that mode until mid-May, 1987. Actually, an effort was made to establish nitrification during December, 1986, but it failed, first because of low temperatures and then because of very high influent flows. The monthly average flows and raw influent characteristics, listed in Table 6.5, show the high flows that

TABLE 6.4. Operating Log: York River STP BNR Demonstration Project.

Operating Objective	Time Period
Evaluation of conventional treatment	June 19-August 3, 1986
BPR start-up	August 4-27, 1987
BPR design flow operation	Sept. 1, 1986-Jan. 16, 1987
BPR high flow operation	Jan. 18-April 30, 1987
Nitrification start-up	June 1-30, 1987
BPR operation with nitrification	July 1-August 3, 1987
BNR start-up (nitrate recycle)	August 4-31, 1987
BNR operation*	Sept. 1-Oct. 31, 1987

*Simultaneous nitrogen and phosphorus removal.

began in late January and continued into May. The flows reached peaks of almost 68 ML/d (18 mgd), and averaged 53 ML/d (14 mgd) for periods as long as two weeks. This resulted in very high organic loading rates and very short HRTs in the biological process.

Table 6.5 shows that the raw influent BOD_5 increased substantially from June through October, 1987. The increase was more than could be explained by flow decreases, and the COD increase was much smaller. Apparently fermentation of the wastewater in the sewers increased substantially during the summer, and this increased the BOD_5 value relative to the COD. The BOD_5:TCOD ratio during February, March and April averaged less than 0.40, whereas in August, September and October it averaged almost 0.73. Biological phosphorus removal (BPR) was enhanced, but the system was still organics limited, illustrating that the COD:TP ratio is a more reliable indicator of the effluent phosphorus concentration achievable than is the BOD_5:TP ratio (see Figure 4.5).

Process Start-Up

The modified basins were put on-line on August 4, 1986, and operated in accordance with the conditions listed in Table 6.6. As indicated by the

TABLE 6.5. Raw Influent: York River STP BNR Demonstration Project.

Month	Flow ML/d	Flow mgd	TBOD mg/L	TCOD mg/L	TP mg/L	TKN mg/L
July '86	23.2	6.12	174	N/S	9.15	30.2
Aug	23.4	6.19	156	N/S	8.64	28.6
Sept	23.6	6.24	172	N/S	9.24	29.9
Oct	23.2	6.13	208	391	9.74	32.0
Nov	23.0	6.09	225	395	9.66	33.0
Dec	25.0	6.61	204	471	10.17	33.0
Jan '87	39.3	10.38	161	334	8.88	25.3
Feb	39.5	10.43	135	318	7.31	24.4
Mar	43.0	11.36	130	344	7.34	23.2
Apr	37.4	9.88	145	368	7.66	24.3
May	27.9	7.38	182	374	8.38	27.2
June	26.6	7.03	207	348	8.99	25.5
July	23.8	6.28	254	377	9.58	29.1
Aug	23.1	6.10	270	376	10.27	29.5
Sept	22.7	6.01	286	395	10.51	30.4
Oct	20.9	5.53	304	413	11.83	30.4
Oct-Oct Averages	28.9	7.63	209	377	9.26	28.4

N/S = No sample.

TABLE 6.6. **Start-Up Operating Conditions August 4-27, 1986: York River STP BNR Demonstration Project.**

Effluent flow (ML/d)	23.0
Target MCRT (d)	4
Average hydraulic retention time (h)	5.25
Number of anaerobic cells in use	4
Number of anoxic cells in use	0
Anaerobic sludge mass fraction (%)	33.3
Aerobic sludge mass fraction (%)	66.7
Return activated sludge (RAS) recycle rate (% of raw influent flow)	70

All influent flow into cell 1.
Number of basins in operation = 2.

anaerobic mass fraction, the first four cells in each basin were operated anaerobically while the remaining two-thirds of the volume was operated aerobically. There was no anoxic zone. The activated sludge was not seeded with bacteria from an operating BPR plant or from any other source.

Phosphorus removal by the process began increasing almost immediately upon start-up and the effluent reached steady-state phosphorus levels for the operating conditions in only twenty-three days, as illustrated by Figure 6.4. The system was operated with start-up conditions for an additional two weeks after steady state was achieved, and produced average effluent phosphorus concentrations of 2.95 mg/L total phosphorus (TP), and 2.15 mg/L soluble phosphorus (SP), based on early morning grab sampling procedures. It was later determined that the grab sampling procedure substantially underestimated the effluent phosphorus concentrations, as measured by composite samples, but all data in Figure 6.4 were collected by consistent grab sampling and the relative changes are accurate, i.e., the process accomplished the removals shown.

The percent phosphorus in the activated sludge increased as the phosphorus removal increased and this is shown in Figure 6.5. Before BPR start-up, the phosphorus in the VSS was between 3 and 4%. By mid-September, it was between 10 and 11%.

BPR Operation Results

Several changes occurred in the treatment plant flows with the establishment of BPR. Two of these changes are illustrated by Figure 6.6. It can be seen there was a substantial decrease in the effluent phosphorus concentration, but there was also a significant increase in the phosphorus concentration influent to the biological process, i.e., in the primary clarifier effluent. The changes in the primary effluent over the entire project are shown by the monthly average values listed in Table 6.7. It can be seen that

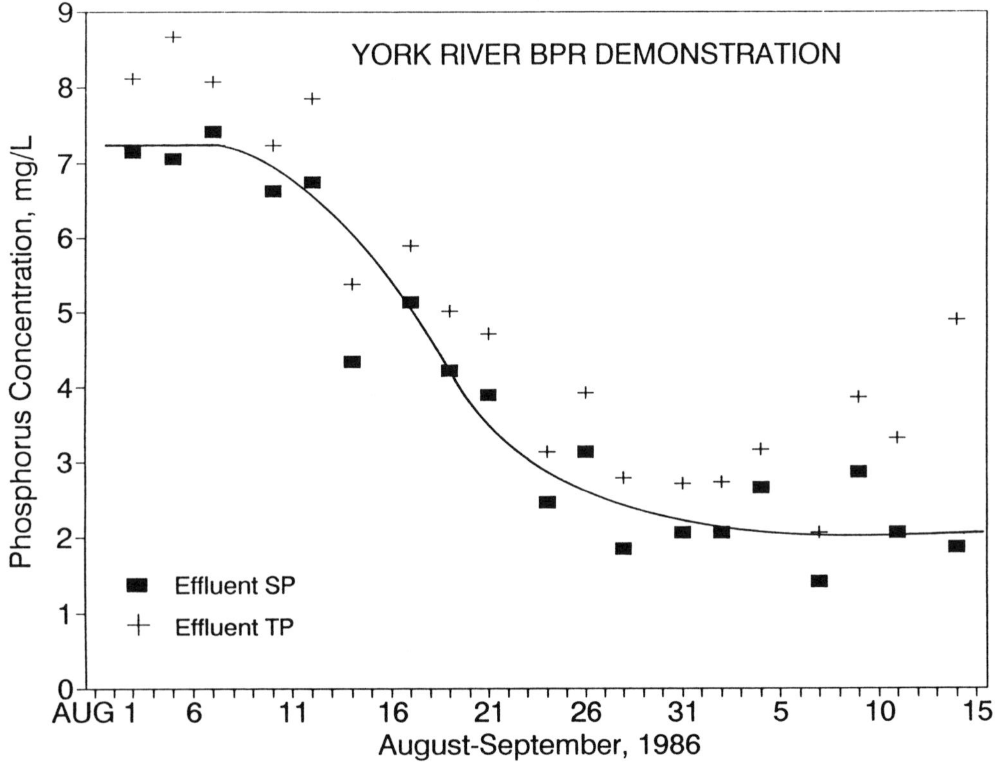

Figure 6.4 Reduction of effluent phosphorus concentrations during start-up.

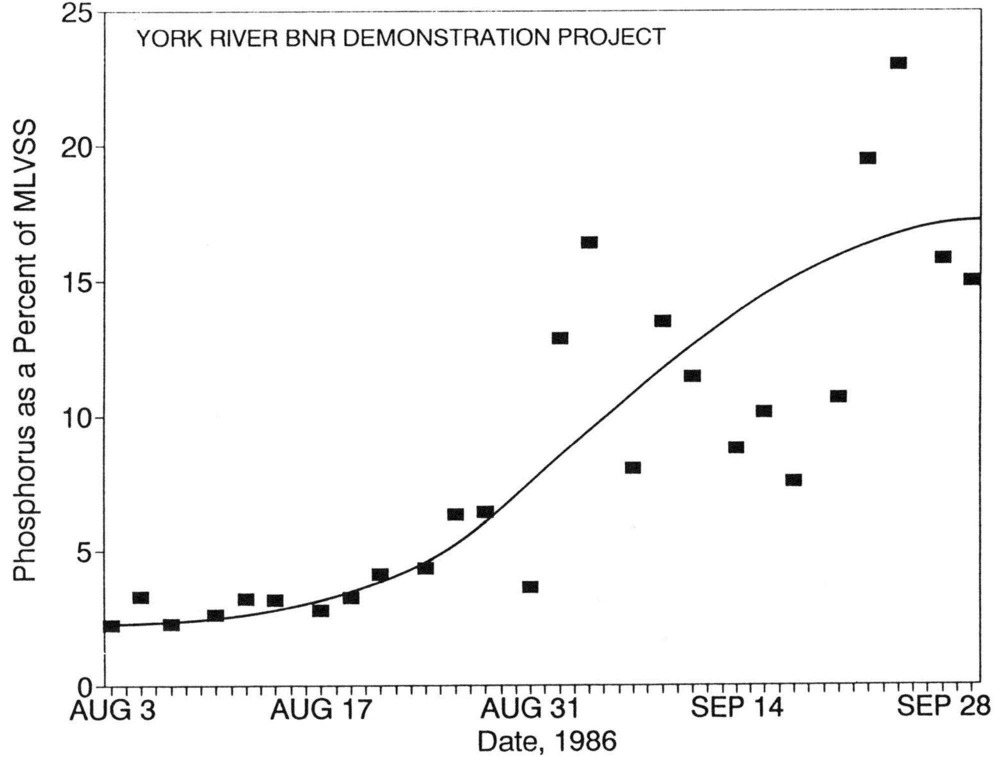

Figure 6.5 Increase in activated sludge phosphorus content during BPR start-up.

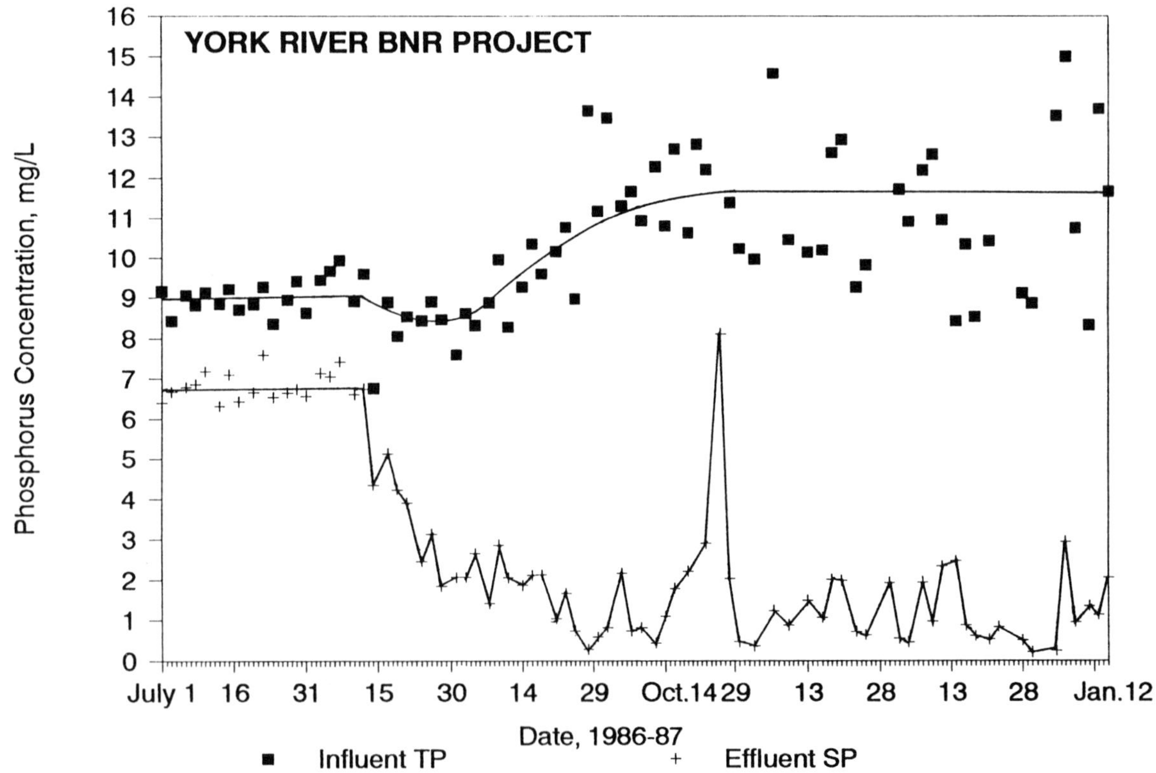

Figure 6.6 Changes in influent and effluent phosphorus during BPR operation.

TABLE 6.7. Primary Effluent: York River STP BNR Demonstration Project.

	*Flow		$TBOD_5$		$SBOD_5$		TCOD		TP		TKN	
Month	ML/d	mgd	mg/L	lb/d	mg/L	lb/d	mg/L	lb/d	mg/L	lb/d	mg/L	lb/d
July '86	30.3	8.01	95	6346	N/S	N/S	N/S	N/S	8.92	596	25.6	1707
Aug	30.6	8.09	94	6342	N/S	N/S	N/S	N/S	8.71	588	27.4	1851
Sept	30.4	8.03	119	7969	58	3884	N/S	N/S	9.84	659	28.2	1891
Oct	28.6	7.55	128	8060	N/S	N/S	244	15364	11.71	737	31.4	1977
Nov	27.2	7.20	122	7326	67	4023	261	15673	12.05	724	36.0	2161
Dec	29.2	7.71	138	8874	52	3344	303	19483	12.59	810	32.0	2054
Jan '87	45.0	11.89	140	13883	39	3867	249	24691	10.92	1083	28.7	2844
Feb	45.2	11.95	109	10863	33	3289	246	24517	9.78	975	30.9	3084
Mar	46.7	12.34	75	7719	37	3808	224	23053	7.88	811	24.5	2523
Apr	42.4	11.21	84	7853	40	3740	258	24121	9.02	843	28.4	2650
May	32.6	8.62	91	6542	N/S	N/S	235	16894	10.58	761	27.1	1950
June	30.7	8.11	123	8319	53	3585	208	14069	11.98	810	26.0	1760
July	29.0	7.65	183	11676	68	4338	247	15759	11.70	746	29.0	1849
Aug	29.0	7.65	173	11038	60	3838	291	18566	10.67	681	29.2	1865
Sept	27.5	7.27	219	13278	63	3820	228	13824	12.12	735	26.9	1630
Oct	26.7	7.06	225	13248	69	4063	240	14131	13.37	787	29.0	1708
Oct-Oct Averages	33.8	8.94	139	9898	53	3791	249	18473	11.11	808	29.2	2158

*Combined flow = influent plus recycles.
N/S = No samples.

BPR operation caused a substantial increase in the phosphorus concentration influent to the process and that high flows diluted these concentrations.

The cause of the increased phosphorus load to the biological process was recycle from sludge processing, particularly from belt filtration of the anaerobically digested sludge. The increase in the phosphorus content of the belt filter press recycle back to the headworks over the entire project is illustrated by Figure 6.7. The filtrate phosphorus concentration steadily increased from September l, 1986 through early January, 1987. This was because the anaerobic digesters were considerably underloaded and had a HRT, based on sludge flow, of nearly 120 days. After one HRT, the filtrate TP concentration had increased from an average of less than 20 mg/L to an average of about 250 mg/L. There was also a large amount of variation in the concentration after one HRT, and the changes appeared to be cyclic. The reason for the cycle could not be established.

Although it took four months to reach equilibrium with the anaerobic digesters and the TP concentrations in the filtrate, the primary clarifier effluent TP reached apparent equilibrium much sooner. Figure 6.6 indicates the TP concentration had reached what became the primary effluent average value for design flow BPR operation by early October, 1986. After that there was a lot of cyclic variation but little change in the average value. Apparently a substantial fraction of the phosphorus returned with the recycles spontaneously precipitated, and was removed from the primary clarifier with the primary sludge. This limited the concentration discharged by the clarifier, although the phosphorus inputs continued to increase.

The large effluent spike shown in Figure 6.6 occurred for two reasons. The basins adjacent to the modified basin had been full of water, but were drained in early October. This caused hydrostatic pressure to force MLSS from the operating basins through wall joints into the empty adjacent basins. Because of this, the MLSS concentration decreased from an average of 1600 mg/L down to 850 mg/L on October 26, the day of the spike. In addition, there was a large return of suspended solids and other pollutants back to the headworks from the gravity thickener on the same day. The primary clarifier effluent concentrations on October 16 were 118 mg/L TP and 15 mg/L SP. Very high TSS and TKN were observed in the same sample. Refilling the adjacent basins with water resulted in an increase in the MLSS back to normal levels, thickener excursions were minimized, and efficient phosphorus removal was restored in a very short period of time.

The MCRT was decreased before start-up to reduce or eliminate nitrification, because it was believed that true anaerobic conditions were needed in the anaerobic zone to induce excess phosphorus removal. Low MCRT operation reduced the effluent nitrate concentration to less than 2 mg/L on August 10. Increased phosphorus removal was not observed on August 12,

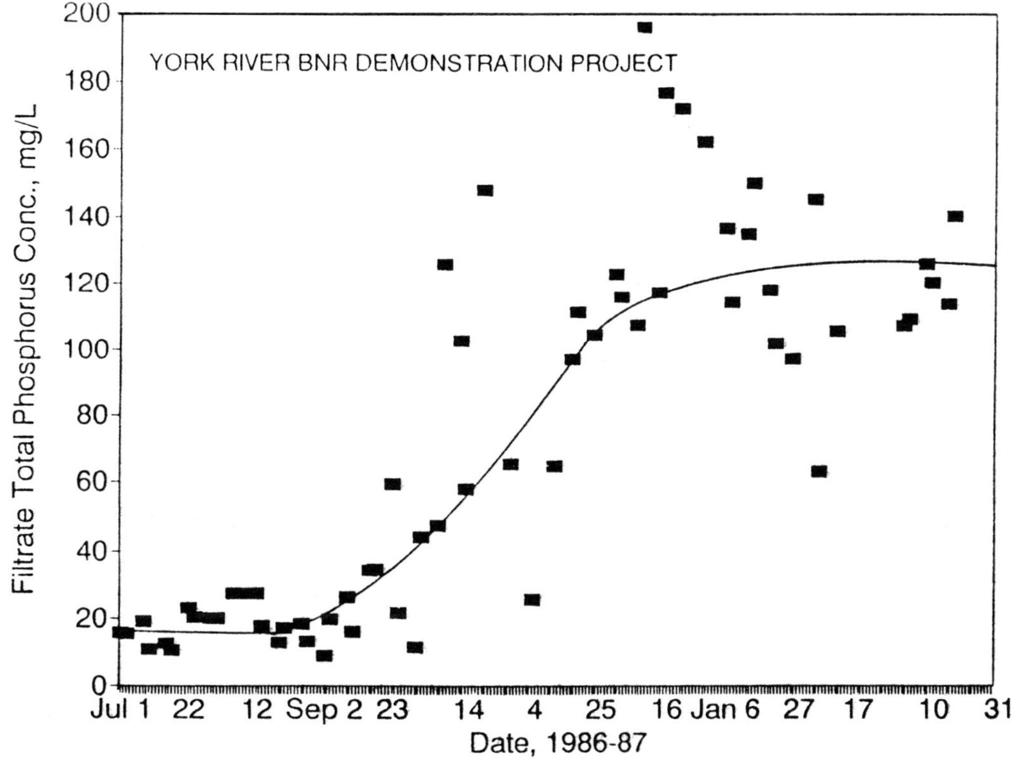

Figure 6.7 Increase in phosphorus concentration of belt filter press filtrate.

but was observed on the 14th. Comparison of the data indicates that zero nitrate recycle back to the anaerobic zone was needed to induce enhanced BPR.

The sludge volume index (SVI) varied as shown in Figure 6.8 during design flow BPR operation. Settleability worsened for a short period after start-up, but had improved considerably by the time steady-state BPR was obtained. The average was less than 70 when the process was operated with a 33% anaerobic mass fraction. The SVI increased to values generally between 70 and 100 when the anaerobic mass fraction was increased to 50%. Settleability remained good and relatively constant until the MCRT was increased in an effort to re-establish nitrification. This resulted in the substantial increases in SVI shown in Figure 6.8 during December and January. Throughout this project, increases in MCRT resulted in increases in SVI, and decreases in MCRT produced decreases in SVI. This was probably because high MCRTs increased the impacts of oxidized materials in the anaerobic zone and made it possible for low DO filamentous forms to grow. The worst settling occurred in September and October, 1987 after complete nitrification and nitrogen removal were established. However, treatment performance was excellent during this period even though the sludge settleability was the worst of the study.

Treatment Plant Performance

The monthly average final effluent concentrations observed throughout the project are given in Table 6.8. The October, 1986 through October, 1987 averages are given to indicate how the system performed under all operating conditions once BPR was established. Composite phosphorus samples were not collected from July through November, 1986, so the values listed were estimated from grab sample results. It is likely the adjustment relationship used overestimated the effluent concentrations, and lower values than those shown were actually achieved during these months.

The data show that excellent effluent quality in terms of organic matter and TSS was generally achieved in spite of high flows and experimental operational changes. It is interesting to note that lower BOD_5 values were observed while nitrification was being accomplished by the process, and that even the extremely high winter flows did not cause the effluent TSS to exceed acceptable levels. The latter was achieved because it was possible to put a third secondary clarifier on-line and because RAS was increased to increase clarifier solids throughput.

After BPR was established, the only time the process was operated with complete nitrification was from mid-June through October, 1987. The data show that low effluent nitrogen concentrations were obtained during the nitrification period because nitrates were recycled back to non-oxic zones,

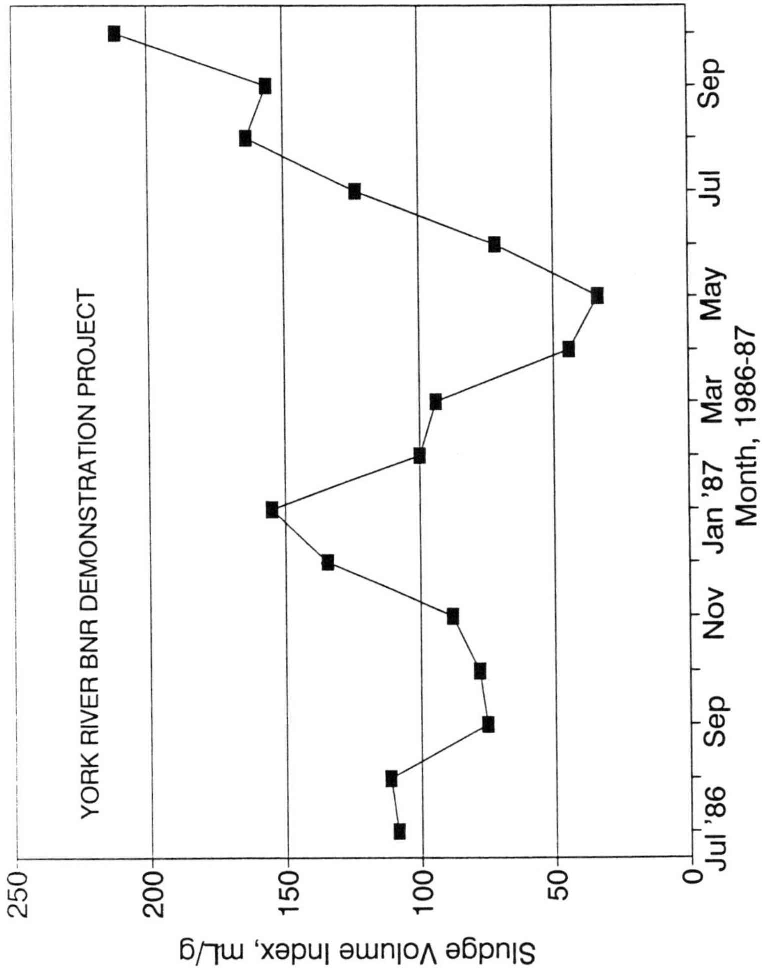

Figure 6.8 Change in sludge volume index during BPR operation.

TABLE 6.8. Final Effluent Concentrations—Monthly Average Values*: York River STP BNR Demonstration Project.

Month	$TBOD_5$	TCOD	SCOD	TSS	COMP (TP)	**COMP (SP)	TN	TSN	NH_3—N	NO_X—N
July '86	3	N/S	N/S	7	†10.18	8.96	14.0	13.2	0.2	11.9
Aug	9	N/S	N/S	9	† 7.78	6.67	19.1	18.0	12.9	3.5
Sept	17	N/S	N/S	8	† 4.27	2.74	19.0	18.1	11.2	5.2
Oct	11	55	45	7	† 4.42	3.21	19.1	18.2	14.0	3.6
Nov	11	51	44	7	† 3.40	2.28	22.8	21.9	19.4	2.3
Dec	28	50	37	7	2.94	1.80	21.7	20.9	17.7	3.1
Jan '87	14	45	45	13	3.63	2.32	17.5	16.0	14.5	1.2
Feb	10	58	34	10	3.71	2.62	17.3	16.1	15.2	0.2
Mar	8	48	36	6	3.34	2.58	16.5	15.8	14.8	0.1
Apr	13	67	41	13	3.45	2.22	19.4	17.4	17.0	0.2
May	10	58	48	5	2.86	2.18	20.9	20.3	18.8	0.3
June	6	48	32	4	3.55	3.34	13.6	12.2	7.4	4.8
July	4	30	30	3	3.98	3.64	8.6	8.2	0.3	6.7
Aug	3	63	35	5	5.97	5.39	7.2	6.6	0.6	5.5
Sept	4	46	41	7	3.51	2.35	5.7	4.8	0.5	4.4
Oct	4	34	26	7	3.27	2.30	5.6	4.8	0.7	3.6
Oct-Oct Averages	10	50	38	7	3.69	2.79	15.1	14.1	10.8	2.8

*All values given in mg/L.
**Calculated using SP/TP ratio grab samples.
†Calculated from grab samples.
N/S = No sample.

either through RAS recycle or a combination of RAS and mixed liquor recycle. Mixed liquor recycle was started up on August 4, 1987 at a constant rate of 27.0 ML/d (7.14 mgd) and operated through October, 1987. During the last two months of this project, simultaneous nitrogen and phosphorus removal produced average effluent concentrations of 5.6 mg/L TN, 4.8 mg/L total soluble nitrogen (TSN), 3.4 mg/L TP, and 2.3 mg/L SP. Stated another way, the biological process removed an average of 10.4 mg/L phosphorus (influent TP-effluent SP) and 23.2 mg/L nitrogen (influent TN-effluent TSN) removal during this two-month period. This was accomplished with process removals of 218 mg/L BOD_5 and 200 mg/L COD.

The near equivalency of total BOD_5 and total COD in the primary effluent during the last two months of the project was particularly interesting. It probably indicates that virtually all of the organics were readily available for storage by the phosphorus-removing bacteria. Yet, the soluble BOD_5 was a relatively small fraction (30%) of the total BOD_5 during the same period. BOD removal was greater during treatment than COD removal because during growth bacteria produce soluble organics that are difficult to degrade, and these were discharged in the effluent, thus increasing the COD.

Phosphorus profiles through the treatment train for the entire project, based on monthly average values, are presented in Figure 6.9. This figure illustrates the various changes that took place in the treatment system with establishment of BPR and with changes in operating conditions.

Figure 6.9 shows the changes in primary clarifier effluent phosphorus that occurred with BPR start-up. Before BPR, primary treatment removed slightly more than 2 mg/L phosphorus whereas after equilibrium with the recycle streams was achieved, the primary effluent phosphorus was approximately 2 mg/L more than the raw influent concentration. The figure also shows that the high influent flows substantially diluted the raw influent and primary effluent phosphorus concentrations. In spite of the high flows, there were only modest increases in the effluent phosphorus concentrations. However, the combination meant the mg/L phosphorus reductions accomplished by both the plant and the biological process were much smaller during high flows although the mass of phosphorus removed did not decrease. The monthly averages for both are given in Table 6.9.

The mass of phosphorus removed by the biological process increased during the high flow period, but the high organic loadings caused a rapid washout of the small nitrifier population (see NO_X-N, Table 6.8) and a substantial decrease in the relative population of the phosphorus-storing bacteria. Calculations showed that the fraction of phosphorus removal that could be attributed to normal activated sludge growth increased considerably at the expense of enhanced BPR during this period.

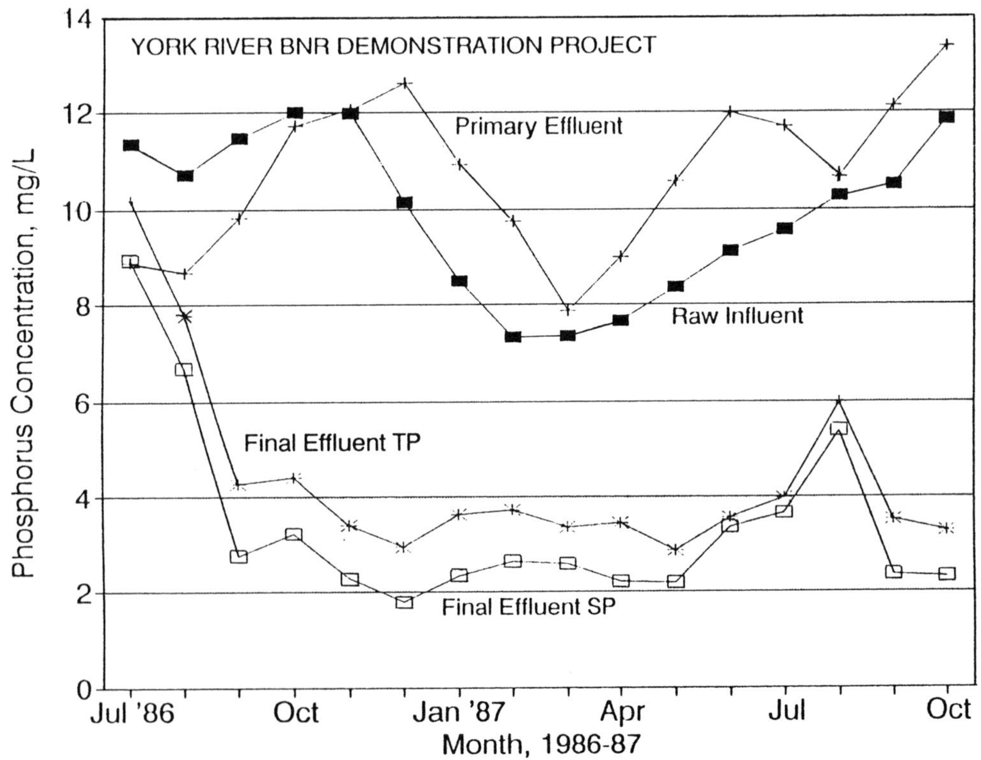

Figure 6.9 Monthly average phosphorus concentration profiles through the plant.

TABLE 6.9. Plant and Process Phosphorus Removal: York River STP BNR Demonstration Project.

	Removal through Plant				Removal through Biol. Process			
Month	TP-TP (mg/L)	TP-TP (kg/d)	TP-SP (mg/L)	TP-SP (kg/d)	TP-TP (mg/L)	TP-TP (kg/d)	TP-SP (mg/L)	TP-SP (kg/d)
July '86	1.17	27	2.39	55	−1.26	−38	−0.04	−1
Aug	2.93	69	4.04	95	0.93	28	2.04	62
Sept	7.19	170	8.72	206	5.57	169	7.10	216
Oct	7.58	176	8.79	204	7.29	208	8.50	243
Nov	8.58	198	9.70	224	8.65	236	9.77	266
Dec	7.23	181	8.37	209	9.65	282	10.79	315
Jan '87	4.89	192	6.20	244	7.29	328	8.60	387
Feb	3.60	142	4.69	185	6.07	275	7.16	324
Mar	4.00	172	4.76	205	4.54	212	5.30	248
Apr	4.21	157	5.44	204	5.57	236	6.80	289
May	5.52	154	6.20	173	7.72	252	8.40	274
June	5.59	149	5.80	154	8.43	259	8.64	265
July	5.60	133	5.94	141	7.72	224	8.06	233
Aug	4.30	99	4.88	113	4.70	136	5.28	153
Sept	7.00	159	8.16	186	8.61	237	9.77	269
Oct	8.56	179	9.53	200	10.10	270	11.07	296
Oct-Oct Averages	5.90	161	6.80	188	7.41	243	8.32	274

During February, 1987, phosphorus removal by normal bacterial growth was slightly greater than that accomplished by BPR. But, when design flow prevailed, BPR accounted for from 69 to 74% of the phosphorus removal, with an average of 71%. The results of these calculations are given in Table 6.10. BPR returned to the normal range in May, 1987 when the flows returned to near design level.

Figure 6.8 also shows that the effluent phosphorus concentrations increased when a program of nitrification without mixed liquor recycle to an anoxic zone was started in June, 1987. It continued to increase during July and reached the highest effluent concentrations observed during the entire project when mixed liquor recycle was started in August. However, once denitrification in the anoxic zone was established, the effluent phosphorus concentrations returned to values similar to those observed during BPR-only operation.

The high phosphorus concentrations measured during June, July, and August, 1987 can be easily explained using Figure 6.10, which is a graph of the mass of nitrate returned to the influent of the biological process by RAS. The graph shows that once nitrification was reduced to establish BPR in August, 1986, very little nitrate was recycled to the biological process until complete nitrification was established in June, 1987. The recycle of nitrates with the RAS accomplished a 70% removal of total nitrogen during July, but it reduced the process phosphorus removal by an estimated 20%. This occurred because bacteria will use nitrate as a DO substitute to metabolize organic matter rather than store it for later use to enhance phosphorus storage, if nitrates are available. The results indicate that substantial quantities of nitrates were removed by denitrification in the secondary clarifiers and sludge return lines. This adequately protected the anaerobic zone during complete nitrification without mixed liquor recycle to the anoxic zone, or during mixed-liquor recycle start-up.

There was a sharp decrease in RAS nitrates one week after recycle to the anoxic zone was started, but it was nearly forty days before the returned quantities were similar to those before nitrification start-up.

Sludge Production

Waste sludge production information is summarized in Table 6.11. The data show the system was typically operated at high food-to-microorganism (*F/M*) values and it was not surprising that such an *F/M* range would generate large quantities of waste sludge. However, relatively low yields were observed except during the high flow period. When operated for BPR-only at design flow and with an MCRT between 3.2 to 7.9 days, the biological process generated 0.48 kg TSS (0.31 kg VSS) per kg BOD_5

TABLE 6.10. Waste Sludge Production: York River STP BNR Demonstration Project.

Month	$TBOD_5$ *F/M*	TCOD *F/M*	WAS TSS kg/d	WAS VSS kg/d	Biomass in Reactors kg TSS	kg VSS	Actual MCRT (days)	VSS COD	VSS $TBOD_5$	TSS $TBOD_5$
July '86	0.41	N/S	1279	1023	8753	7002	4.6	N/S	0.37	0.46
Aug	0.60	N/S	957	756	6124	4838	6.4	N/S	0.29	0.37
Sept	0.80	N/S	1559	1045	6765	4532	4.3	N/S	0.34	0.51
Oct	0.73	1.39	1383	871	7940	5002	5.7	0.16	0.26	0.41
Nov	0.47	1.00	1568	1035	10783	7116	6.9	0.18	0.34	0.52
Dec	0.45	0.98	1060	700	13611	8983	12.8	0.09	0.22	0.33
Jan '87	0.50	0.89	3349	2311	18214	12568	5.4	0.25	0.41	0.59
Feb	0.37	0.84	3656	2632	18342	13206	5.0	0.31	0.59	0.82
Mar	0.32	0.96	2903	2148	14787	10942	5.1	0.26	0.69	0.93
Apr	0.36	1.12	2237	1611	13629	9813	6.1	0.20	0.53	0.74
May	0.35	0.90	1947	1324	12594	8564	6.5	0.23	0.50	0.74
June	0.36	0.60	1544	926	17697	10618	11.5	0.19	0.26	0.43
July	0.47	0.63	1481	904	18664	11385	12.6	0.14	0.17	0.28
Aug	0.53	0.90	1140	753	14251	9406	12.5	0.11	0.15	0.23
Sept	0.63	0.66	1501	916	15640	9541	10.4	0.18	0.15	0.25
Oct	0.61	0.65	1240	694	17661	9890	14.2	0.13	0.12	0.21
Oct-Oct Averages	0.47	0.89	1924	1294	14909	9772	8.8	0.19	0.34	0.50

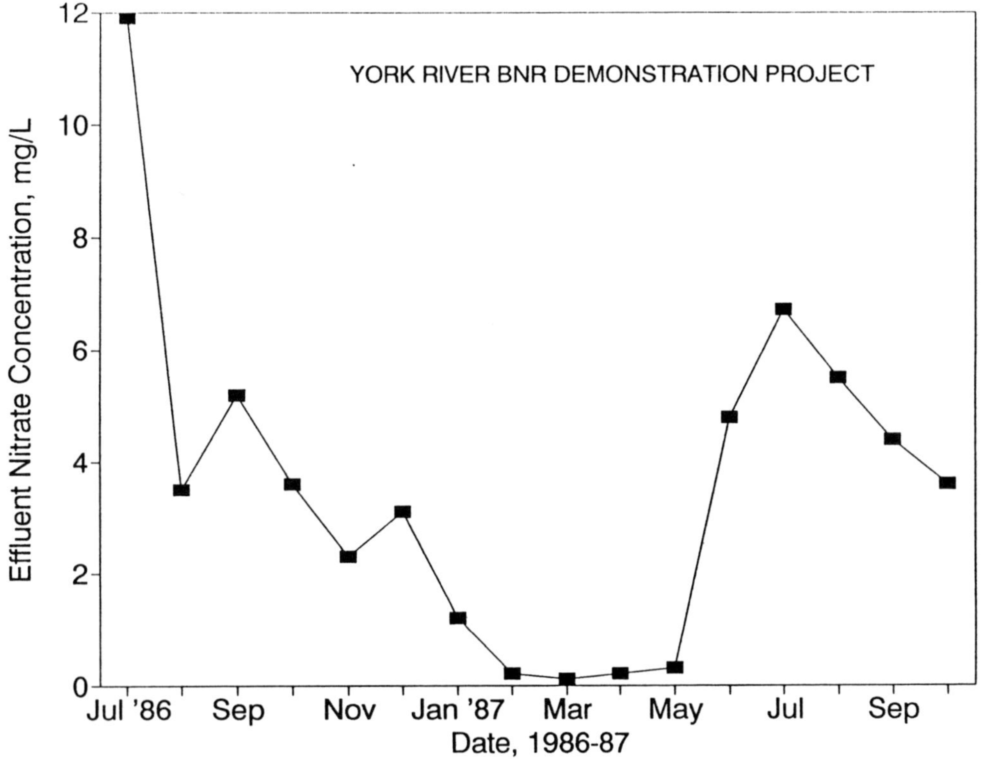

Figure 6.10 Changes in effluent nitrate concentrations during BPR operation.

TABLE 6.11. Phosphorus Content of the Biological Sludge during Excess Phosphorus Removal: York River STP BNR Demonstration Project.

	Waste Sludge Characteristics					Phosphorus Removal			
								Excess	
Month	Flow ML/d	TSS mg/L	VSS %	TP mg/L	TP/VSS %	Actual k/d	*Aer Proc kg/d	kg/d	%
July '86	0.567	2255	80	68	3.79	39	39	0	0
Aug	0.360	2661	79	103	4.92	37	29	9	23
Sept	0.545	2860	67	245	12.81	134	40	94	70
Oct	0.401	3445	63	339	15.62	136	33	103	76
Nov	0.337	4653	66	382	12.42	129	39	89	69
Dec	0.163	6512	66	525	12.21	85	27	59	69
Jan '87	0.426	7851	69	547	10.10	233	88	146	62
Feb	0.466	7850	72	419	7.41	195	100	95	49
Mar	0.447	6497	74	425	8.83	190	81	108	57
Apr	0.386	5792	72	357	8.57	138	61	77	56
May	0.379	5142	68	467	13.34	177	50	126	72
June	0.261	5908	60	453	12.79	118	35	83	70
July	0.249	5938	61	462	12.87	115	34	81	70
Aug	0.225	5061	66	316	9.46	71	29	43	60
Sept	0.227	6607	61	547	13.56	124	35	89	72
Oct	0.174	7120	56	571	14.32	99	26	73	74
Oct-Oct Averages	0.319	6029	66	447	11.65	139	49	90	66

*Assumed growth phosphorus removal for totally aerobic activated sludge operation. July 1986 operation was totally aerobic.

removed. By comparison, a nutrient removal pilot plant at the HRSD Lambert's Point STP site generated 0.77 kg TSS/kg BOD_5 removed.

When the operating MCRT was increased to the range of ten to fourteen days, the waste sludge production decreased, as expected, and averaged 0.26 kg TSS/kg BOD_5 removed. VSS production was about 65% of TSS production.

The amount of phosphorus incorporated into the waste sludge determines the amount of phosphorus removed by a BPR system. The percent phosphorus in the York River STP was very high throughout the project as shown in Table 6.11. Excluding the high flow period and August when upset conditions prevailed, the percent TP/VSS was never below 12% based on monthly averages, and averaged 13.3%. On a TP/TSS basis the average was 8.4%. This value is considerably higher than the 4 to 5%, or lower, typically reported in the literature for BNR systems, because the percent phosphorus in the sludge is a function of the process influent TCOD:TP ratio, as discussed in Chapter 4.

Analysis of the data indicated that 22 mg/L of BOD_5 was consumed for each mg/L of phosphorus removed by the biological process when the anaerobic mass fraction was 33%. During BNR operation the ratio was only 21 mg/L although the anaerobic mass fraction was only 25%.

Summary

Two aeration basins of a six-basin step-aeration design, high-rate activated sludge plant were modified for A/O and A^2/O operation, for a cost of approximately $155,000. This provided a rated BNR capacity of about 30 ML/d (6.5 mgd). The process was started up for BPR without seeding, and reached steady-state effluent phosphorus concentrations within twenty-three days.

Operation for BPR caused a large increase in the phosphorus content of the waste sludge. This increased the amount of phosphorus recycled back to the headworks from sludge processing, particularly from belt filtration of the anaerobically digested sludge. Before BPR operation, the primary clarifier effluent contained 2 mg/L less phosphorus than the raw influent. After BPR was established, the primary effluent contained an average of 2 mg/L more phosphorus than the raw influent.

The biological process removed both nitrogen and phosphorus very efficiently. During BNR operation, the process simultaneously removed 10.4 mg/L of phosphorus and 23.2 mg/L of nitrogen while utilizing 218 mg/L of BOD_5. The average amount of BOD_5 used to remove one mg/L of phosphorus was 22 mg/L, when the system was operated near design flow and the anaerobic mass fraction as 33%. However, it was not possible to reduce the effluent phosphorus concentration to less than 1.0 mg/L because the BOD_5:TP ratio was too low after primary treatment.

The York River BNR system attained very high concentrations of phosphorus in the waste sludge compared to values reported in the literature. When operated near design flow, the system averaged 8.6% TP/TSS. The TP/VSS ratio was 13.3%, with a range of 12.21 to 15.62. Sludge production rates were low compared to literature values and pilot plant experience.

Biological phosphorus removal was substantially decreased by both high influent flows and the recycle of large quantities of nitrates back to the anaerobic zone. The high flows produced high organic loading rates and it is believed the high flows and *F/M* ratios caused a significant fraction of the phosphorus-storing bacteria to wash out of the system because their competitors thrived under the high flow conditions.

Increases in MCRT to obtain nitrification consistently resulted in a worsening of sludge settleability.

Acknowledgements

This project was financially supported by both the Commonwealth of Virginia through the State Water Control Board Chesapeake Bay Office, and the Hampton Roads Sanitation District. All laboratory analysis was performed either on-site by Bob Jones or at the HRSD peninsula laboratory. Special recognition is due to Bob Jones for performing the day-to-day monitoring of the process and for compiling the data. Milind Wable and Vikram Pattarkine spent many hours performing computer manipulation of the data, and their contributions are gratefully recognized. The Air Products and Chemical Company, Inc. permitted the use of the patented A/O and A^2/O processes to facilitate the investigative objectives of this project. Also, Dr. Sun Nan Hong of Air Products made significant contributions to the successful completion of the project through data review and operating advice.

Bonnybrook WWTP, Calgary, Alberta, Canada

The retrofit of this plant has been described by Wilson et al. (1990). The city of Calgary's Bonnybrook Wastewater Treatment Plant was a conventional activated sludge process plant with a design capacity of 450 ML/d. The treatment train consists of screening, grit removal, primary sedimentation, and plug-flow activated sludge, with gravity thickening of waste activated sludge, DAF thickening of waste activated sludge, and anaerobic digestion of the combined, thickened sludges. Digested sludge is pumped to off-site lagoons for dewatering and ultimate disposal to land. Supernatant from the sludge lagoons is returned to the plant headworks approximately three days per week. The effluent from the plant is discharged to the Bow River.

In contrast to the York River WWTP, which was BOD limited for phosphorus removal, the biological process of the Bonnybrook plant was receiving an average of 145 mg/L BOD_5 and 5.0 mg/L TP at the time of start-up, for a BOD_5:TP ratio of 29:1. This ratio indicates that effluent TP concentrations well below 1.0 mg/L should have been achievable. The permit issued by Alberta Environment mandates monthly average effluent concentrations of:

20 mg/L	BOD_5
20 mg/L	Total suspended solids
1.0 mg/L	Total phosphorus (summer months)
1.25 mg/L	Total phosphorus (winter months)

The plant has historically used liquid alum for the removal of phosphorus, at an annual cost of $2 million (Canadian). The decision was made to investigate the feasibility of using BPR to remove phosphorus and reduce or eliminate the chemical cost. It was decided to use the two-stage (A/O) configuration for the investigation. Only a part of the total plant was used for the demonstration. The "test" section received a flow of 105 ML/d compared to a total plant flow of 450 ML/d.

The design parameters for the Bonnybrook Plant and those for the section that was modified for BPR are compared in Table 6.12. The modified portion of the plant consisted of two of the four mechanically aerated plug-flow aeration basins and six of the twelve square final clarifiers of the secondary section of the plant that was commissioned in 1970. Each

TABLE 6.12. Bonnybrook Wastewater Treatment Plant Secondary Section Design Parameters.

Parameter	"Total" Plant	"Test" Section
Nominal rated flow (ML/d)	450	105
Aeration tank volume (m^3)	4 @ 13,200 4 @ 17,500	2 @ 13,200
Nominal detention time (h)	4 @ 6.0 4 @ 7.0	2 @ 6.0
Avg. BOD_5 load (kg/d)	27,500	6400
Max. BOD_5 load (kg/d)	106,000	24,600
Final clarifier sizes (m)	12 @ 32 m square 12 @ 39 m round	6 @ 32 m square
Avg. overflow rate ($m^3/m^2/d$)	17	17
Max. overflow rate ($m^3/m^2/d$)	34	34

aeration basin in this part of the plant was equipped with ten fixed-mounted 45 kw surface mechanical aerators with draft tubes. There were five aerators in each pass and there were two passes per activated sludge basin. The modifications made were very simple. A baffle was installed between the third and fourth aerators in the first pass of each basin, and the first three aerators and draft tubes were removed and replaced with submerged turbine mixers. A schematic of a modified basin is given in Figure 6.11.

The test section was initially operated with an SRT of 4.5 to 5.5 days to minimize the potential for nitrification. Because start-up was in November, nitrification was very effectively eliminated.

Process Performance

BPR operation commenced on November 1, 1989. The liquid alum addition was maintained during the first six weeks of operation to ensure phosphorus removal in compliance with Alberta Environment requirements. The alum metering rate was diminished in stages over the six week period, and was finally discontinued on December 21. It was assumed that all of the alum would be flushed out of the system via sludge wasting by January 1, 1990. The phosphorus removal data from January 1 through September 12 is illustrated by Figure 6.12, and the monthly average performance data for the same period is listed in Table 6.13.

The data show that the effluent TP concentration was above 1.0 mg/L for the first two months of operation. This was because the baffles installed were not rigid baffles, but instead consisted of two beams, one high and one low, from which were suspended flexible partitions that moved with the flow. However, because the sections immediately after the baffles were being aerated, the water surfaces in those sections were higher than in the upstream sections, and backflow of aerated mixed liquor into the anaerobic section occurred. The DO in the backflow stream reduced the amount of organic matter stored by the polyP bacteria, and consequently reduced the amount of phosphorus removal accomplished. The baffle design was corrected in late February to prevent backflow, and the effluent TP concentration immediately decreased to less than 0.5 mg/L, and remained at or near that level until partial nitrification began occurring during the summer months.

The recycle of nitrates to the anaerobic section during July and August resulted in effluent TP spikes, as shown in Figure 6.12. The spikes were caused by increases in SP rather than higher suspended solids carryover. However, the monthly averages were only 0.6 mg/L during both of those months in spite of the spikes. The effluent TP values above 1.0 were observed on Mondays and Tuesdays of most weeks, but for the balance of

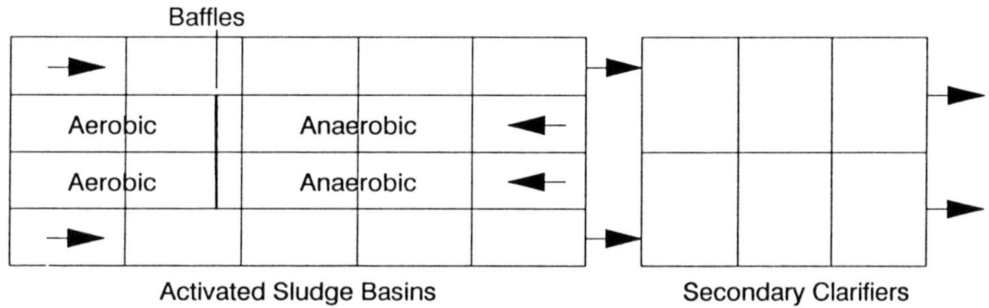

Figure 6.11 Retrofit of Bonnybrook WWTP for biological phosphorus removal.

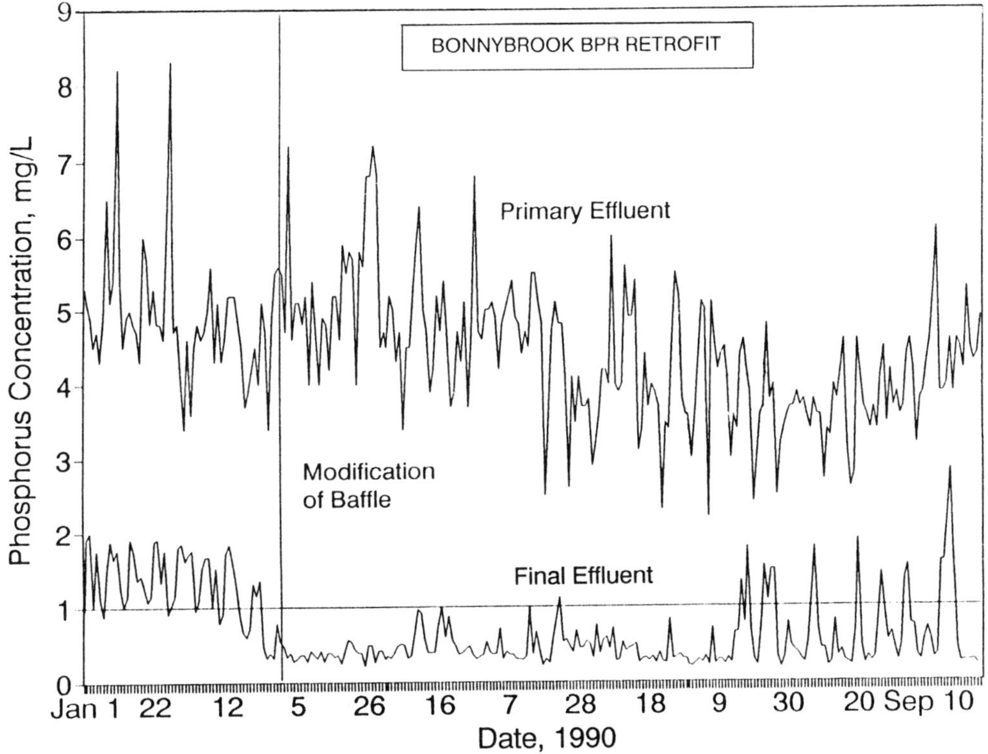

Figure 6.12 Biological phosphorus removal performance at Bonnybrook.

TABLE 6.13. Summary of Monthly Average Performance.

Month, 1990	BOD_5 (mg/L)		Suspended Solids (mg/L)		Total Phosphorus (mg/L)		Volatile Fatty Acids in PE (mg/L)	Total Daily Flow (ML/d)
	PE	FE	PE	FE	PE	FE		
January	154	7	117	11	5.3	1.4	21	80
February	141	7	89	12	4.6	1.1	5	80
March	144	5	95	8	5.3	0.4	12	78
April	141	5	79	7	4.8	0.5	11	87
May	128	4	79	7	4.5	0.5	10	96
June	104	3	66	6	4.1	0.4	8	109
July	101	3	63	7	3.8	0.6	7	92
August	100	2	60	6	3.7	0.6	11	88
September (1 to 12)	111	2	65	7	4.1	0.7	13	84

PE indicates primary effluent.
FE indicates final effluent.

the week the values were generally within the 1.0 mg/L limit. This occurred because loadings were lower during the weekends than during the weekdays, and more nitrification occurred as a result.

A review of the nitrification data showed that some nitrification occurred in the modified system even during the winter months when the mixed liquor temperatures were typically 10 to 11°C, with drops to 7 to 8°C during cold spells. Nitrate concentrations of 3 to 5 mg/L occurred during January, February, and March, but after the baffle was fixed there was sufficient BOD to remove this quantity of nitrates and still accomplish excellent phosphorus removal. When the temperatures warmed up in the spring, the effluent nitrate concentration increased to 5 to 6 mg/L, and varied from 7 to 10 mg/L during the summer months. The increased nitrate loads caused reductions in BPR.

Summary

The demonstration project showed that BPR could be used at the Bonnybrook WWTP to reliably achieve effluent TP concentrations of less than 0.5 mg/L without chemical addition. The operating data clearly showed the detrimental effects of both dissolved oxygen and nitrates on phosphorus removal, when they are permitted to enter the anaerobic zone. The DO problem was easily solved by installing a rigid baffle that prevented surface backflow. Because nitrates reduce phosphorus removal, it was decided to further modify the plant to include an anoxic section for the removal of nitrogen. That modification has since been made and the plant is currently removing both nitrogen and phosphorus.

MODIFICATIONS FOR BIOLOGICAL NITROGEN REMOVAL

Approaches to converting an existing wastewater treatment facility to a BNR plant for nitrogen removal are dependent on site conditions and the level of treatment required. In many cases, major plant modifications and additions may be required, while in other cases minor efforts may be needed. For existing systems that accomplish only BOD removal, a higher SRT will have to be provided for nitrification by increasing the aeration tank volume and/or the MLSS concentration. A greater quantity of oxygen will also have to be transferred to the mixed liquor. If the system is already designed for nitrification, additional volume may be required to provide anoxic zones for denitrification. The anoxic volume in an activated sludge nitrification-denitrification system may account for 20 to 40% of the total tank volume. Where denitrification is added, the energy required for oxygen supply may be reduced; but the mixing energy required in the anoxic zone and the pumping energy for the internal recycle will offset some of this energy saving.

The plant retrofit procedure must first consider the effluent nitrogen requirements and the treatment alternatives that can meet such requirements. The next step is to determine if the existing facility precludes the use of any of the particular alternatives. Preliminary designs are then prepared for the remaining alternatives and the ability to incorporate these designs into the existing facility is evaluated. Once this is done, the final design can be selected based on capital and operating costs, treatment reliability, and operational needs.

FACTORS AFFECTING RETROFIT DESIGNS

The following are considerations that must be taken into account when retrofitting an existing activated sludge facility for nitrogen removal.

Compatibility with Existing Treatment Processes

Some plants will be easier and less costly to convert to biological nitrogen removal than others. For example, if nitrification is already accomplished at the plant, then perhaps only simple modifications are needed to create anoxic zones for denitrification. The tank configuration and type of aeration in the existing plant is important. A plant with plug-flow type basins and diffused aeration may be more amenable to the creation of separate anoxic and aerobic zones by using simple baffle walls. On the other hand, a completely mixed basin with mechanical aeration would be more difficult to section into anoxic and aerobic zones. A plant with existing fixed film processes may be more expensive to modify since the fixed film process is not compatible with internal recycle flow schemes.

Site Constraints

Site space limitations might preclude the use of suspended growth processes for nitrogen removal where substantial tank capacity increases are needed. Such conditions may favor the use of fixed film nitrification and fixed film denitrification with methanol addition.

Wastewater Characteristics

Successful biological nitrogen removal requires a complete and efficient nitrification step. Influent toxicity can in some cases jeopardize biological nitrification or decrease the nitrification rates.

Other important wastewater characteristics include the wastewater temperature, alkalinity, and BOD_5:nitrogen ratio. Lower temperatures will require greater tank volumes for nitrification and denitrification. Sufficient alkalinity must be available to maintain desirable pH levels during nitrification, or provisions must be made for lime or soda ash addition. The influent BOD_5:N ratio should be well above 4.0 to assure that there is sufficient oxygen demand from the carbonaceous material to utilize the nitrate produced.

Sludge Characteristics and Processing

The thickening characteristics of the activated sludge produced in the biological nitrogen removal process can have operational and economic impacts. The use of a pre-anoxic zone often results in sludges with low SVIs and good thickening characteristics. This feature can affect the secondary clarification requirements and the sludge thickening and dewatering rates prior to disposal.

Permit Compliance during Plant Modifications

Construction activities necessary for plant retrofitting may require that some treatment units be temporarily removed from service. If the existing plant does not have sufficient capacity and redundancy, the retrofit implementation may require temporary treatment facilities. This will increase the retrofit costs.

CASE HISTORIES

Phoenix, Arizona

The city of Phoenix activated sludge treatment facility was modified to upgrade the plant performance from conventional secondary treatment levels to advanced wastewater treatment requirements with nitrogen removal. Table 6.14 lists the effluent quality required for the retrofit design. The retrofit was accomplished with minimal plant modifications and costs (about 2 cents/liter of installed capacity) to meet these needs (Stensel et al., 1991). Improvements in sludge thickening characteristics and aeration efficiency following the retrofit design helped to minimize the retrofit capital and operating costs.

The city of Phoenix was being served by two major wastewater treatment facilities – the 23rd Avenue wastewater treatment plant with a design capacity of 1630 L/s (37.2 mgd) and the 91st Avenue wastewater treatment plant with a design capacity of 6733 L/s (153.7 mgd). The 23rd Avenue

TABLE 6.14. Phoenix 23rd Avenue Wastewater Facility Nitrogen Removal Effluent Requirement.

Parameter	Concentration, mg/L
BOD_5	10.0
TSS	5.0
NH_4—N	0.5
NO_3—N	8.0

plant had an unusual option of accepting only the amount of flow that it could adequately treat and bypassed remaining flow to the 91st Avenue plant along with recycle streams. Due to excessive sludge bulking and limited clarifier capacity, the 23rd Avenue plant was typically operated to treat 657-876 L/s (15-20 mgd) instead of the design flow. Intermittent *Nocardia* foaming also was occurring.

Plans were made to expand the 23rd Avenue plant to an average flow of 2500 L/s (57 mgd) and a future peak flow of 4120 L/s (94 mgd) and to also provide advanced wastewater treatment by the summer of 1994. Evaluation of the existing plant to improve current capacity led to several plant modifications that were carried out by the plant personnel and provided a prototype model for the plant expansion. The modifications included conversion of the plant to an anoxic-aerobic process and internal changes in the secondary control.

The plant consisted of primary treatment, two activated sludge aeration basins, four secondary clarifiers, and chlorination. Primary sludge was anaerobically digested and discharged for ultimate disposal on agricultural lands. The two aeration basins were serpentine four-pass units: 378 m (1240 ft) long; 7.6 m (25 ft) wide; and a water depth of 4.6 m (15.2 ft). Each aeration basin was equipped with 7270 17.8-cm (7 in.) ceramic diffusers, and the system could be operated as a plug-flow, contact stabilization, or step feed system. Normal operating practice was to feed pass 1 and 2 in the step feed mode due to oxygen transfer limitations in pass 1. The two 13.360 m^3 (3.53 MG) aeration basins provided for 4.6 hours retention time at average design flow. The firm recycle activated sludge capacity was 1250 L/s (28.5 mgd) or 77% of the average design flow rate.

Secondary clarification consisted of one unit with a 48.8 m (160 ft) diameter and a 2.7 m (9.0 ft) SWD, two units with 42.7 m (140 ft) diameter and 2.7 m (9.0 ft) SWD, and one unit with 42.7 m (140 ft) diameter and 2.68 m (8.8 ft) SWD. The clarifiers employed center feedwells and were equipped with a double inboard weir and lauder. The four clarifiers provided a retention time of 2.86 hrs at an average design overflow rate of 1.1 m/hr (560 gal/ft^2d).

Modifications to the full-scale system occurred between March and May, 1990. Biological process modifications were first performed on the west aeration basin, and after improvements were demonstrated, the east aeration basin was similarly modified. The modified east basin is shown in Figure 6.13. Baffles were installed at three locations in the first pass and at the first 30% length of the second pass. The location of the first pass baffles were 11.0 m (36 ft), 29.3 m (96 ft) and 66.1 m (217 ft) from the inlet along the first pass, which has a total length of 94.5 m (310 ft). Fine bubble diffusers were replaced by coarse bubble diffusers operated at low air application rates to mix the first three zones created. Over 1400 fine bubble diffusers were moved from pass 1 to pass 2 to better meet the high oxygen demand at the beginning of pass 2. These changes, along with the installation of an internal recycle pump to move mixed liquor from the fourth pass to the first stage in the first pass, provided a high *F/M* filamentous control selector zone and anoxic zones for nitrate removal. The internal recycle ratio, based on average influent flow, was 1.6 for the east basin and 2.4 for the west basin. Later, flexible membrane diffusers were added to the last zone of pass 1 along with coarse bubble diffusers to allow operation under either anoxic or aerobic conditions.

Effect of Plant Retrofit

After the modifications were made in March, 1990 the SVI gradually decreased to 134 mL/g by April and 100 mL/g by May (Figure 6.14). Table 6.15 summarizes the treatment performance before and after the retrofit. The wastewater temperature ranges from 20-24°C for the November-February period and from 24-30°C for the March-July period. Excellent

TABLE 6.15. Comparison of Average Treatment Performance before and after Retrofit to Anoxic-Aerobic Operation, Phoenix, Arizona 23rd Avenue WWTP.

Parameter	Before Nov. 89-Feb. 90	After July 90-Dec. 90
Avg. flow, m^3/d	87,000	95,000
Infl. BOD, mg/L	211	274
SRT, d	2	10
MLSS, mg/L	500	2600
SVI, mL/g	310	70
Effluent		
BOD, mg/L	11	4
TSS, mg/L	10	5
NH_4—N, mg/L	—	0.6
NO_3—N, mg/L	—	4.8

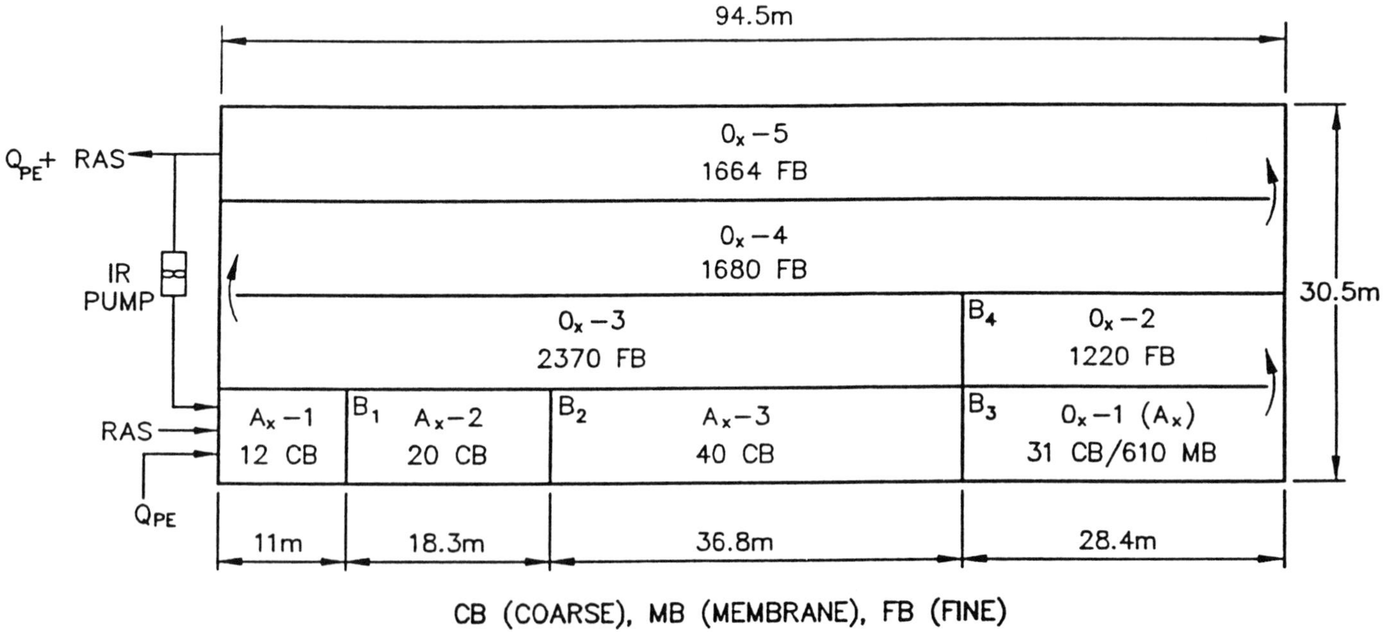

Figure 6.13 (East) aeration basin zones, no. and type diffuser and baffle (B) arrangement.

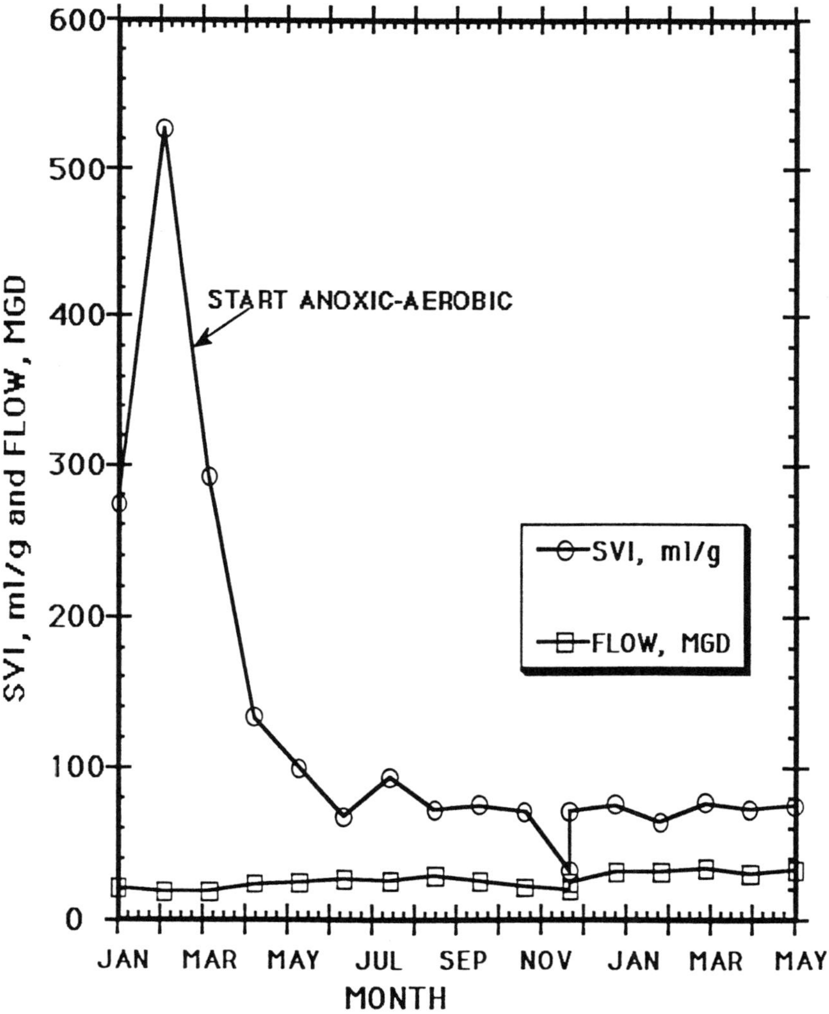

Figure 6.14 Change in SVI and flow with anoxic-aerobic operation.

nitrification and over 85% nitrogen removal was achieved. The improved SVI resulted in increased solids loadings to the clarifiers from 0.8 kg/m^2-hr (4 lb/ft^2-d) to 4.1 kg/m^2-hr (20 lb/ft^2-d) with a minimal sludge blanket. Due to a stringent effluent requirement of less than 0.15 ml/L settleable solids, the influent flow had normally been within 113,000 m^3/d (30 mgd), but during the latter part of December, 1990 and January, 1991 flows of up to 151,000 m^3/d (40 mgd) were treated with production of single digit effluent suspended solids concentrations. The MLSS concentration and SRT were increased to about 2000 mg/L and ten days, respectively, to provide for full nitrification (Figure 6.15).

Nocardia developed when the SRT was being increased and chlorine dosing at 1360-1800 kg/d (3000 to 4000 lb/d) to the return sludge was unsuccessful and caused a turbid effluent. Two boxes that were provided with an adjustable flap to allow foam to enter were constructed across the aeration passes. Chlorine was sprayed over the mixed liquor surface within the boxes. Portable chlorinators were used to spray a low volume of high-concentration chlorine (2000-3000 mg/L) onto the *Nocardia* foam surface. The *Nocardia* was controlled within two days using about 70 kg (150 lb/d) of chlorine per aeration tank without an increase in effluent turbidity.

Off-gas testing was performed in August, 1986 on the fine bubble aeration system during the low SRT operation and later during the higher SRT, anoxic-aerobic operation in May, 1990. The average tank alpha value between those periods increased from 0.27 to 0.47, and the average actual standard oxygen transfer efficiency increased from 6.9% to about 12.0%. The oxygen transfer efficiency increase was attributed to the reduced substrate concentration in the aeration mixed liquor due to the removal of substrate in the pre-anoxic zones and the higher operating SRT. The improved oxygen transfer efficiency offsets the increased oxygen demand for nitrification. The plant operating data showed that no increase in air supply was required after the modifications to nitrification, and the oxic basin dissolved oxygen concentrations were typically between 2.0-4.0 mg/L. The average amount of air supplied was 51 m^3/kg (656 ft^3/lb) BOD applied before the modification and 42 m^3/kg (543 ft^3/lb) BOD applied after.

In summary, sufficient aeration tank capacity was available for nitrification, and it could be used with a higher SRT operating condition as a result of an improved sludge SVI. The improved SVI was the result of incorporating a pre-anoxic zone into the basin, which served as a selector for non-bulking sludge and provided for nitrogen removal. The pre-anoxic zone and longer SRT operation also resulted in improved aeration oxygen transfer efficiencies. Thus, for a modest retrofit capital cost of only $730,000 or about 2¢/liter (7¢/gal) of average influent flow capacity, the existing facility was easily converted for nitrogen removal.

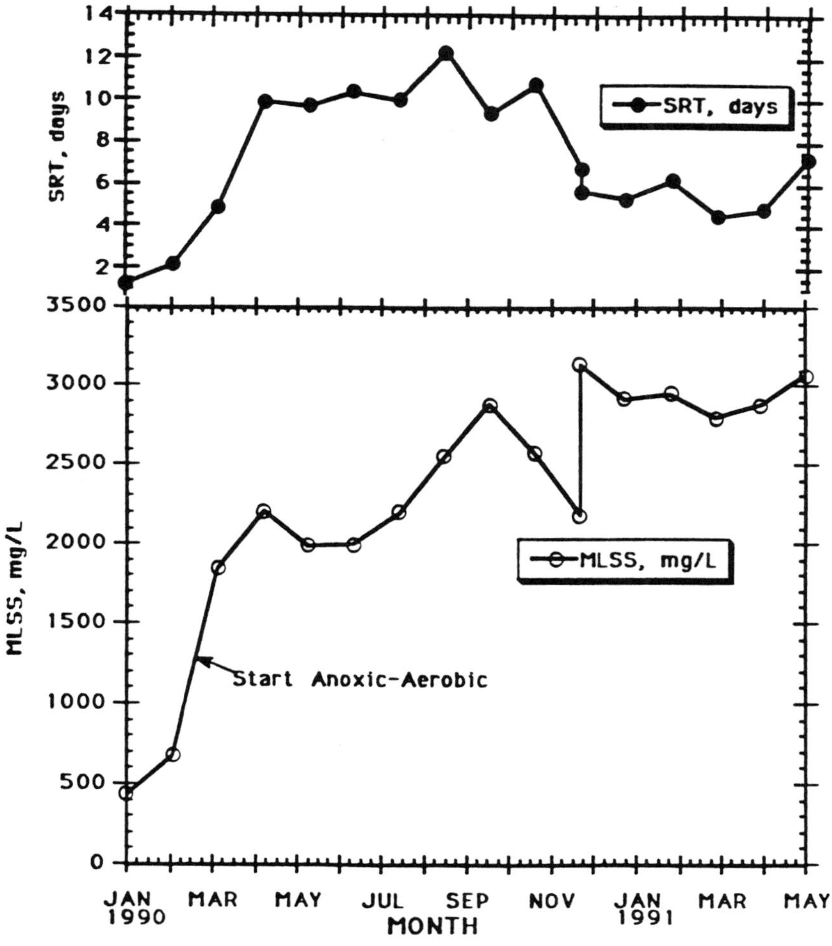

Figure 6.15 Average MLSS and SRT versus time.

Columbus, Ohio

Columbus, the state capital of Ohio, is a major commercial and industrial center located in the geographic center of the state. It is located on the Scioto River, a tributary of the Ohio River originating in northwestern Ohio and flowing 217 km southeast to the Ohio River. The Scioto River at low flow in the Columbus area is characterized as a small pool and riffle stream with much of the slope gradient in the range of 0.31 m/km. Approximately 65% of the population in the Scioto River basin live in the Columbus area. Several water supply, flood control, and reaeration impoundments have been constructed on the Scioto River and its major tributaries, altering the flow characteristics.

Columbus operates two major wastewater treatment facilities. The Jackson Pike Wastewater Treatment Plant (WWTP) in southwest Columbus was constructed in the late 1930s with a hydraulic capacity of 4030 L/s (92 mgd). It was modified in 1988 and currently has an advanced wastewater treatment (AWT) capacity of 2630 L/s (60 mgd). The Southerly WWTP was constructed in 1967 for a 4380 L/s (100 mgd) capacity, expanded to AWT in 1988, and now provides a capacity of 4990 L/s (114 mgd). The plants serve a population of 872,000, including twenty adjacent small communities besides the city of Columbus as well as some major industries. The plants are interconnected by a 3.9 m diameter sewer to transfer flow from the larger Jackson Pike service area to Southerly.

During dry weather and low river flow periods, effluent discharged to the Scioto River from both plants has a significant impact on the water quality of the receiving stream. Columbus is authorized, by a 1913 statute, to divert all upstream flows of the Scioto River for the purpose of maintaining the public water supply. This authority has resulted in an occasional "no flow" condition at the Dublin Dam water intake. The only assured water sources during low flow periods are a guaranteed 140 L/s flow from an upstream reservoir located on a major tributary and the city's two treatment plants. Jackson Pike WWTP routinely contributes as much as 95% of the extreme low flow discharge in the river stretch between the two plants.

The regulatory agency designated the reach of the Scioto River receiving the effluent from the treatment plants as capable of supporting warm water habitat. The agency also issued use designations of agricultural water supply, industrial water supply, and primary contact recreation.

The original effluent permit values of 5 mg/L carbonaceous BODs ($CBOD_5$), 10 mg/L TSS, and 1.5 mg/L NH_4-N were renegotiated to yield about the same ultimate or total oxygen demand (TOD) value, i.e., to allow $CBOD_5$, TSS, and NH_4-N limitations of 8, 16, and 1.0 mg/L. This change resulted in the elimination of effluent sand filters at a savings of $50,000,000 to Columbus.

TABLE 6.16. Description of Southerly WWTP Primary and Secondary Treatment Basins before and after Retrofit.

Primary Clarifiers	Before	After (New)
No. of units and sizes	4—51.8 m × 24.4 m × 4.57 m 4—51.8 m × 24.4 m × 3.0 m	2—53.9 m dia. × 4.57 m
Avg. overflow rate		
m/hr	1.6	1.2
gpd/ft^2	940	730
Aeration basins	—	—
No. of units and sizes	10—274 m × 7.92 m × 4.57 m	6—274 m × 792 m × 4.57 m
Detention time (hrs)	6.3	8.8
Secondary Clarifiers		
No. of units and sizes	Demolished in 1988	6—61 m dia. × 4.57 m
Avg. overflow rate		
m/hr		1.0
gpd/ft^2		600
No. of units and sizes	10—274 m × 7.92 m × 4.57 m	6—274 m × 7.92 m × 4.57 m

Descriptions of Southerly Plant Facility and Loadings

The Southerly plant, prior to its modification and expansion, consisted of conventional pretreatment, pre-aeration, rectangular primary clarification, two-pass plug-flow or step feed aeration basins, and rectangular final clarifiers. Sludge disposal was accomplished by centrifugal dewatering and combustion or land disposal. Table 6.16 summarizes the design of the major liquid processing unit operations before and after the retrofit design modification. Table 6.17 summarizes the design loadings for year 2008 for the retrofit.

TABLE 6.17. Influent Design Loadings for Southerly Plant Retrofit—Year 2008.

Parameter	Value
Flow	4,900 L/s (114 mgd)
BOD_5	84,700 kg/d 196 mg/L
TSS	85,140 kg/d 197 mg/L
TKN	10,800 kg/d 40 mg/L
Total phosphorus	3,500 kg/d 8 mg/L

The system design SRT and MLSS concentration were 8.7 days and 3500 mg/L, respectively. Besides increasing the system SRT and aeration basin detention time to assure nitrification, the aeration basin configuration was also selected to prevent bulking sludge. By providing for bulking sludge control a higher aeration mixed liquor concentration could be maintained to minimize the amount of additional aeration volume needed to convert the plant to nitrification. Figure 6.16 shows a schematic of the aeration basin configuration, termed a semi-aerobic system because it features either aeration or anoxic conditions in the initial contact zone of return sludge and primary effluent. The 274 m long basins were baffled to provide three bulking sludge control selector zones producing total *F/M* gradients of 5.5, 2.8, and 1.4 kg BOD/kg MLSS-d at design loadings. The remaining basin volume was baffled to provide seven additional aeration zones of equal size. Zones 1a, 1b, 1c, and 2 were equipped with jet aerators so that either mixing or aeration can be employed. Nitrified mixed liquor can be recycled from Zone 8 to Zones 1a and 2. The design recycle ratio was selected to produce a total effluent nitrogen concentration of less than 10 mg/L. Depending on the mixed liquor recycle ratio used, Zones 1b, 1c, and 2 could also be operated under anaerobic conditions following the removal of nitrate in Zone 1a. Tubular membrane diffusers were installed in Zones 3 through 8 for continuous aerobic operation.

Laboratory Testing

The process design for Southerly used a high MLSS of 3500 mg/L and initially the first six of eight clarifiers would have a high solids loading. Since Southerly had a history of bulking, successful operation was dependent on the capability of the selectors to control bulking. Though it was necessary to complete engineering and start construction immediately, laboratory studies were initiated to evaluate the design selected. The studies were also necessary due to the low SRTs used to minimize aeration volume requirements and the fact that nitrification inhibition was detected at the existing plant.

Laboratory sequencing batch reactors (SBRs) were selected to simulate the planned staged selector/aeration basin. With nitrification occurring in the SBRs, the initial mixing period of influent and mixed liquor was under anoxic conditions as in the full-scale design. Two eight-liter SBR units were used to operate simultaneously at temperatures equal to the full-scale facility winter and summer temperatures. A total of eleven individual study phases were conducted at winter and summer mixed liquor temperatures with SRTs ranging from seven to twelve days.

The initial studies quickly determined the existence of high sludge yields, which would result in a much lower SRT than the original design basis of

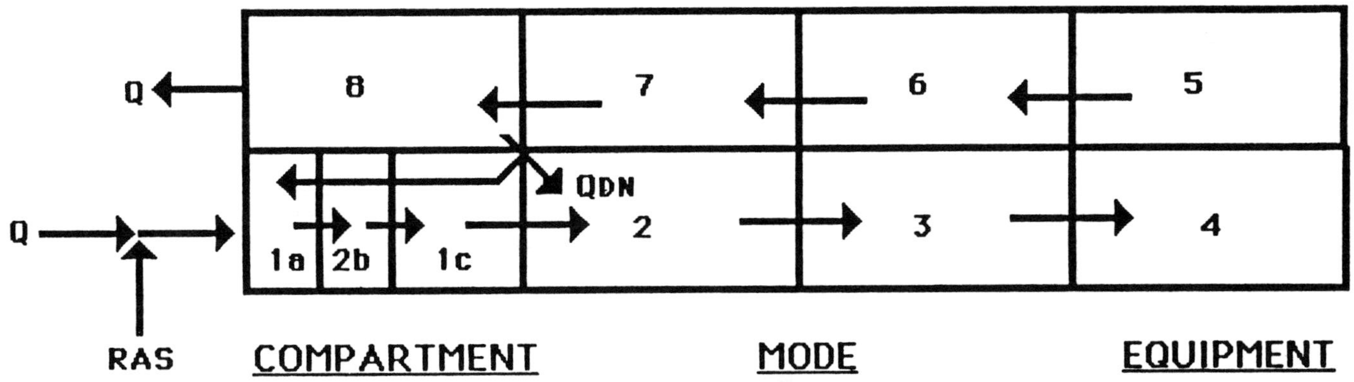

COMPARTMENT	MODE	EQUIPMENT
1a	ANOXIC	JET AERATORS
1b & 1c	AEROBIC/ANOXIC	JET AERATORS
2	AEROBIC/ANOXIC/OXIC	JET AERATORS
3–8	OXIC	MEMBRANE-DIFFUSERS

Figure 6.16 Southerly aeration basin configuration retrofit.

twelve days at 12°C. In order to determine if the existing design was operable, the SRT was reduced to nine days and then to seven days to evaluate the range of full-scale operation.

Although some data were not available until the end of the construction phase, the results of the 1 1/2-year SBR studies still provided insights to the wastewater treatment capabilities. Specifically,

(1) Excess sludge production was 0.85-0.95 kg/kg BOD_5 removed, in part due to the high influent 2.4-1.7 COD/BOD_5 ratio and the use of selectors.
(2) Nitrification toxicity was both chronic and acute for the Southerly wastewater.
(3) Early results permitted revision of the full-scale plant selector baffling arrangement to increase the *F/M* gradient.
(4) Oxygen utilization profiles could be established as well as denitrification and nitrification rates.
(5) An excellent profile of the wastewater characteristics of the Southerly plant primary effluent was produced.

The testing basis and average results for the laboratory study are shown in Table 6.18. Most importantly, the tests indicated that the existing design, while using a plant-scale SRT with a minimal safety factor, would produce the required effluent quality and provide SVI control. Low ammonia-nitrogen concentrations were consistently maintained. There was evidence of biological phosphorus removal, but the removal was not repeated for similar operating conditions. Effluent suspended solids concentrations increased at the seven-day SRT operation.

Retrofit Plant Effluent Quality

A summary of the Southerly plant performance over a 2.5 year period is shown in Table 6.19. The plant had not reached full design capacity and was operating at 90, 64, and 87% of the design flow, BOD loading, and TKN loading, respectively. The resultant average operating SRT was 13.8 days with an average MLSS concentration of 3200 mg/L. Excellent BOD, TSS, and ammonium removal was achieved. Though the effluent nitrate nitrogen was not routinely measured, values less than or equal to 8 mg/L were observed. The average effluent total phosphorus concentration indicates the occurrence of biological phosphorus removal. The plant receives wastewater from a large local brewery, which provides more readily fermentable substrates that are important for promoting biological phosphorus removal.

TABLE 6.18. Laboratory SBR Tests Operating Phases and Results.

			Average Effluent Quality (mg/L)					
Phase	Temp °C	SRT_{ox} (days)	$CBOD_5$	TSS	TKN	NH_4—N	TP	SVI mL/g
I	12	12	1.5	2.9	1.2	<0.2	0.1	83
II	12	9	2.0	3.7	1.4	<0.2	0.1	89
III	15	9	2.0	3.6	1.2	<0.2	0.2	101
IV A	12	9	3.0	5.5	1.3	<0.2	0.5	94
IV B	21	9	4.1	9.9	1.8	<0.2	3.1	77
V A	12	9	2.6	9.2	1.6	<0.2	3.5	99
V B	21	9	2.0	4.9	1.4	<0.2	3.3	79
VI A-1	12	7	3.9	11.4	2.1	<0.2	4.3	86
VI B-1	21	7	4.0	9.8	2.0	<0.2	5.1	84

TABLE 6.19. Performance of Retrofit Facility Southerly WWTP, Columbus, Ohio.

Parameter	Average	80th Percentile
$CBOD_5$, mg/L	2.0	2.0
TSS, mg/L	5.5	7.0
TKN, mg/L	1.5	1.7
NH_4—N, mg/L	0.13	0.18
TP, mg/L	1.1	2.0
SVI, mL/g	77	90

Comparisons of Tables 6.18 and 6.19 show that the SBR laboratory study effectively described the full-scale plant performance. Average effluent TKN and NH_4-N concentrations were very close. The SVIs were within the same range. The average effluent TSS concentration for the full-scale facility was a little higher. Overall, with the exception of biological phosphorus removal, the laboratory SBR system effectively predicted the full-scale performance.

RETROFITTING PLANTS FOR NITROGEN AND PHOSPHORUS REMOVAL

In retrofitting plants for BNR, the designer should be aware of a number of factors that may influence his/her decision, many of which have been discussed in preceding sections. Each plant is different and some offer unique opportunities at minimal changes or cost.

RETROFITTING SURFACE AERATION PLANTS

Retrofitting of many plants for biological nitrogen and phosphorus removal usually, but not always, entails reducing the nitrates in the RAS. Exceptions are where there is sufficient VFA in the feed, as discussed before.

Surface aeration plants usually offer better opportunities for retrofit or even simple changes in operation to effect nitrogen and sometimes phosphate removal. While the channel type systems were not originally designed for denitrification, it was soon discovered that they could be easily manipulated for good nitrogen removal. Controlling the oxygen input has been successful in improving nitrogen removal to greater than 80% in many plants. The multi-channel Orbal plants were especially successful in this regard due to the fairly high load or *F/M* ratio on the first channel. Lower

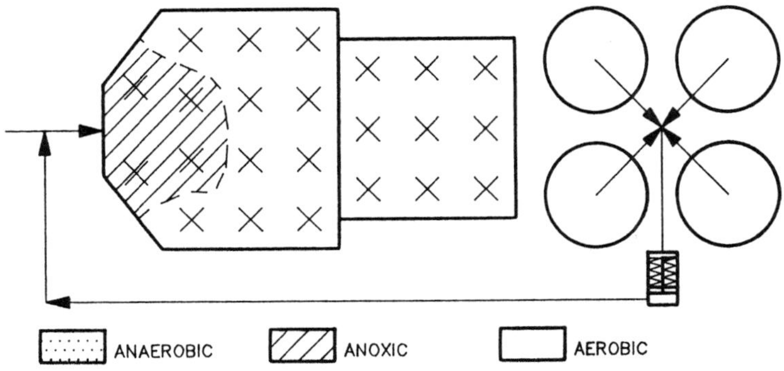

Figure 6.17 Alexandra plant with aerators off.

MCRT plants were more successful than high MCRT plants, mainly due to the higher active fraction of the mixed liquor, resulting in a more rapid uptake of the available oxygen and leaving more pronounced aerobic and anoxic zones. Higher MLSS concentrations had the same effect. Stronger influent wastewater would also improve denitrification due to the higher oxygen demand where introduced.

Good denitrification can be achieved in a plug-flow type aeration basin using surface aerators or in a basin having a large number of surface aerators. By switching off four aerators in the Alexandra plant in Johannesburg, as shown in Figure 6.17, the improvement in effluent quality shown in Table 6.20 was achieved. The pumping action of the aerators would recycle nitrified mixed liquor back to the influent area where denitrification took place. Further denitrification took place in anoxic zones formed inadvertently near the bottom of the aeration basin. This has led to the development of the unbalanced aeration tank shown in Figure 6.17 for nitrification and denitrification. This type of nitrogen removal system has been very successful in Zimbabwe and has also been applied to many smaller unsophisticated BNR plants. Excellent phosphate removal was

TABLE 6.20. Alexandra Performance before and after Switching Aerators Off.

	Full Nitrification		Partial Denitrification	
	Influent	Effluent	Influent	Effluent
BOD	420	10	340	7
COD	600	38	610	50
TKN	38	5.2	38	5.4
NO_3—N	—	22	—	0.9
Alkalinity	290	44	190	150

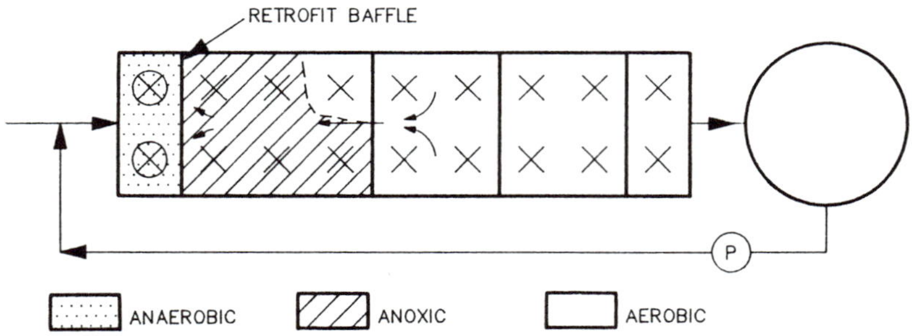

Figure 6.18 Retrofitting at Olifantsfontein.

made possible by the addition of an anaerobic zone. In this case some P removal was achieved by switching off more aerators, but the open arrangement of the aerators and the relative freshness of the wastewater was not conducive to phosphate removal.

The Olifantsfontein plant of the city of Johannesburg was converted from an extended aeration nitrification plant to a BNR plant by the removal of the first two aerators, replacing them with stirrers and partitioning this section to form an anaerobic basin. The remainder of the aeration basin was operated as a combined nitrification/denitrification basin similar to the Alexandra plant (see Figure 6.18).

A Carrousel plant in the town of Heidelberg, Transvaal, in South Africa, was converted to BioP removal by the addition of an anaerobic zone as shown in Figure 6.19. The anaerobic zone was provided with a selector of about five minutes retention to prevent bulking by low *F/M* filaments. The wastewater is strong and is pumped to the plant providing SCFA development. The primary sedimentation tank was provided with a recycle for

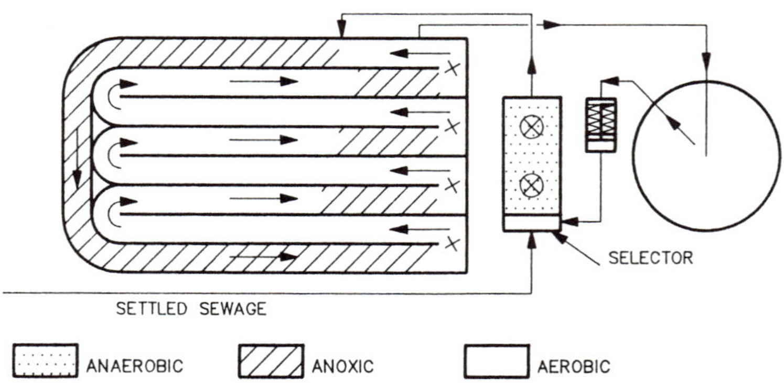

Figure 6.19 Heidelberg Carrousel.

further SCFA production. The effluent ortho-P averaged 0.5 mg/L over a one-year period.

RETROFITTING MEDIUM BUBBLE AERATION PLANTS

While the target may be only to remove phosphates, retrofitting for biological P removal must be cognizant of the role of nitrates. In this sense, the standard plug-flow multi-pass aeration plants found in North America can be converted to BNR simply by using the first sections of the first pass as anaerobic and anoxic basins, and installing a mixed liquor recycle pump. If nitrates in the RAS were to present a problem, the first partition could be for denitrifying of the RAS as shown in Figure 6.20 by either or all of convenient methods. This will ensure reliable P removal. If the VFA in the influent is not sufficient, the primary sludge may be recycled to elutriate more VFA.

Where the need for nitrogen removal is not that important but nitrate production cannot be avoided, there are several options. One would be a modified UCT configuration as shown in Figure 4.14 with recycle from the end of the anaerobic basin to the first anoxic basin for removal of nitrates from the RAS but with no secondary recycle. In warmer climates and with sufficient VFA generation in the sewerage pipework, the expedient of switching off a few aerators may be sufficient. At the Walt Disney World, Florida, plant, the first eight downers in the first pass of the aeration basin were switched off with the results shown in Figure 2.12. Because of the high temperatures, the effluent was well nitrified. However, even though nitrates were present in the RAS, the number of forced mains discharging to the plant generated sufficient short-chain volatile fatty acids in the form of acetates and propionates that nitrates did not interfere with the process,

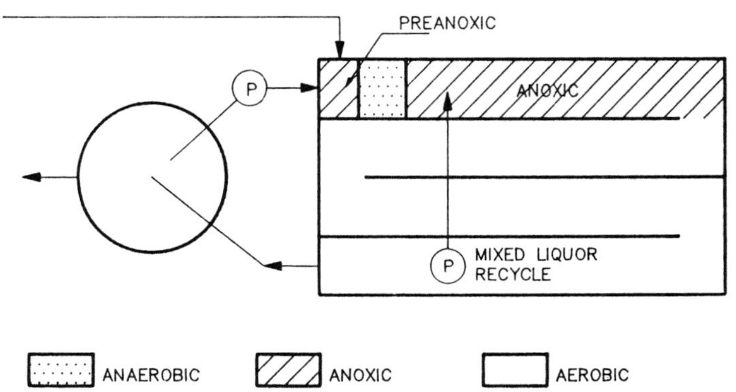

Figure 6.20 Coarse bubble multipass aeration basin.

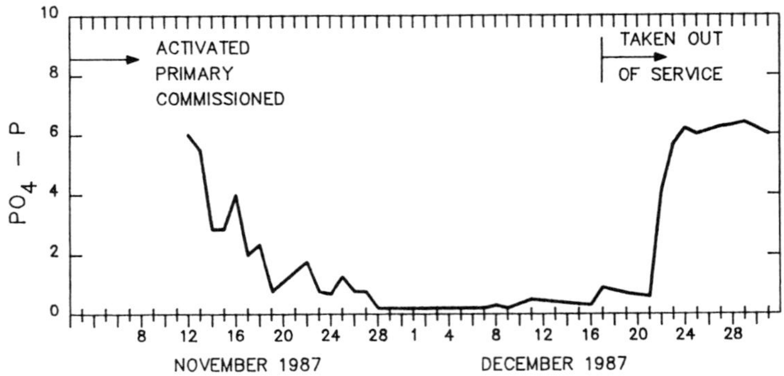

Figure 6.21 Effect of activated PST at Olifantsfontein.

as explained in Chapters 2 and 3. When sufficient acetates are introduced, acetate uptake by the *Acinetobacters* occurs in the presence of nitrates while they are being denitrified, leading to good P removal without the need for any further recycles to remove nitrates. The effluent is then passed through a wetland system where most of the remaining nitrates are removed.

MISCELLANEOUS RETROFITTING

Instead of actually retrofitting trickling filters for BNR, the designer could incorporate them into extensions of the plant. Trickling filters may be designed as high rate without nitrification or low rate units with partial nitrification. In the former case, trickling filters may be used for treating a portion of the wastewater while combining the effluent in an activated sludge plant for N and P removal. This was done at a number of plants in South Africa and is especially useful where the COD:TKN ratio is in excess of 12.

At the Olifantsfontein plant near Pretoria, 45% of the flow was passed through the old trickling filter plant while 55% of the flow was discharged

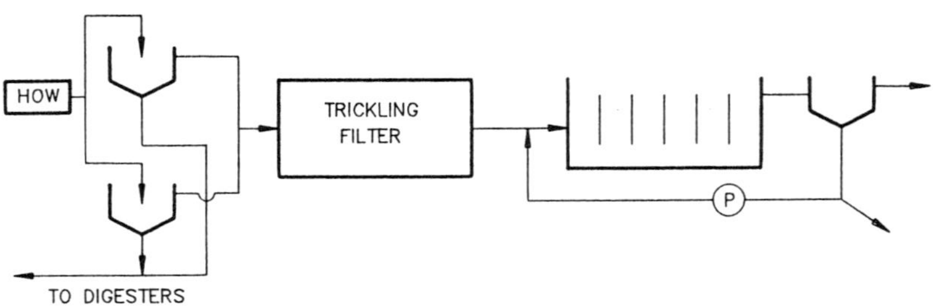

Figure 6.22 Flow diagram of OWASA before.

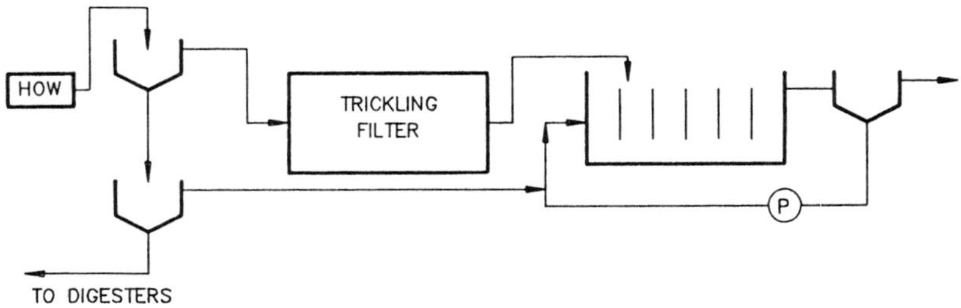

Figure 6.23 Flow diagram of OWASA after.

to the activated sludge plant. An activated primary sedimentation tank was provided for generating VFA. The trickling filter effluent was discharged to the aeration zone of the BNR plant, which consisted of a three-stage anaerobic/anoxic/aerobic process. Aeration was by slow speed vertical spindle, surface aerators. Excellent results were obtained as long as the recycle of sludge to the PST was continued. The graph in Figure 6.21 shows the results of taking the tank out of service.

At the Orange County Water and Sewerage Association plant in North Carolina, the flow diagram shown in Figure 6.22 was changed to that shown in Figure 6.23 resulting in excellent BioP removal. In the second flow diagram, only one PST was used while the primary sludge was stored and elutriated in the other tank. The elutriated supernatant was passed to the first activated sludge compartment and mixed with the RAS. The trickling filter effluent was added to the third compartment. Jet aerators provided oxygen for the remainder of the compartments. The oxygen was turned down in some of the compartments to effect denitrification, while the remainder of the nitrates returned in the RAS were removed partly by endogenous respiration and partly by the VFA added. Sufficient VFA remained to ensure good P removal.

REFERENCES

Albertson, O. E. et al. 1991. "Expansion and Upgrading of Columbus, OH WWTPs to Advanced Wastewater Treatment," *Proceedings of the 6th IAWPRC Conference on Wastewater Treatment Facilities, Prague, Czechoslovakia, August, 1991.*

Randall, C. W., G. D. Waltrip and M. V. Wable. 1990. "Upgrading a Municipal Activated Sludge Plant for High-Rate Biological Nutrient Removal," *Wat. Sc. Tech.*, 22:7/8:21-33. Printed in Great Britain.

Randall, C. W., K. P. Brannan, S. A. McClintock and V. M. Pattarkine. 1992. "The Case for Anaerobic Reduction of Oxygen Requirements in Biological Phosphorus Removal Systems," submitted to the *Journal of the Water Pollution Control Federation*. Department of Civil Engineering, Virginia Tech, Blacksburg, VA.

Stensel, H. D. et al. 1991. ''Anoxic-Aerobic Activated Sludge Treatment,'' *Amer. Soc. of Civil Eng. Environ. Eng. Spec. Conf., Reno, NV.*

Tetreault, M. J., A. H. Benedict, C. Kaempfer and E. F. Barth. 1985. ''Biological Phosphorus Removal: A Technology Evaluation,'' Presented at the *1985 Water Pollution Control Federation Conference*, Pleasant Hill, CA: Brown and Caldwell.

Tetreault, M. J., A. H. Benedict, C. Kaempfer and E. F. Barth. 1985. ''Biological Phosphorus Removal: a Technology Evaluation,'' Los Angeles, California, 1985. Pleasant Hill, California: Brown and Caldwell.

Waltrip, G. D. 1991. ''Full-Scale Nutrient Removal Demonstration Project at the York River Treatment Plant with the VIP Process,'' Presented at the *45th Annual Conference, Virginia Water Pollution Control Association, Williamsburg, April 29-May 1, 1991*. Hampton Roads Sanitation District, Virginia Beach, VA.

Wilson, A. W., E. I. Meckelborg and P. Do. 1990. ''Full-Scale Biological Phosphorus Removal Trails at Bonnybrook,'' Presented at the *42nd Annual Conference of the Western Canada Water and Wastewater Association, Regina, Saskatchewan, October 16-19, 1990.* Reid Crowther & Partners Ltd., Calgary, Alberta, Canada.

CHAPTER 7

General Activated Sludge Model for Biological Nitrogen and Excess Phosphorus Removal

INTRODUCTION

THE activated sludge system was designed primarily for the removal of carbonaceous biodegradable material using a mixed culture of heterotrophic bacteria operating in an aerobic environment. Subsequently, the design was modified for the growth of autotrophic bacteria to oxidize ammonia to nitrates. Unaerated zones were included at a later date to develop configurations that would remove nitrogen (denitrification). Thereafter, it was observed that the unaerated zones also encouraged the growth of organisms that demonstrated the phenomenon of biological excess phosphorus removal (BEPR). This chapter discusses an activated sludge model based on the IAWPRC model for nitrogen removal (Henze et al., 1987a, 1987b). The model is then extended for the growth of Bio-P (Comeau et al., 1985) or polyP (Wentzel et al., 1986) organisms and is based on the models presented by Wentzel et al., (1991, 1990) and Dold (1991).

MATRIX FORMAT PRESENTATION OF THE ACTIVATED SLUDGE MODEL

A matrix format, similar to that used to present the IAWPRC Activated Sludge Model 1, is used here. The matrix notation is introduced through an example in Table 7.1.

The aerobic growth of heterotrophic bacteria is modeled in Table 7.1 in accordance with Monod kinetics and decay is modeled according to first order kinetics. The effect of growth and decay on three components is examined. Those three components are:

(1) Dissolved oxygen concentration S_o[g DO or g ($-$COD) m^{-3}]

(2) Organic substrate concentration S_s[g COD m^{-3}]

(3) Heterotrophic biomass concentration Z_{bh}[g COD m^{-3}]

TABLE 7.1. Example of a Simple Biokinetic Model in Matrix Format.

Component, i	1 S_O	2 S_S	3 Z_{bh}	Process Rate Equation ϱ_j
List of processes, j Aerobic growth	$-(1-Y_H)/Y_H$	$-1/Y_H$	$+1$	$\mu_H\,[S_S/K_{SH}+S_S)]Z_{bh}$
Decay		$+1$	-1	$b_H\,Z_{bh}$
Observed transformation rate, r_i	$r_i = \Sigma\nu_{j,i} * \varrho_j\,[M_iL^{-3}T^{-1}]$			
Definition of stoichiometric parameters Y_H = heterotrophic yield [g cell COD (g COD util)$^{-1}$]	Dissolved oxygen [g(−COD)/m^3]	Dissolved organic substrate [g COD/m^3]	Biomass (heterotrophs) [g COD/m^3]	Definition of kinetic parameters μ_H = Max specific growth rate (d^{-1}) K_{SH} = Half saturation constant (g COD/m^3) b_H = Rate constant for decay (d^{-1})

These parameters have an index, i, and are introduced as an array of concentrations in the first row along with short definitions. The manner in which they are measured (COD, DO) is also indicated.

The two transformation processes have an index, j. Their names are listed in column 1.

Stoichiometric coefficients, $\nu_{j,i}$, form the matrix at the center of Table 7.1. The growth of heterotrophs consumes dissolved oxygen and substrate. This leads to negative coefficients for $\nu_{1,1}$ and $\nu_{1,2}$. It generates biomass, which results in a positive coefficient for $\nu_{1,3}$. Decay consumes biomass and liberates biodegradable substrates, which are subsequently available for the growth process. The corresponding coefficients for biomass consumption ($\nu_{2,3}$) and substrate generation ($\nu_{2,2}$) are shown in Table 7.1. To simplify this table, it is assumed that all the biomass is destroyed through decay, and the entire COD content of the biomass is available for growth. It is assumed that decay does not affect and is not affected by the dissolved oxygen concentration.

Process rate equations (ϱ_j) are listed in the extreme right hand column. The Monod equation is used to describe the growth of heterotrophs and a first order rate is used for decay.

Stoichiometric parameters are defined in the lower left hand corner and kinetic parameters are defined in the lower right hand corner.

For each process, one stoichiometric coefficient, $\nu_{j,i}$, is selected freely, preferably as a dimensionless unit with the value of +1 or −1. For aerobic growth, $\nu_{1,3} = +1$, and for decay, $\nu_{2,3} = -1$. Since these coefficients are fixed, the dimensions of the process rate equations are also fixed. In Table 7.1, the dimensions of the process rate equations are $[M_{\text{biomass COD}}L^{-3}T^{-1}]$.

In the horizontal direction, the stoichiometric matrix shows a conservation of mass for individual elements (COD in this example). This is clearly indicated for decay where the sum of all stoichiometric coefficients add up to zero

$$\Sigma\ \nu_{2,1} = 0$$

Since dissolved oxygen is equivalent to negative chemical oxygen demand (−COD), the COD balance for the growth process can be written as

$$\nu_{1,1} + \nu_{1,2} + \nu_{1,3} = 0$$

In the vertical direction, the stoichiometric matrix relates to materials balance equations. Any materials balance is written for one component, i. Each process, j, may affect this component. Therefore, the observed transformation rate, r_j, as entered into the materials balance equation, is the sum of the transformation rates of this material in the processes:

$$r_j = \Sigma(\nu_{j,i} * \varrho_j)$$

ACTIVATED SLUDGE MODEL FOR ORGANIC CARBON REMOVAL, NITRIFICATION, AND DENITRIFICATION

A mathematical model for the removal of carbonaceous biodegradable material, nitrification, and denitrification, based on the Activated Sludge Model No. 1 (ASM1) of the IAWPRC Task Group (Henze et al.; 1987a,b), and subsequent modifications by Wentzel (1991) and Dold (1991), is presented in Table 7.2. The definitions for the switching functions for this model plus those for the phosphorus removal model are given in Table 7.3.

A total of ten dissolved and seven particulate components are used to characterize the wastewater and the activated sludge. These include:

- dissolved oxygen, bicarbonate alkalinity, and soluble phosphorus
- three forms of biomass [heterotrophs and two types of autotrophs—*Nitrosomonas* and *Nitrobacter* spp. (all represented in terms of COD)]
- five forms of nitrogen (particulate and soluble biodegradable organic nitrogen, ammonia, nitrite, and nitrate)
- six forms of COD (inert soluble and particulate in feed, two forms of biodegradable soluble, enmeshed slowly degradable particulate, and inert particulate COD from endogenous decay)

TABLE 7.2. Model Matrix for Organic Carbon and Nitrogen Removal.

Component i	(1) Z_{bh}	(2) Z_{ba1}	(3) Z_{ba2}	(4) Z_e	(5) S_{enm}	(6) X_i
List of processes, j						
Aerobic growth of heterotrophs on S_{bsc} with NH_3	1					
Anoxic growth of heterotrophs on S_{bsc} with NH_3	1					
Aerobic growth of heterotrophs on S_{bsc} with NO_3	1					
Anoxic growth of heterotrophs on S_{bsc} with NO_3	1					
Aerobic growth of heterotrophs on S_{bsa} with NH_3	1					
Anoxic growth of heterotrophs on S_{bsa} with NH_3	1					
Aerobic growth of heterotrophs on S_{bsa} with NO_3	1					
Anoxic growth of heterotrophs on S_{bsa} with NO_3	1					
Decay of heterotrophs	−1			$f_{Ep,H}$	$1 - f_{Ep,H}$	
Hydrolysis of stored COD						−1
Hydrolysis of org-N						
Ammonification						
Conversion of complex S_{bsc} to acetate						
Nitrosomonas growth		1				
Nitrobacter growth				1		
Nitrosomonas decay			−1		$f_{Ep,A1}$	$1 - f_{Ep,A1}$
Nitrobacter decay				−1	$f_{Ep,A2}$	$1 - f_{Ep,A2}$

TABLE 7.2. (continued).

Component i	(7) X_{on}	(8) S_{bsc}	(9) S_{bsa}	(10) NH_3	(11) N_{os}	(12) NO_2
List of processes, j						
Aerobic growth of heterotrophs on S_{bsc} with NH_3		$-1/Y_H$		$-f_{ZBH,N}$		
Anoxic growth of heterotrophs on S_{bsc} with NH_3		$-1/Y_{DN}$		$-f_{ZBH,N}$		$-(1 - Y_H)/(1.71\ Y_H)$
Aerobic growth of heterotrophs on S_{bsc} with NO_3		$-1/Y_H$				
Anoxic growth of heterotrophs on S_{bsc} with NO_3		$-1/Y_{DN}$				
Aerobic growth of heterotrophs on S_{bsa} with NH_3			$-1/Y_H$	$-f_{ZBH,N}$		
Anoxic growth of heterotrophs on S_{bsa} with NH_3			$-1/Y_{DN}$	$-f_{ZBH,N}$		$-(1 - Y_H)/(1.71\ Y_H)$
Aerobic growth of heterotrophs on S_{bsa} with NO_3			$-1/Y_H$			
Anoxic growth of heterotrophs on S_{bsa} with NO_3			$-1/Y_{DN}$			
Decay of heterotrophs	$f_{ZBH,N} - f_{Ep,H} * f_{ZEH,N}$					
Hydrolysis of stored COD		1				
Hydrolysis of org-N	−1				1	
Ammonification				1	−1	
Conversion of complex S_{bsc} to acetate		−1	1			
Nitrosomonas growth				$-1/Y_{A1}$		$(1/Y_{A1} - f_{ZBA1,N})$
Nitrobacter growth				$-f_{ZBA2,N}$		$1/Y_{A2}$
Nitrosomonas decay				$f_{ZBA1,N} - f_{Ep,A1} * f_{ZEA1,N}$		
Nitrobacter decay				$f_{ZBA2,N} - f_{Ep,A2} * f_{ZEA2,N}$		

(continued)

TABLE 7.2. (continued).

Component i	(13) NO_3	(14) Alk	(15) P_s	(16) S_{us}	(17) S_o
List of processes, j					
Aerobic growth of heterotrophs on S_{bsc} with NH_3		$-f_{ZBH,N}/14$	$-f_{ZBH,P}$		$-(1 - Y_H)/Y_H$
Anoxic growth of heterotrophs on S_{bsc} with NH_3	$-(1 - Y_H)/(2.86\ Y_H)$	$(1 - Y_H)/(14 * 2.86\ Y_H) - f_{ZBH,N}/14$	$-f_{ZBH,P}$		
Aerobic growth of heterotrophs on S_{bsc} with NO_3	$-f_{ZBH,N}$	$f_{ZBH,N}/7 - f_{ZBH,N}/14$	$-f_{ZBH,P}$		$-(1 - Y_H)/Y_H$
Anoxic growth of heterotrophs on S_{bsc} with NO_3	$-f_{ZBH,N}$ $-(1 - Y_H)/(2.86\ Y_H)$	$(1 - Y_H)/(14 * 2.86\ Y_H) + f_{ZBH,N}/14$	$-f_{ZBH,P}$		
Aerobic growth of heterotrophs on S_{bsa} with NH_3		$-f_{ZBH,N}/14$	$-f_{ZBH,P}$		$-(1 - Y_H)/Y_H$
Anoxic growth of heterotrophs on S_{bsa} with NH_3	$-(1 - Y_H)/(2.86\ Y_H)$	$(1 - Y_H)/(14 * 2.86\ Y_H) - f_{ZBH,N}/14$	$-f_{ZBH,P}$		
Aerobic growth of heterotrophs on S_{bsa} with NO_3	$-f_{ZBH,N}$	$f_{ZBH,N}/7 - f_{ZBH,N}/14$	$-f_{ZBH,P}$		$-(1 - Y_H)/Y_H$
Anoxic growth of heterotrophs on S_{bsa} with NO_3	$-f_{ZBH,N}$ $-(1 - Y_H)/(2.86\ Y_H)$	$(1 - Y_H)/(14 * 2.86\ Y_H) + f_{ZBH,N}/14$	$-f_{ZBH,P}$		
Decay of heterotrophs					
Hydrolysis of stored COD					
Hydrolysis of org-N					
Ammonification		$+1/14$			
Conversion of complex S_{bsc} to acetate					
Nitrosomonas growth		$-1/7(1/Y_{A1} - f_{ZBA1,N}) - f_{ZBZA1,N}/14$	$-f_{ZBA1,P}$		$-[3.43(1/Y_{A1} - f_{ZBA1,N}) - 1]$
Nitrobacter growth	$1/Y_{A2}$	$-f_{ZBA2,N}/14$	$-ZBA2,P$		$-[1.14/Y_{A2} - 1]$
Nitrosomonas decay			$f_{ZBA1,P} - f_{Ep,A1} * f_{ZEA1,P}$		
Nitrobacter decay			$f_{BA2,P} - f_{Ep,A2} * f_{ZEA2,P}$		

TABLE 7.2. (continued).

Process	Process Rate Equations, ϱ_i
Aerobic growth of heterotrophs on S_{bsc} with NH_3	$\mu_H * \frac{S_{bsc}}{K_{SH} + S_{bsc}} * Z_{bh} * AirYes * NH_3Yes$
Anoxic growth of heterotrophs on S_{bsc} with NH_3	$\eta_{grow} * \mu_H * \frac{S_{bsc}}{K_{SH} + S_{bsc}} * Z_{bh} * AirNo * NO_3Yes * NH_3Yes$
Aerobic growth of heterotrophs on S_{bsc} with NO_3	$\mu_H * \frac{S_{bsc}}{K_{SH} + S_{bsc}} * Z_{bh} * AirYes * NO_3Yes * NH_3No$
Anoxic growth of heterotrophs on S_{bsc} with NO_3	$\eta_{grow} * \mu_H * \frac{S_{bsc}}{K_{SH} + S_{bsc}} * Z_{bh} * AirNo * NO_3Yes * NH_3No$
Aerobic growth of heterotrophs on S_{bsa} with NH_3	$\mu_H * \frac{S_{bsa}}{K_{SH} + S_{bsa}} * Z_{bh} * AirYes * NH_3Yes$
Anoxic growth of heterotrophs on S_{bsa} with NH_3	$\eta_{grow} * \mu_H * \frac{S_{bsa}}{K_{SH} + S_{bsa}} * Z_{bh} * AirNo * NO_3Yes * NH_3Yes$
Aerobic growth of heterotrophs on S_{bsa} with NO_3	$\mu_H * \frac{S_{bsa}}{K_{SH} + S_{bsa}} * Z_{bh} * AirYes * NO_3Yes * NH_3No$
Anoxic growth of heterotrophs on S_{bsa} with NO_3	$\eta_{grow} * \mu_H * \frac{S_{bsa}}{K_{SH} + S_{bsa}} * Z_{bh} * AirNo * NO_3Yes * NH_3No$
Decay of heterotrophs	$b_H * Z_{bh}$
Hydrolysis of stored COD	$K_H * \frac{S_{enm}/Z_{bh}}{K_X + S_{enm}/Z_{bh}} * Z_{bh} * (AirYes + \eta_{sol} * AirNo * NO_3Yes)$
Hydrolysis of organic N	$\varrho_{10} * \frac{X_{on}}{S_{enm}}$
Ammonification	$K_R * N_{os} * Z_{bh}$
Conversion of complex S_{bsc} to SCFA	$K_C * S_{bsc} * Z_{bh} * AirNo * NO_3No$
Nitrosomonas growth	$\mu_{A1} * [NH_3/(K_{NH1} + NH_3)] * Z_{ba1} * AirYes * NH_3Yes$
Nitrobacter growth	$\mu_{A1} * [NH_3/(K_{NH2} + NH_3)] * Z_{ba1} * AirYes * NO_2Yes$
Autotrophic decay	$b_{A1} * Z_{ba1}$ and $b_{A2} * Z_{ba2}$

TABLE 7.3. **Definition of Switching Functions.**

Switching Function	Equivalent	Definition
AirYes	1—AirNo	$\frac{S_O}{K_{OH} + S_O}$
AirNo	1—AirYes	$\frac{K_{OH}}{K_{OH} + S_O}$
NH_3Yes	1—NH_3No	$\frac{NH_3}{K_{NA} + NH_3}$
NH_3No	1—NH_3Yes	$\frac{K_{NA}}{K_{NA} + NH_3}$
NO_3Yes	1—NO_3No	$\frac{NO_3}{K_{NO} + NO_3}$
NO_3No	1—NO_3Yes	$\frac{K_{NO}}{K_{NO} + NO_3}$
Switching functions for the biological excess P removal model		
PO_4Yes	1—PO_4No	$\frac{P_S}{K_{LP} + P_S}$
PO_4No	1—PO_4Yes	$\frac{K_{LP}}{K_{LP} + P_S}$
S_{bsa}Yes	1—S_{bsa}No	$\frac{S_{bsa}}{K_{SSEQ} + S_{bsa}}$
S_{bsa}No	1—S_{bsa}Yes	$\frac{K_{SSEQ}}{K_{SSEQ} + S_{bsa}}$
PolyPYes	1—PolyPNo	$\frac{P_{polyP}}{K_{XP} + P_{polyP}}$
PolyPNo	1—PolyPYes	$\frac{K_{XP}}{K_{XP} + P_{polyP}}$

The model is an extension of the IAWPRC ASM1 model in the following areas:

- the nitrogen source for cell synthesis
- denitrification and heterotrophic growth on biodegradable COD and fermented products (short chain fatty acids, SCFA)
- formation of nitrites and nitrates by two species of autotrophs, and their denitrification

A modification that formulates matrix coefficients for conversion of ammonia to nitrites and nitrites to nitrates in a manner that is slightly different from the ASM1 model is discussed later.

THE NITROGEN SOURCE FOR CELL SYNTHESIS

Dold and Marais (1986) postulated that under certain circumstances, nitrate, instead of ammonia nitrogen, can serve as the nitrogen source for cell synthesis. This may be observed in reactors operating at high mean cell residence times (MCRTs) and in aerobic digesters. The use of nitrate nitrogen as a nitrogen source for polyP organisms was also observed for polyP organisms when ammonia concentrations dropped to low levels (Wentzel et al., 1989a). On the basis of this information, two additional processes were incorporated in the model by Dold (1991) to give four growth processes: aerobic and anoxic growth of heterotrophs with either ammonia or nitrate serving as the nitrogen source for synthesis.

CONVERSION OF SOLUBLE SUBSTRATE TO SCFA

Different vessels may be used to convert part of the influent soluble COD to readily biodegradable complex forms and short chain fatty acids (SCFA). This may be accomplished in the sewer system itself under anaerobic conditions, in gravity thickeners whose supernatants may contain SCFAs, in fermentation tanks, or in the anaerobic zone of a biological phosphorus removal system. Denitrification rates can be enhanced if the concentration of SCFAs (and therefore, the readily degradable simple substrate) entering the anoxic tank can be increased (Sen et al.; 1990, 1991a). It may be modeled as an increase in maximum specific growth rate (μ_H) and reduction in half saturation constant (K_{SH}). Additional discussion on denitrification rates observed for polyP and non-polyP organisms is presented with the model for biological excess phosphorus removal.

As far as possible, the nomenclature for nitrogen removal follows that of the IAWPRC task group for nitrification-denitrification. The equivalents for those changed are listed in Table 7.4. Suggested values for the stoichiometric parameters, the kinetic parameters, and the switching function parameters are given in Tables 7.5, 7.6, and 7.7, respectively.

SUGGESTED ALTERATION TO THE IAWPRC MODEL FOR NITRIFICATION

The oxygen coefficient for nitrification in the IAWPRC Activated Sludge Model 1 (ASM1) is expressed as

$$-(4.57 - Y_A)/Y_A$$

where Y_A is the total nitrifier yield (*Nitrosomonas* and *Nitrobacter*), expressed in g cell COD per g NO_3 generated. The negative sign represents utilization of oxygen (i.e., negative of oxygen generation represents utilization).

TABLE 7.4. **Definition of Component Symbols and IAWPRC Activated Sludge Model 1 (ASM1) Equivalents.**

Component *i*	Model Symbol	ASM1 Symbol	Definition	Units
1	Z_{bh}	X_H	Active non-polyP heterotrophs	g cell COD m^{-3}
2	Z_{ba1}	X_A	Active *Nitrosomonas* (autotrophs)	g cell COD m^{-3}
3	Z_{ba2}	X_A	Active *Nitrobacter* (autotrophs)	g cell COD m^{-3}
4	Z_e	X_p	Endogenous biomass	g cell COD m^{-3}
5	S_{enm}	X_S	Enmeshed slowly biodeg. substrate	g cell COD m^{-3}
6	X_i	X_i	Inert particulate COD	g COD m^{-3}
7	X_{on}	X_N	Slowly biodegradable organic nitrogen	g m^{-3}
8	S_{bsc}	S_S	Readily biodeg "complex" substrate	g COD m^{-3}
9	S_{bsa}	S_S	Readily biodegradable SCFA substrate	g COD m^{-3}
10	NH_3	S_{NH}	Ammonia nitrogen	g N m^{-3}
11	N_{os}	X_{12}	Soluble biodegradable organic nitrogen	g N m^{-3}
12	NO_2	S_{NO}	Nitrite nitrogen	g N m^{-3}
13	NO_3	SNO	Nitrate nitrogen	g N m^{-3}
14	Alk	SAlk	Biocarbonate alkalinity	mole HCO_3^- m^{-3}
15	P_s		Soluble biologically available P	g P m^{-3}
16	S_{us}	S_I	Unbiodegradable soluble substrate	g COD m^{-3}
17	S_O	S_O	Dissolved oxygen	g (–COD) m^{-3}
Biological P removal component				
18	Z_{bp}		Active polyP heterotrophs	g cell COD m^{-3}
19	S_{phb}		Stored SCFA	g COD m^{-3}
20	P_{polyP}		Stored polyP	g P m^{-3}

The expression for IAWPRC ASM1 computation is derived as follows:

- Consumption of DO is 4.57 g m^{-3} for every g m^{-3} of nitrate nitrogen formed in nitrification.
- The nitrifier cell COD generation from inorganic oxidized carbon (bicarbonate) reduces the oxygen demand (nitrifier cell COD $= Y_A * N$, where N is the quantity of nitrate nitrogen generated).

TABLE 7.5. Stoichiometric Parameters for the Organic Carbon and Nitrogen Removal Model.

Symbol	Description	Value	Units	Ref.
Heterotrophs				
Y_H	Yield, aerobic	0.666	g cell COD yield (g COD utilized)$^{-1}$	a
Y_{DN}	Yield, anoxic	0.403	g cell COD yield (g COD utilized)$^{-1}$	b
$f_{ZBH,\mathrm{N}}$	N content of active mass	0.068	g N (g COD)$^{-1}$ in active organisms	a
$f_{ZEH,\mathrm{N}}$	N content of endogenous mass	0.068	g N (g COD)$^{-1}$ in endogenous organisms	a
$f_{ZBH,\mathrm{P}}$	P content of active mass	0.015	g P (g COD)$^{-1}$ in active organisms	c
$f_{ZEH,\mathrm{P}}$	P content of endog. mass	0.015	g P (g COD)$^{-1}$ in endogenous organisms	c
$f_{Ep,H}$	Fraction of active mass remaining as endogenous residue	0.08	g COD endogenous mass (g COD active mass)$^{-1}$	a
$f_{cv,H}$	Ratio COD/VSS	1.48	g COD (g VSS)$^{-1}$	a
Autotrophs				
Y_{A1}	Yield, *Nitrosomonas*	0.074	g cell COD yield (g NH_3N utilized)$^{-1}$	d
Y_{A2}	Yield, *Nitrobacter*	0.074	g cell COD yield (g NO_2N utilized)$^{-1}$	d
x in subscripts below represents 1 and 2 for *Nitrosomonas* and *Nitrobacter*, respectively				
$f_{ZBAx,\mathrm{N}}$	N content of active mass	0.068	g N (g COD)$^{-1}$ in active organisms	a
$f_{ZEAx,\mathrm{N}}$	N content of endog. mass	0.068	g N (g COD)$^{-1}$ in endogenous organisms	a
$f_{ZBAx,\mathrm{P}}$	P content of active mass	0.015	g P (g COD)$^{-1}$ in active organisms	c
$f_{ZEAx,\mathrm{P}}$	P content of endog. mass	0.015	g P (g COD)$^{-1}$ in endogenous organisms	c
$f_{Ep,Ax}$	Fraction of active mass remaining as endog. residue	0.08	g COD endogenous mass (g COD active mass)$^{-1}$	a
$f_{cv,Ax}$	Ratio COD/VSS	1.48	g COD (g VSS)$^{-1}$	a

[a]Dold, 1991.
[b]McClintock et al., 1988.
[c]Sen et al., 1990; Wable and Randall, 1989.
[d]Hanaki et al., 1990.

TABLE 7.6. **Kinetic Parameters for the Organic Carbon and Nitrogen Removal Model.**

Symbol	Description	Value	Units	Reference
Heterotrophs				
μ_H	Maximum specific growth rate, aerobic	3.4	d^{-1}	b
μ_H	Maximum specific growth rate, anoxic	3.2	d^{-1}	b
$K_{S,H}$	Half saturation coefficient for growth	5.0	g COD m^{-3}	a
b_H	Decay rate	0.24	d^{-1}	d
n_{sol}	Anoxic solubilization factor	0.37		a
n_{grow}	Anoxic growth factor	1.00		a
K_H	Maximum specific hydrolysis rate	2.81	d^{-1}	a
K_X	Half saturation coefficient for hydrolysis	0.15	g COD (g COD)$^{-1}$	a
K_R	Ammonification rate	0.016	m^3 (g COD·d)$^{-1}$	a
Autotrophs				
μ_{A1}	Max specific growth rate, *N-somonas*	0.66	d^{-1}, 25 Celsius	c
μ_{A2}	Max specific growth rate, *N-bacter*	0.58	d^{-1}, 25 Celsius	c
b_{A1}	Decay rate *N-somonas*	0.12	d^{-1}, 25 Celsius	c
b_{A2}	Decay rate *N-bacter*	0.11	d^{-1}, 25 Celsius	c
K_{NH1}	Half saturation constant for growth, *Nitrosomonas*	1.0	g NH_3—N m^{-3}	c
K_{NH2}	Half saturation constant for growth, *Nitrobacter*	0.35	g NO_2—N m^{-3}	c

[a]Dold, 1991. [c]Hanaki et al., 1990.

TABLE 7.7. **Definition of Parameters in Switching Functions.**

Symbol	Description	Value	Units	Reference
K_{OH}	Aerobic or anoxic growth	0.2	g O m^{-3}	b
K_{NA}	Ammonia limit	0.05	g NH_3—N m^{-3}	a
K_{NO}	Nitrate limit	0.2	g NO_3—N m^{-3}	c
K_{LP}	Soluble phosphorus limit	1.0	g P m^{-3}	a
K_{XP}	PolyP limit	1.0	g P m^{-3}	a
K_{SSEQ}	SCFA limit	1.0	g COD m^{-3}	a

[a]Dold, 1991. [c]Grady and Lim, p. 895, 1980.
[b]Sen, 1991c.

Therefore, the generation of N g of nitrate nitrogen results in the following relationship:

$$DO = -[4.57\,N - YA * N]$$

$$DO = -Y_A * N\,[4.57/Y_A - 1]$$

where N is the g m^{-3} of nitrate formed.

In the matrix form, the value of $Y_A * N$ is fixed at 1. Therefore,

- quantity of DO generated $= -[4.57/Y_A - 1]$
- quantity of nitrate nitrogen generated $= 1/Y_A$
- quantity of ammonia nitrogen utilized $= [-f_{ZBA,N} - 1/Y_A]$

In the revised version presented here, yield for *Nitrosomonas* is expressed in terms of g cell COD per g NH_3 utilized (Y_{A1}) and yield for *Nitrobacter* is expressed in terms of g cell yield per g NO_2 utilized (Y_{A2}). The value of $Y_{A1}N$ and $Y_{A2}N$ are both fixed at 1.0.

For *Nitrosomonas*

- quantity of ammonia nitrogen utilized $= 1/Y_{A1}$
- quantity of nitrite nitrogen generated $= [\,1/Y_{A1} - f_{ZBA1,N}\,]$
- quantity of DO generated $= -[\,3.43\,(1/Y_{A1} - f_{ZBA1,N}) - 1$

For *Nitrobacter*

- quantity of ammonia nitrogen utilized $= -f_{ZBA2,N}$
- quantity of nitrate nitrogen generated $= 1/Y_{A2}$
- quantity of DO generated $= -[\,1.13/Y_{A2} - 1\,]$

ACTIVATED SLUDGE MODEL FOR NITRIFICATION, DENITRIFICATION AND BIOLOGICAL EXCESS PHOSPHORUS REMOVAL (NDBEPR)

The activated sludge model for describing the kinetics of polyP heterotrophic organisms is based on the enhanced culture model presented by Wentzel et al. (1989). Dold (1991) incorporated one additional process for anoxic growth of polyP organisms with the additional stoichiometric constant, $f_{P,upt}$, for P uptake during anoxic growth.

SIMPLIFIED ENHANCED CULTURE STEADY-STATE MODEL

Research on establishing biochemical pathways and mechanisms for *Acinetobacter* spp. and polyP organisms by Marais et al. (1983), Comeau et al. (1985), Matsuo (1985), Wentzel et al. (1986), and Lötter and Dubery (1987, 1989) led to the development of a biochemical model for biological excess phosphorus removal. The model proposed by Wentzel et al. (1986) was for *Acinetobacter* spp., and specifically for organisms not possessing the Embden-Meyerhoff (EM) pathway. Mino et al. (1987) proposed a biochemical model for polyP organisms that possess this pathway, which was in agreement with that of Wentzel et al. (1986) except that under anaerobic conditions, the TCA cycle was not implicated. Protons and electrons were, instead, supplied by utilization of carbohydrate via the EM pathway. This was verified experimentally by Mino et al. (1987) who demonstrated utilization of carbohydrate under anaerobic conditions. The utilization of extracellular carbohydrate under anaerobic conditions has also been observed experimentally by Osborn et al. (1989).

A general kinetic model that describes the behavior of NDBEPR systems was developed by Wentzel et al. (1988, 1989b). It was assumed that in a NDBEPR system, a mixed culture would develop which could be categorized into three groups of organisms:

(1) Heterotrophic organisms able to accumulate polyP
(2) Heterotrophic organisms unable to accumulate polyP
(3) Autotrophic organisms mediating nitrification

The kinetic model to describe the behavior of non-polyP heterotrophs and autotrophs has already been described.

Experimental observations made on enhanced culture systems fed with acetates and batch tests allowed Wentzel et al. (1989a) to observe certain characteristics and kinetic responses of polyP organisms. They observed that:

(1) PolyP organisms showed little propensity to denitrify.
(2) PolyP organisms had an extremely low endogenous mass loss rate of 0.04 mg active VSS/(mg active VSS-day), as compared to an endogenous mass loss rate of 0.24 mg active VSS/(mg active VSS-day) for aerobic activated sludge systems (Marais and Ekama, 1976).

A simplified enhanced culture kinetic model was developed by Wentzel et al. (1990) to simulate the steady-state response of a system under constant flow and load. Two assumptions were made to simplify the development of the steady-state model:

(1) P release for anaerobic maintenance energy requirements was assumed to be small compared to P release for energy requirements for sequestration of SCFA.
(2) All substrate sequestered in the anaerobic zone is subsequently utilized in the aerobic zone.

DEVELOPMENT OF A MIXED CULTURE STEADY-STATE MODEL

Wentzel et al. (1991) and Dold (1991) extended the enhanced culture model to incorporate mixed cultures of polyP and non-polyP organisms present in NDBEPR systems receiving domestic wastewater as the influent. These groups studied the difference in endogenous mass loss rates between non-polyP and polyP organisms. In a conventional aerobic activated sludge system, they argue, the high specific endogenous mass loss rate with non-polyP organisms results from the high rate of predation and regrowth. In a biological excess phosphorus removal system, however, polyP organisms are not preyed upon to the same degree as non-polyP organisms. Simulations with the steady state mixed culture model have shown that if the polyP organisms were subjected to the same rate of predation as the non-polyP organisms, there would be no significant accumulation of these organisms in the system and the biological excess phosphorus removal would be near zero.

Since the polyP organisms have a significantly lower rate of predation, and the polyP and non-polyP organisms essentially do not compete for the same form of substrate, the two populations can be modeled virtually independently of each other in a mixed culture. However, there is some degree of interaction for three scenarios, two of which occur in the anaerobic reactor and one that occurs in the anoxic reactor. They are as follows:

(1) The readily biodegradable complex substrate forms are converted to SCFA by non-polyP organisms (fermenters) in the anaerobic reactor. This makes SCFA available to polyP organisms for sequestration. Since the rate of conversion is much slower than the rate of sequestration, the reaction rate is controlled by the former except in cases where abundant SCFAs are available in the feed to the anaerobic reactor.

(2) If electron acceptors such as oxygen or nitrate are recycled to the anaerobic reactor, the readily biodegradable complex substrate forms are also utilized by the non-polyP organisms, thereby reducing the mass converted to SCFA. If SCFAs are supplied externally, the rate of SCFA sequestration (product of the specific rate of SCFA sequestration, K_p, and polyP organism concentration, Z_{bp}) will be in direct competition with the rate of SCFA utilization under anoxic or aerobic conditions by both polyP and non-polyP organisms. Since the growth and substrate utilization rates of polyP organisms are lower, the fraction of polyP organisms will decrease gradually, reducing the quantity of biological phosphorus removal.

TABLE 7.8. Extension of the Activated Sludge Model for Biological Excess Phosphorus Removal.

Component i	1 Z_{bh}	2 Z_{ba1}	3 Z_{ba2}	**18** $\mathbf{Z_{bp}}$	4 Z_e	5 S_{enm}	6 X_i
List of processes, j							
Aerobic growth of poly P organisms on S_{phb} with NH_3				1			
Aerobic growth of polyP organisms on S_{phb} with NO_3				1			
Process 16 if S_p limited				1			
Process 17 if S_p limited							
Aerobic decay of polyP organisms				−1	$f_{Ep,G}$		
PolyP lysis upon aerobic death of polyP organisms							
S_{phb} lysis upon aerobic death of polyP organisms							
Anaerobic decay of polyP organisms				−1	$f_{Ep,G}$		
PolyP lysis upon anaerobic death							
S_{phb} lysis upon anaerobic death							
Cleavage of polyP for anaerobic maintenance							
Sequestration of SCFA by polyP organisms							
Anoxic growth of polyP organisms on S_{phb} with NH_3				1			

TABLE 7.8. (continued).

Component i	7 X_{on}	19 $\mathbf{S_{phb}}$	20 $\mathbf{P_{polyP}}$	8 S_{bsc}	9 S_{bsa}	10 NH_3	11 N_{os}	12 NO_2
List of processes, j								
Aerobic growth of poly P organisms on S_{phb} with NH_3		$-1/Y_G$	$f_{P,upt1}/Y_G$			$-f_{ZBG,N}$		
Aerobic growth of polyP organisms on S_{phb} with NO_3		$-1/Y_G$	$f_{P,upt1}/Y_G$					
Process 16 if S_p limited		$-1/Y_G$	$-f_{ZBG,P}$			$-f_{ZBG,N}$		
Process 17 if S_p limited		$-1/Y_G$	$-f_{ZBG,P}$					
Aerobic decay of polyP organisms						$f_{ZBG,N} - f_{Ep,G} * f_{ZEG,N} - f_{Es,G} * f_{EsG,N}$	$f_{Es,G} * f_{EsG,N}$	
PolyP lysis on aerobic death of polyP organisms			-1					
S_{phb} lysis on aerobic death of polyP organisms		-1			1			
Anaerobic decay of polyP organisms						$f_{ZBG,N} - f_{Ep,G} * f_{ZEG,N} - f_{Es,G} * f_{EsG,N}$	$_{Es,G} * f_{EsG,N}$	
PolyP lysis on anaerobic death			-1					
S_{phb} lysis on anaerobic death		-1			1			
Cleavage of polyP for anaerobic maintenance			-1					
Sequestration of SCFA by polyP organisms		1	$-f_{P,rel}$		-1			
Anoxic growth of polyP organisms on S_{phb} with NH_3		$-1/Y_G$	$f_{P,upt2}/Y_G$			$-f_{ZBG,N}$		

(continued)

TABLE 7.8. (continued).

Component i	13 NO_3	14 Alk	15 P_s	16 S_{us}	17 S_o
List of Processes, j					
Aerobic growth of poly P organisms on S_{phb} with NH_3			$-f_{P,upt1}/Y_G - f_{ZBG,P}$		$-(1 - Y_G)/Y_G$
Aerobic growth of polyP organisms on S_{phb} with NO_3	$-f_{ZBG,N}$		$-f_{P,upt1}/Y_G - f_{ZBG,P}$		$-(1 - Y_G)/Y_G$
Process 16 if S_p limited					$-(1 - Y_G)/Y_G$
Process 17 if S_p limited	$-f_{ZBG,N}$				$-(1 - Y_G)/Y_G$
Aerobic decay of polyP organisms			$-f_{ZBG,P} - f_{Ep,G} * f_{ZEG,P}$	$f_{Es,G}$	$-(1 - f_{Ep,G} - f_{Es,G})$
PolyP lysis on aerobic death of polyP organisms			1		
S_{phb} lysis on aerobic death of polyP organisms					
Anaerobic decay of polyP organisms			$-f_{ZBG,P} - f_{Ep,G} * f_{ZBG,P}$	$1 - f_{Ep,G}$	
PolyP lysis on anaerobic death			1		
S_{phb} lysis on anaerobic death					
Cleavage of polyP for anaerobic maintenance			1		
Sequestration of SCFA by polyP organisms			$f_{P,rel}$		
Anoxic growth of polyP organisms on S_{phb} with NH_3	$-(1 - Y_G)/(2.86\ Y_G)$		$-f_{P,upt2}/Y_{G,DN} - f_{ZBG,P}$		

TABLE 7.8. (continued).

	Process Rate Equations, ϱ_i
Aerobic growth of polyP organisms on S_{phb} with NH_3	$\mu_{G1} * \frac{S_{phb}/Z_{bp}}{K_{SG1} + S_{phb}/Z_{bp}} * Z_{bp} * AirYes * NH_3Yes * PO_4Yes$
Aerobic growth of polyP organisms on S_{phb} with NO_3	$\mu_{G1} * \frac{S_{phb}/Z_{bp}}{K_{SG1} + S_{phb}/Z_{bp}} * Z_{bp} * AirYes * NH_3No * NO_3Yes * PO_4Yes$
Process 16 if S_p limited	$\mu_{G2} * \frac{S_{phb}/Z_{bp}}{K_{SG2} + S_{phb}/Z_{bp}} * Z_{bp} * AirYes * NH_3Yes * PO_4No$
Process 17 if S_p limited	$\mu_{G2} * \frac{S_{phb}/Z_{bp}}{K_{SG2} + S_{phb}/Z_{bp}} * Z_{bp} * AirYes * NH_3No * NO_3Yes * PO_4No$
Aerobic decay of polyP organisms	$b_G * Z_{bp} * Air\ Yes$
PolyP lysis on aerobic death of polyP organisms	$\varrho_{20} * \frac{P_{polyP}}{Z_{bp}}$
S_{phb} lysis on aerobic death of polyP organisms	$\varrho_{20} * \frac{S_{phb}}{Z_{bp}}$
Anaerobic decay of polyP organisms	$b_G * Z_{bp} * AirNo$
PolyP lysis on anaerobic death	$\varrho_{23} * \frac{P_{polyP}}{Z_{bp}}$
S_{phb} lysis on anaerobic death	$\varrho_{23} * \frac{S_{phb}}{Z_{bp}}$
Cleavage of polyP for anaerobic maintenance	$b_{PP} * Z_{bp} * AirNo * PolyPYes$
Sequestration of SCFA by polyP organisms	$K_P * Z_{bp} * S_{bsa}Yes * PolyPYes$
Anoxic growth of polyP organisms on S_{phb} with NH_3	$\eta_P * \mu_{G1} * \frac{S_{phb}/Z_{bp}}{K_{SG1} + S_{phb}/Z_{bp}} * Z_{bp} * AirNo * NO_3Yes * NH_3Yes * PO_4Yes$

TABLE 7.9. Stoichiometric Parameters for the Extension of the Activated Sludge Model for Biological Excess Phosphorus Removal.

Symbol	Description	Value	Units	Ref.
Y_G	Aerobic Yield	0.639	g cell COD yield (g COD utilized)$^{-1}$	a
$Y_{G,DN}$	Anoxic Yield	0.387	g cell COD yield (g COD utilized)$^{-1}$	b
$f_{P,upt1}$	P uptake/COD utilized in aerobic growth	1.00	g P (g stored COD)$^{-1}$	a
$f_{P,upt2}$	P uptake/COD utilized in anoxic growth	0.20	g P (g stored COD)$^{-1}$	a
$f_{ZBG,N}$	N content of active mass	0.068	g N (g COD)$^{-1}$ in active organisms	a
$f_{ZEG,N}$	N content of endog. mass	0.068	g N (g COD)$^{-1}$ in endogenous residue	a
$f_{EsG,N}$	N content of soluble unbiodegradable COD	0.068	g N (g COD)$^{-1}$ in endogenous residue	a
$f_{ZBG,P}$	P content of active mass (excluding polyP content)	0.015	g P (g COD)$^{-1}$ in active organisms	c
$f_{ZEG,P}$	P content of endogenous residue	0.015	g P (g COD)$^{-1}$ in endogenous residue	c
$f_{Ep,G}$	Fraction of active mass remaining as particulate unbiodeg. endogenous residue	0.25	g COD endog. mass (g COD active mass)$^{-1}$	a
$f_{Es,G}$	Fraction of active mass remaining as soluble unbiodeg. residue	0.20	g COD (g COD active mass)$^{-1}$	a
$f_{P,rel}$	P release/SCFA COD uptake for sequestration	0.48	g P (g SCFA COD)$^{-1}$	a
$f_{cv,G}$	Ratio COD/VSS	1.42	g COD (g VSS)$^{-1}$	a

[a]Dold, 1991.
[b]McClintock et al., 1988.
[c]Sen et al., 1990; Wable and Randall, 1989.

TABLE 7.10. Kinetic Parameters for the Extension of the Activated Sludge Model for Biological Excess P Removal.

Symbol	Description	Value	Units	Ref.
μ_{G1}	Max. specific growth rate (no P limit)	1.0	d^{-1}	a
μ_{G2}	Max. specific growth rate (P limit)	0.42	d^{-1}	a
K_{SG1}	Half sat. coefficient (no P limit)	0.18	g COD (g COD active mass·d)$^{-1}$	a
K_{SG2}	Half sat. coefficient (P limit)	0.18	g COD (g COD active mass·d)$^{-1}$	a
n_p	Anoxic growth factor	1.00		a
b_G	Organism endogenous mass loss rate	0.04	d^{-1}	a
b_{pp}	PolyP cleavage rate for anaerobic "maintenance" energy generation	0.03	d^{-1}	a
K_p	Specific rate of SCFA sequestration	6.00	g COD (g COD active mass·d)$^{-1}$	a
K_C	Specific rate of conversion of "complex" readily biodegradable COD to SCFA	0.04	g COD (g COD active mass·d)$^{-1}$	a

[a] Dold, 1991.

(3) PolyP organisms can grow in the anoxic reactor in the presence of sufficient ammonia and phosphorus. The reduction in phosphorus uptake per mg of COD utilized ($f_{P,upt}$) from 1.0 to 0.2 mg P/mg COD utilized has yielded satisfactory results in modeling the observed phosphorus response (Dold, 1991).

There is considerable discrepancy in denitrification rates observed in full-scale NDBEPR systems in the U.S. (Stearns and Wheler, 1991; Sen et al., 1991b). Additional research needs to be conducted to determine factors that influence denitrification rates for polyP and non-polyP organisms in NDBEPR systems.

The activated sludge model presented earlier is extended for biological excess phosphorus removal in Table 7.8. Stoichiometric and kinetic parameters are presented in Tables 7.9 and 7.10.

REFERENCES

Comeau, Y., K. J. Hall, R. E. W. Hancock and W. K. Oldham. 1985. "Biochemical Model for Enhanced Biological Phosphorus Removal." *Proc. BC Conference on New Methods and Research in Waste Treatment and Residuals Management, June, Vancouver, Canada.* [Pub. in *Water Research*, 20(12):1511-1521 (1986)].

Dold, P. L. 1991. "Incorporation of Biological Excess Phosphorus Removal in a General Activated Sludge Model," Paper, Dept. of Civil Engr. and Engr. Mechanics, McMaster University, Ontario L8S 4L7.

Ekama, G.A. and G. v. R. Marais. 1984. "Carbonaceous Material Removal," in *Theory, Design and Operation of Nutrient Removal Activated Sludge Processes*, Pretoria, SA: Water Research Commission, pp. 3-4.

Grady, C. P. L. and H. C. Lim. 1980. "Performance of Denitrification Systems," in *Biological Wastewater Treatment, Theory and Applications*. New York: Marcel Dekker, Inc., p. 895.

Hanaki, K., C. Wantawin and S. Ohgaki. 1990. "Effects of Heterotrophs on Nitrification in Suspended-Growth Reactor," *Water Research*, 24(3):289-296.

Henze, M., C. P. L. Grady, W. Gujer, G. v. R. Marais and T. Matsuo. 1987a. "Activated Sludge Model No. 1," *IAWPRC Scientific and Technical Report No. 1*. London: IAWPRC.

Henze, M. C., P. L. Grady, W. Gujer, G. v. R. Marais and T. Matsuo. 1987b. "A General Model for Single Sludge Wastewater Systems," *Water Research*, 21:505-515.

Lötter, L.H. and I. A. Dubery. 1989. "Metabolic Behavior of β-hydroxybutyrate dehydrogenase in *Acinetobacter calcoaceticus* var. *lwoffi*," *Water SA*, 15(2):65-70.

Lötter, L. H. and I. A. Duber. 1987. "Metabolic Control of Phosphate Accumulating Bacteria and Its Role in Enhanced Biological Phosphate Removal," in *Advances in Water Pollution Control 4: Biological Phosphate Removal from Wastewaters*, R. Ramadori, ed. Oxford: Pergamon Press, pp. 7-14.

Marais, G. v. R., R. E. Loewenthal and I. P. Siebritz. 1983. "Review: Observations Supporting Phosphate Removal by Biological Excess Uptake," *Wat. Sci. Tech.*, 15(3/4):15-41.

Marais, G. v. R. and G. A. Ekama. 1976. "The Activated Sludge Process: Part I—Steady State Behavior," *Water SA*, 2:164-200.

Matsuo, Y. 1985. "Functioning of the TCA Cycle under Anaerobic Conditions in the Anaerobic Aerobic Acclimitization Activated Sludge," (in Japanese), *Proc. of 40th Annual Conf. of the Japan Soc. of Civ. Eng.*, 40(2):989-990.

McClintock, S.A., J. H. Sherrard, J. T. Novak and C. W. Randall. 1988. "Nitrate versus Oxygen Respiration in the Activated Sludge Process," *Journal WPCF*, 60(3):342-350.

Mino, T., V. Arun, Y. Tsuzuki and T. Matsuo. 1987. "Effect of Phosphorus Accumulation on Acetate Metabolism in the Biological Phosphorus Removal Process," in *Advances in Water Pollution Control 4: Biological Phosphate Removal from Wastewaters*, R. Ramadori, editor. Oxford: Pergamon Press, pp. 27-38.

Osborn, D.W., L. H. Lötter, A. R. Pitman and H. A. Nicholls. 1989. "Two Year Study on the Enhancement of Biological Phosphate Removal by Altering Process Feed Composition (Plant and Laboratory Studies)," Report to Water Research Commission, WRC 137/2/89, P.O. Box 824, Pretoria 0001, SA.

Sen, D., C. W. Randall and T. J. Grizzard. 1991a. "Process Design and Operational Modifications of Oxidation Ditches for Biological Nutrient Removal," *Proceedings 6th IAWPRC Conf. on Design and Operation of Large Wastewater Treatment Plants, Prague, Czechoslovakia*. pp. 253-260.

Sen, D., C. W. Randall and T. J. Grizzard. 1991b. "Design and Process Control Features to Optimize Biological Nutrient Removal in Temperate Climates,"

Chesapeake Water Pollution Control Association. (submitted to *Water Quality International '92, 16th Biennial Conference, IAWPRC, Washington D.C.).*

Sen, D. (1991c). Results from Plant Performance and Denitrification Rate Simulations for Different D.O. Levels in the Anoxic Zone at the Broadneck WRF, Little Patuxent WWTP, Old Maryland City WWTP and Patuxent WRF, Maryland, USA. U.S.EPA, County and University Studies, Civil Engr. Research Division, VPI&SU, Blacksburg, USA.

Sen, D., C. W. Randall and T. J. Grizzard. 1990. "Biological Nitrogen and Phosphorus Removal in Oxidation Ditches and High Nitrate Recycle Systems," EPA Chesapeake Bay Program Tech. Publ., CBP/TRS 47/90, 120 pp.

Stearns and Wheler. 1991. "Single Sludge Systems," *USEPA Nitrogen Control Manual Revision*. In press.

Wable, M. and C. W. Randall. 1989. Evaluation of VIP/UCT Process at York River WWTP, Final Report. Chesapeake Bay Project, State Water Control Board, VA.

Wentzel, M.C., P. L. Dold, G. A. Ekama and G.v.R. Marais. 1989a. "Enhanced Polyphosphate Organism Cultures in Activated Sludge Systems. Part II: Experimental Behavior," *Water SA*, 15(2):71-88.

Wentzel, M. C., P. L. Dold, G. A. Ekama and G. v. R. Marais. 1989b. "Enhanced Polyphosphate Organism Cultures in Activated Sludge Systems. Part III: Kinetic Model," *Water SA*, 15(2):89-102.

Wentzel, M. C., G. A. Ekama and G. v. R. Marais. 1991. "Kinetics of Nitrification Denitrification Biological Excess Phosphorus Removal Systems – A Review," *Wat. Sci. Tech.*, 23, Kyoto, 555-565.

Wentzel, M. C., Loewenthal, R.E., Ekama, G. A. and Marais, G. v. R. 1988. "Enhanced Polyphosphate Organism Cultures in Activated Sludge Systems – Part I: Enhanced Culture Development," *Water SA*, 14(2):81-92.

Wentzel, M. C., L. H. Lötter, R. E. Loewenthal and G. v. R. Marais. 1986. "Metabolic Behavior of *Acinetobacter* spp. in Enhanced Biological Phosphorus Removal – A Biochemical Model," *Water SA*, 12(4):209-224.

CHAPTER 8

Control of Bulking and Foaming Organisms

OVERVIEW

THE discussion will be directed to the causes and cures of problems associated with bulking and foaming organisms. The procedures for identifying the organisms are not presented, but the literature citations are provided at the end of this chapter. While methods for controlling filamentous bulking are proving quite successful, a second bulking phenomenon, hydrous or viscous bulking, is less well understood and may only be obvious in the lack of filamentous bulking. Case histories are provided to demonstrate that significant progress has been made in bulking and foaming control.

Bulking sludges and other types of solids-liquid separation problems have often been confused. For the purpose of this manual, the types of problems are segregated as shown in Table 8.1. It is possible for the procedures used to control bulking to induce problems such as dispersed growth or pin floc. Thus, this analysis will be concerned with all aspects that can affect the biological and clarification effectiveness.

THE PROBLEM

Bulking sludges have been the major nemesis of activated sludge designers and operators for seventy years, since the original fill-and-draw systems were converted to continuous flow. Many site-specific remedies have been evaluated and proven successful, but with little general success until recently. Until recent years, chlorine or peroxide was the only widely effective means of suppressing the presence of filamentous organisms that cause bulking.

The BNR processes that include nitrogen removal require much longer SRTs than are necessary for carbonaceous BOD_5 and biological phosphorus

TABLE 8.1. Activated Sludge Solids—Liquid Separation Problems.

Name of Problem	Nature of Problem	Characterization of Problem
Dispersed growth	Flocs are dispersed, forming only small clumps or single cells	Turbid effluent. Poor floc formation. Lack of zone settling of sludge
Pin floc	Small, compact, dense flocs which settle rapidly, leaving lighter flocs in suspension	Low SVI. Cloudy, turbid effluents. Low quantity of filamentous organisms
Filamentous bulking	An excessive filamentous population which interferes with compaction and settling of activated sludge	High SVI, clear supernatant. Low RAS concentration. Increasing sludge blanket depth. Poor sludge handling characteristics
Hydrous zoogleal or viscous bulking (non-filamentous)	The sludge flocs become more hydrated and lose density	High SVI. Lower RAS concentration. High sludge blankets
Rising sludge	Gas entrainment and/or gas release caused by denitrification	Flocs or clumps of sludge rise rapidly to the surface. Surface effervescence solids carryover
Foaming and scum	Aeration basin and clarifier accumulation of floating materials, froth and scum of activated sludge	Lightly colored frothy foam forms on aeration. Persistent dark heavy foam with high solids content
Excessive sludge blankets: non-bulking sludge	Diffuse sludge blankets approaching weirs, excessive solids loading or transport problems (see Chapter 5)	Loss of solids and unstable operation. Solids accumulation in clarifier

removal. In order to process the maximum quantity of flow in existing and new facilities, it is desirable to operate at the highest possible MLSS. This mode of operation necessitates the control of bulking organisms. With good control of bulking sludge, high MLSS of 3500-4500 mg/L can provide reliable, stable performance in properly designed biological basins and clarifiers.

While the presence of nuisance foams in activated sludge plants has always been a problem, the severity has increased in those facilities dedicated to advanced wastewater treatment. Two general types of foaming problems exist: one is a low-density, billowy foam that is caused by surface active agents and the second is a dense heavy foam that is associated with specific organisms in the presence of surface active agents. The first type of foam can be controlled by water sprays, and it is generally not a serious problem. The discussion in this chapter will be primarily directed to the types of foam that cannot be suppressed by water sprays—i.e., *Nocardia* type foaming. Recently, progress has been made on controlling *Nocardia* type foam and case histories will be provided. Tankage construction features to assist in foam management are discussed.

DEFINITIONS AND MEASUREMENTS

It is to be recognized that the term "sludge bulking" lacks scientific clarity and specific reports of bulking may not satisfy the general consensus of the term "bulking." Furthermore, an operator used to coping regularly with a Sludge Volume Index (SVI) of 300-400 mL/g, would find an SVI of 150 mg/L a refreshing non-bulking situation. Plant operators who have been able to maintain stable operations at SVIs of 60-80 mL/g will become concerned when the SVI exceeds 100 mL/g.

Some reports of "bulking sludge" are founded on the formation of microscopic bubbles of nitrogen gas causing flotation of sludge solids while others come from rising sludge blankets. These situations may result from exceeding the sludge transport capacity of the clarifier scrapers (see Chapter 5), inadequate RAS rate, excessive floor loading, or various combinations of these factors. These observations inflate the reported level of bulking sludge problems.

Even the term SVI has many different meanings in the international technical community since data has been reported as SVI_{30} (and SVI_{60}), $SSVI_{30}$, $SSVI_{2.0}$, $SSVI_{3.5}$, $DSVI_{30}$ and perhaps other definitions. The relationships between these various SVI tests are not constant, and plant-to-plant comparisons are difficult or impossible when the SVI exceeds 125 to 150 mL/g.

The universal need to standardize the SVI procedure is obvious. The question is what procedure? A brief description and comment on each follows:

(1) SVI_{30}. The MLSS settling test is conducted in a 1 L graduated cylinder or a 2 L settlometer for 30 minutes. The recorded 30-minute volume is converted to SVI as a function of the MLSS. When the settled sludge volume (SSV_{30}) exceeds 250-300 mL/L, the relative value of this test is limited. At SSV above 500 mL/L, the values produced may be of little value, and thus misleading. Sometimes this test, and the following tests, are continued for 60 minutes and reported as SVI_{60}, etc.

(2) $SSVI_{30}$ (stirred SVI). This test is conducted in a 1 L graduated cylinder or a 2 L settlometer. A stirrer operating at about 1 rpm is used to disrupt floc particle bridging and assist consolidation of the solids. This test will generally produce a smaller 30-min sludge volume than

the SVI and it is considered more representative of full-scale sludge compaction, but not solids-liquid separation rates.

(3) $SVI_{2.0}$ and $SSVI_{2.0}$. These tests are conducted for 30 minutes at a MLSS of 2000 mg/L (2.0 g/L) and represent an effort to standardize the procedure for comparison. The $SVI_{2.0}$ has the same limitations as the SVI_{30} tests, and the $SSVI_{2.0}$ is an attempt to eliminate the effects of varying MLSS on both the settleability rate as well as the solids-liquid separation rate. It has that advantage over the $SSVI_{30}$ test for plant-to-plant comparisons.

(4) $SSVI_{3.5}$. This test is similar to the $SSVI_{2.0}$ procedure except that it is conducted at 3500 mg/L MLSS. This procedure is more directed to the final concentration characteristics of the sludge than the solids-liquid separation rates at the interface. The $SSVI_{3.5}$ will be increasingly higher than the $SSVI_{2.0}$ as the SVI increases.

(5) $DSVI_{30}$ (Dilute SVI). This procedure involves dilution of the MLSS until the resulting 30-minute settled volume is 200 mL/L. A tolerance of 40 mL/L does not have a large effect on the resulting DSVI. This procedure will provide a more representative value of the solids-liquid separation rate and a better projection of the potential concentration characteristics of the settled solids due to minimization of bridging and MLSS concentration effects.

While there may be some merits to a $DSSVI_{30}$ procedure, data have not been reported. The results of Koopman and Cadee (1983) indicate that the DSVI and SSVI procedures produce reasonably comparable results. Since there are no added encumbrances (stirrers), the $DSVI_{30}$ procedure is recommended for determination of the sludge volume index. Chapter 5 (Figure 5.6) presents a portion of Keinath and Wahlberg (1990) data showing the comparison of SVI_{30} and $SSVI_{30}$ in 1 L cylinders and 2 L settlometers. Additional results of these studies are presented in Figure 8.1 where the SVI and SSVI are compared showing the power correlations.

Sezgin et al. (1978), Palm et al. (1980), and Lee et al. (1982), using the total extended filament length (TEFL) procedure of Sezgin et al., found good correlations between the various bulking sludge indexes and the TEFL. When the TEFL increased beyond 10^7 μm/mL, the SVI increased rapidly. The test results of Palm et al. are shown in Figure 8.2.

Lee et al. (1983) correlated TEFL with several bulking indexes: SVI, $SVI_{1.5}$, $SVI_{2.5}$, $SVI_{3.5}$, DSVI, and $SSVI_{2.5}$. He considered that the DSVI test would be the most quantifiable procedure to use for judging clarifier settling and compaction characteristics. The studies showed that the DSVI was the most consistent correlation obtained between DSVI and TEFL. This conclusion was supported by Rachwal et al. (1982), Koopman and Cadee (1983), Pitman (1984), and Yamamoto and Matsui (1988).

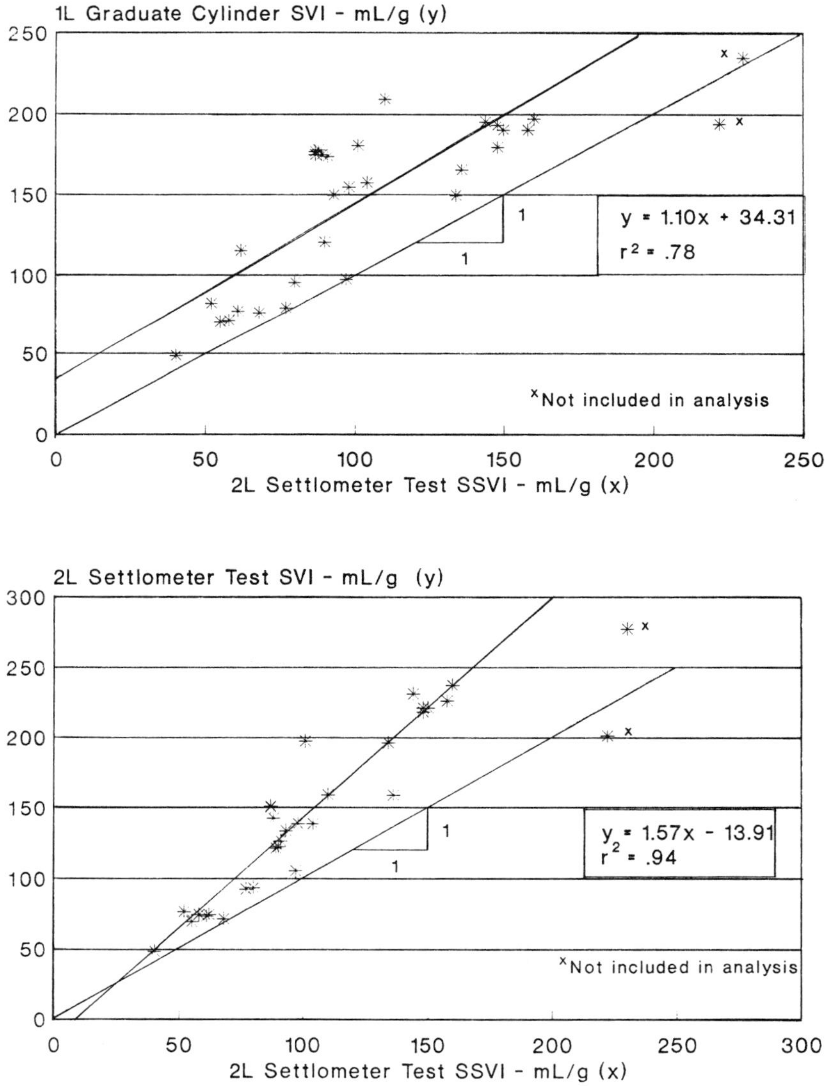

Figure 8.1 Comparison of SVI and SSVI in 1 L and 2 L vessels (see also Figure 5.6).

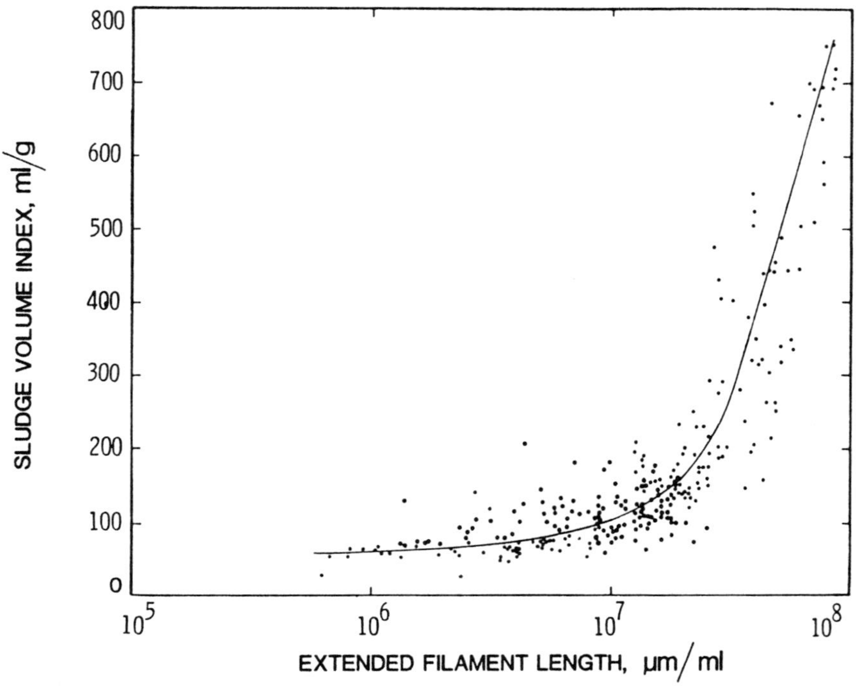

Figure 8.2 Effect of extended filament length on sludge volume index (Palm et al., 1980).

Rachwal et al. (1982) demonstrated the relationship of the settled sludge volume (SSV) in 1 L cylinders using the SVI and SSVI tests from MLSS of 1000-10,000 mL/L at an SSVI of 80-85 mL/g. The results are shown in Figure 8.3 in which the SVI is both less reliable and of no value above 4000 mg/L even at a low SSVI.

The average SSVI predicted in Figure 8.3 at 2000 mg/L would be 75 mL/g, and at 5000 mg/L MLSS it would be 80 mL/g. The SVI tests would project values of 100 mL/g and 180 mL/g, respectively, for these MLSS values. At 10,000 mg/L the SVI is only 100 mL/g (by definition) and the SSVI is about 85 mL/g. The value of SSVI versus the SVI test is obvious and the DSVI tests would have shown comparable settleability results.

The results of the DSVI tests versus the SVI procedure reported in the re-rating study of the Orange County, Florida ESAWWTP (CDM, 1989), provided added criticism of the inadequate standard SVI procedure. The DSVI was about 60% of the SVI when the SVI was 160-220 mL/g. Dick and Vesilind (1969) demonstrated that two sludges with similar SVIs can have significantly different settling characteristics. The designer must be cognizant of the unpredictability and lack of reliability of the standard SVI test. When the SSV_{30} in the SVI tests exceeds 250 mL/L, the procedure will produce inflated values of the actual bulking sludge condition.

Only the DSVI and SSVI procedures are recommended for 1500 to 5000 mg/L MLSS. When the SVI values are conducted, same-time DSVI tests should be run.

Since settled sludge volume (SSV) in either 1 L cylinders or 2 L settlometers is the precursor to the determination of SVI, all comments above are directly applicable to this determination.

The initial settling velocity (ISV) is often determined using the first 5 minutes of subsidence in 1 L or 2 L settling units. The settling distance is translated to m/hr and reported in these terms. While sometimes considered as a measure of the separation velocity, hence an indication of the allowable rise rate, this is not always the case.

The standard test procedure is not analogous to the field conditions for flocculent solids generally in hindered settling conditions. It should not be

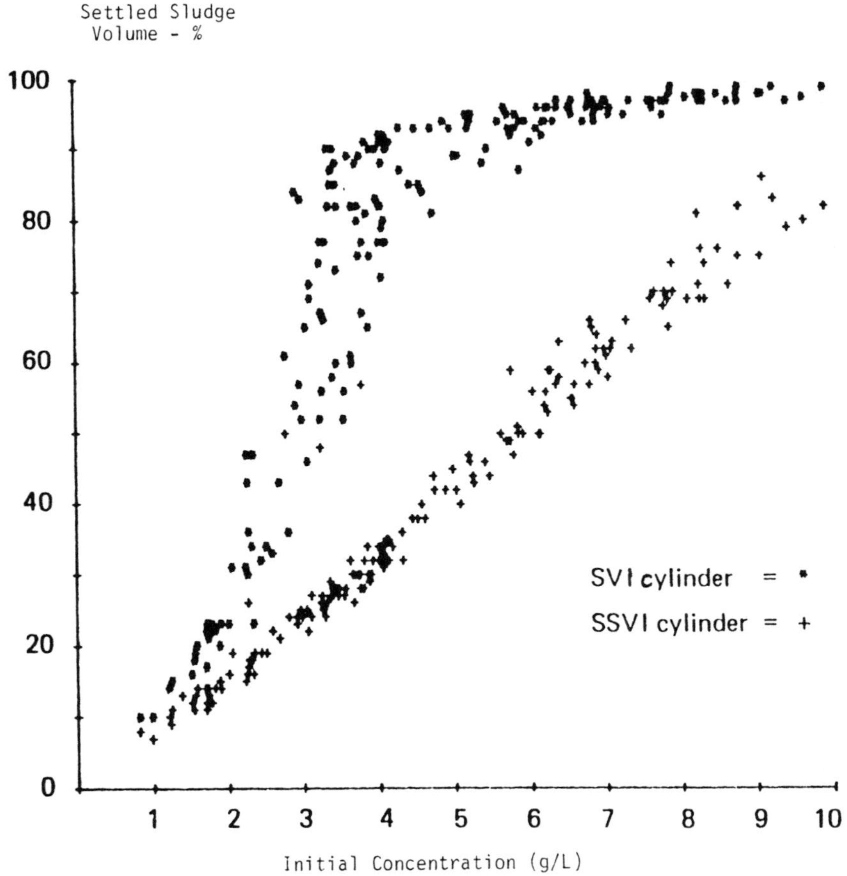

Figure 8.3 Settled volume vs. initial MLSS (SSVI 80-85 mL/g).

interpreted as the effect of SVI on the clarification capacity when conducted in shallow vessels. An example is that long-tube (1.2 m) tests will produce ISV values that are 200-300% higher than the 2 L settlometer tests, even using the DSVI procedure. The long-tube tests should be used for determining the ISV at concentrations that produce 20% sludge volume at 30 minutes in order to properly define the effects of SVI.

HISTORY OF THE BULKING SLUDGE PROBLEM

Sludge bulking in activated sludge systems has plagued designers and users since the early stages of the continuous mode process. It is now evident that the problem of bulking sludge emerged concurrently with the evolution of the activated sludge reactor design from draw-and-fill or batch reactors to continuous flow designs. Lacking a comparative measure of the sludge settling characteristics, such as sludge volume index or sludge density index (SDI), early practitioners and researchers referred instead to what is now recognized as bulking sludge, as the instability of the process or the unsettleability of the sludge.

In spite of much progress, bulking sludges still cause serious problems in wastewater treatment plants. However, technology has slowly crept up on the organisms responsible for bulking, and control theory and practice have evolved from chlorination (kill them) to environmental modifications (discourage them) in the treatment process steps. While the literature in the 1980s contains numerous references to the environmental modification concept, a look back at the original innovators of these concepts reveals an interesting evolution of sludge bulking control technology and practices.

The concepts developed by two individuals, Wellington Donaldson and Alex B. Davidson, predate our current practice of SVI control by thirty to fifty years. Three to five decades earlier, these engineers envisioned, tested, and defined the concepts that we are now using in our advanced wastewater treatment (AWT) plant designs, and ironically, some of these plants are today classified as innovative and alternate technology. Their ideas included compartments in secondary treatment reactors, alternative anaerobic and oxic zones, semi-continuous or batch-fed reactors, and oxygen control.

Wellington Donaldson (1932a, 1932b) recognized the advantages of compartmentalizing aeration basins to eliminate back-mixing. His contributions include the Donaldson Sludge Volume Index (SVI) still used today to define sludge compaction characteristics.

As early as 1932, Donaldson suspected that back-mixing was the cause of the instability of the activated sludge process. However, it was not until thirty to thirty-five years later that the difference in the biological kinetics of PFR versus CMAS reactors was well defined in the wastewater industry. Even so, the U.S. design engineer's preference in reactor design changed from PFR to CMAS in the 1960-1970s and back again to PFR in the 1980s.

As a practicing civil engineer, Donaldson observed and reported that so-called plug-flow aeration tanks were not plug-flow, but in fact, were subjected to a high degree of back-mixing. The first quantification of this observation appears to have been made by Thomas and McKee (1944) although others (Freese, 1938; Milbury et al., 1965) had noted that short-circuiting occurred in spiral flow aeration basins.

Donaldson conjectured: ". . . but the main points presented for discussion are the importance of microscopic growth of filamentous characteristics as an immediate cause of bulking, and the fact that this causative element is seemingly a result of conditions which favor food supply for such growth, which we may refer to as the 'weeds of the activated sludge.' " Fifty years later the causes of the bulking sludge probably cannot be better expressed, especially if all substrate required for growth, including oxygen, are defined as the "food supply."

Alex Davidson (1949) evaluated several processes to treat distillery wastes, which are notorious for producing abundant and uncontrollable levels of filamentous organisms. He tested a process (Davidson, 1957, 1959) that conflicted with all known process designs of the time and with general practice for at least thirty years thereafter. He demonstrated that the sludge bulking in an activated sludge process treating these wastes could be eliminated by alternating anaerobic and aerobic treatment. This process required several hours of anaerobic contact of the aerobic return activated sludge and raw wastewater, followed by aerobic treatment.

Davidson defined his process, shown in Figure 8.4, in the following terms: "Another objective of the invention is to provide a novel method of sewage treatment in which the raw sewage is first subjected to anaerobic digestion and then to aerobic decomposition with a return of the aerobic sludge to the raw sewage being subjected to anaerobic digestion." Davidson refuted the general belief that the return sludge must be kept in an aerobic condition or it will become "sick" and will not perform its purification function, but unfortunately this precious knowledge was lost for twenty-five years.

Before the mid-1960s, this statement reflected a common belief of the sanitary engineering profession. However, at about this same time, a number of practitioners, including advocates of pure oxygen activated sludge, found that the activated sludge was not harmed by anaerobic periods of up to at least twenty-four hours duration. Plant operators controlled bulking by turning off aerators during low flow periods at night. Indeed, Davidson's postulate was confirmed by a later study (Westgarth et al., 1964), but the fully oxic premise died very slowly.

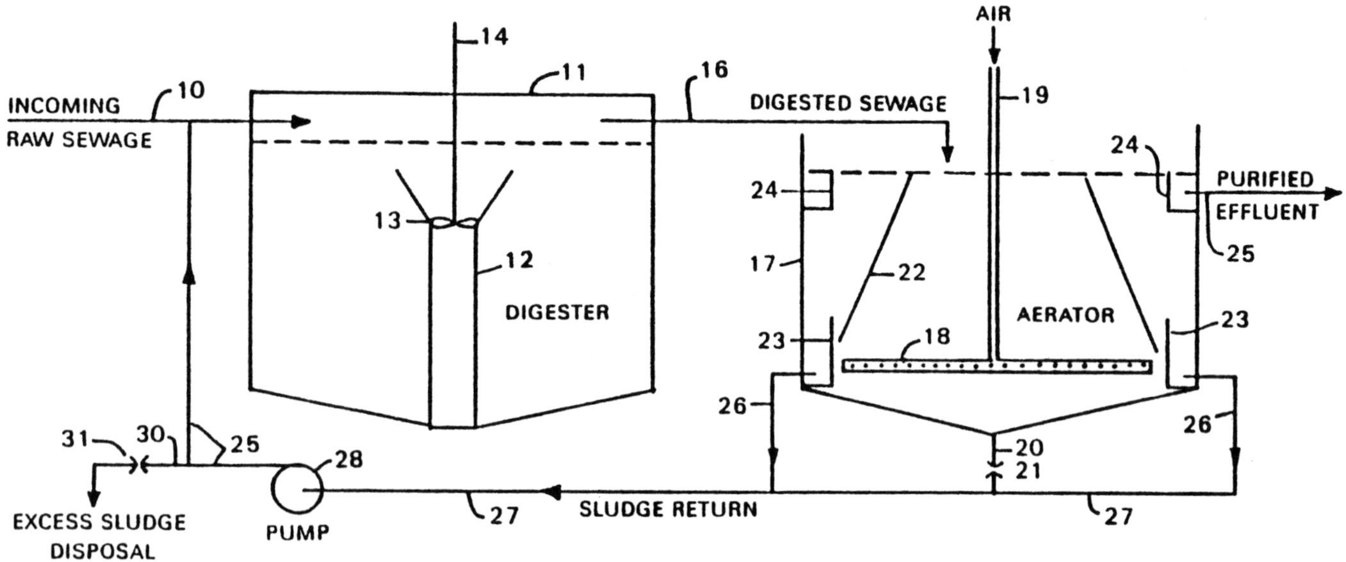

Figure 8.4 The Davidson anaerobic-oxic bulking sludge control flowsheet.

Heide and Pasveer (1974) tested several reactor configurations and waste feed techniques. These tests, a continuation of Pasveer's work (1969) in which the control of SVI in oxidation ditches was accomplished by intermittent feeding, defined several methods of bulking sludge control as noted in Table 8.2.

Heide and Pasveer tested two flowsheets for control of SVI. The first flowsheet was similar to that of Davidson. An alternate flowsheet in a mode similar to later BioP processes recycled MLSS from the aeration basin where it was mixed with the raw wastewater. Both flowsheets shown in Figure 8.5 effectively controlled SVI. The initial zone or mixing tank in Heide and Pasveer's studies was either anoxic and anaerobic depending on the process design and wastewater characteristics. The pump pit *F/M* ratio would have been from 6 to 10 kg/kg MLSSd.

Heide and Pasveer, commenting on the work of Chudoba et al. (1974) which demonstrated SVI control with compartmentalized reactors, were the first wastewater technologists to recognize the now-accepted fact that the organic matter can be taken up by the aerobic bacteria in a zone devoid of oxygen. They reported that the rate of adsorption of organic matter is rapid and at this rate a high floc load level is attained in only a few minutes. They concluded that the bacteria are in and on the floc and the organic matter is to a large extent adsorbed on the floc surface and that, by enzymatic action, is even assimilated within the floc itself.

The Sancta Maria oxidation ditch in Holland experienced a severe bulking condition from startup in 1961 until 1969 when the feeding mode was changed by Pasveer (1969). The SVI had ranged from 150 to over 500 mL/g. As a result of changing from a 15 m^3/h continuous feed to a 60 m^3/h rate for one hour out of every four hours, the SVI was decreased to the range of 100 to 150 mL/g. Pasveer reported that the bulking sludge condition in the Ter Aar oxidation ditch in Holland was also cured by semi-continuous feeding of the unit.

Other investigators had established earlier than 1969 that semi-continuous batch culturing of activated sludge would control the growth of filaments while completely mixed reactors treating the same wastewater

TABLE 8.2. Effect of Reactor Mode of Feeding on Activated Sludge SVI.

Mode of Feeding	SVI, mL/g
Continuous	500-600
Discontinuous	100-150
Continuous via mixing tank	70
Batch reactor	40-50

Source: Heide and Pasveer, 1974.

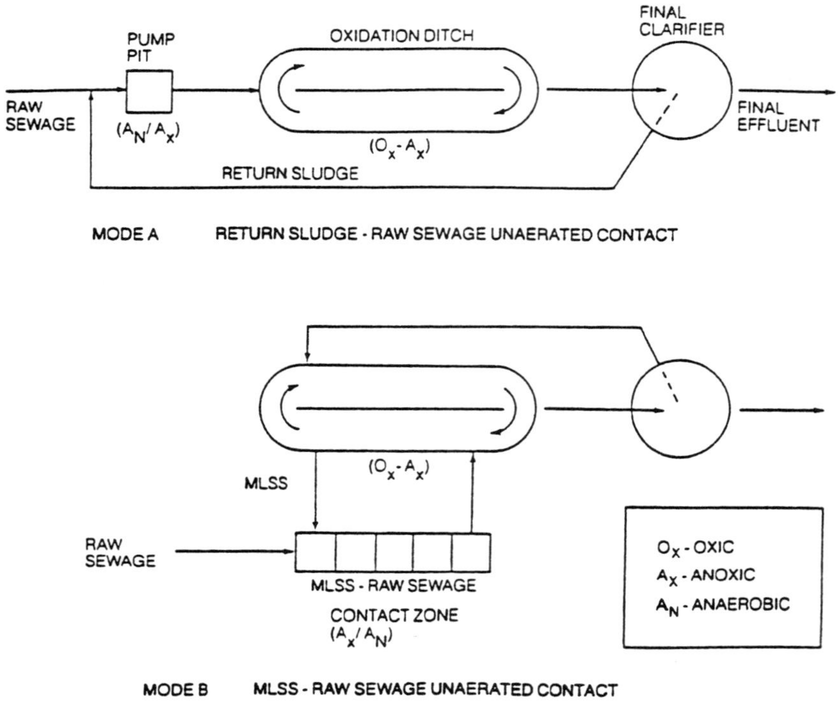

Figure 8.5 Pasveer and Heide's anaerobic/anoxic-oxic flowsheets for SVI control (1974).

contained abundant amounts of filaments. Koller (1966) found that semi-continuous feed (batch) reactors with detention periods of 3, 6, and 12 hours produced an average SVI of 51, 72, and 58 mL/g, respectively, while continuous feed, completely mixed reactors with detention times of 2.4, 3.7, and 6.4 hours resulted in SVIs of 280, 512, and 440 mL/g. Grau and Dohanyos' (1971) tests produced activated sludge with SVIs of 50 to 250 mL/g in semi-continuous feed systems as compared to 450 to 850 mL/g in continuous feed, completely mixed reactors.

The credit for understanding that limiting the quantity of available oxygen in the initial high *F:M* zone would cause filament suppression is more difficult to assign to any specific individual. Certainly, several people contributed to modifying the generally held concept that it was necessary to maintain an aerobic state in the entire aeration basin.

Bhatla's (1967) observations were at odds with those of previous researchers who indicated that low DO concentrations could cause bulking. Using a full-scale, eight-stage activated sludge system, Bhatla was able to control SVI in the basin by maintaining zero DO in the first stages of the aeration basin. When the first-stage DO was increased by increasing aeration power input, the SVI increased from a range of 91-102 to 235-299 mL/g. Bhatla's aeration basin certainly would have had a high *F/M* gradient because it was an eight-stage reactor and the oxygen demand varied from 60-80 mg/g VSS·h in the initial stage to 5-15 mg/g·h in the last stage.

Bhatla's conclusion that the persistent low oxygen tension rather than complete oxygen deprivation probably generated an unsuitable environment for filamentous organisms and retarded their growth had merit.

Westgarth et al. (1964), and later Chambers (1982) and Wagner (1982) reported on pilot plant studies that exposed the sludge to anaerobic or anoxic conditions, producing an improved SVI. These studies provided added support to the Ryder (1969) and Price (1982) full-scale studies, which employed limited uncombined oxygen availability to control SVI.

Researchers (Vacker et al., 1967; Wells, 1969; Scalf et al., 1969; Witherow, 1969) described full-scale plant performances where low to zero initial zone DO levels resulted in 80 to 95% biological phosphorus removal. Garber (1972) summarized the findings from these plants including his own experience at the Hyperion Los Angeles plant. He noted that the sludge density index (SDI) was normally greater than 1.0 (SVI $<$ 100 mL/g), a value that is indicative of a non-bulking sludge. These investigators all used reverse tapered aeration or inadequate aeration to obtain the low to zero DO in the initial zone of treatment.

Contributions by Chudoba et al. (1973a, 1973b, 1974) are extremely important in that the need to employ the plug flow characteristics of feeding the biological zone was firmly established. The bulking problems originating from process designs for complete mix, or those aeration configurations or feeding patterns that approximate complete mix, could be understood and eliminated as by process design.

Confirming the Chudoba et al. work is the English study of bulking sludge distribution characteristics as a function of the process design, plant size, and the plug flow characteristics of the plant. Tomlinson's (1976) full-scale data from sixty-five plants strongly supports the plug flow mode of feeding and compartmentalization of the biological reactor, whether aerated or not.

Albertson (1987a, 1987b) in his review of the past bulking sludge control methodology concluded that the reason the compartmentalization or selection zones were operable was that the organisms that could store soluble substrates in the absence of uncombined oxygen would dominate a high *F/M* ($>$3 kg BOD_5/kg MLSS·d) system whether aerated or anoxic. The rationale was that the unaerated interstitial DO would not penetrate the floc and soluble substrate sorption-conversion reactions rather than oxidation would still be dominant.

Wanner et al. (1987a) added further substance to this premise in their findings regarding the relative utilization of combined oxygen (NO_X) by filamentous and non-filamentous organisms. They concluded that some filamentous organisms had about 10% of the capability of non-filamentous organics in NO_X utilization and would be adversely affected in an anoxic environment. Further, the anoxic removal of $SBOD_5$ in the pre-oxic zones would deprive filaments of needed substrates in the oxic zones.

CAUSES OF NUISANCE ORGANISM GROWTH

FILAMENTOUS BULKING

Filamentous growths have proliferated in many types of environments: high and low *F/M*, high and low DO, high and low pH, fresh and septic wastewater, and perhaps every combination of these conditions. Possibly, there is a filamentous organism growth potential for all environments and the objective must be to reduce, not necessarily eliminate, the growth potential.

The fundamental cause of filamentous bulking is rather simple—the specific environment provides an improved opportunity for filamentous organism growth versus that of the preferred organisms. However, it should be noted that the filamentous organisms do not need to be the dominant mass in the sludge to cause bulking. A change of 5-10% of the sludge mass from non-filamentous to filamentous species can produce a bulking condition.

In the presence of adequate mineral nutrients, the cause of bulking is linked to the design of the biological reactor and the environment maintained within the biological reactor. For this reason, the control of bulking organisms is primarily accomplished by specific design procedures that provide an adverse growth environment for bulking organisms.

Growth of bulking sludges is linked to the presence of soluble substrates in the oxic reactor. More specifically, it is related to the presence of low molecular weight soluble substrate. Heide and Pasveer (1974) observed that 75% of the organic matter is adsorbed in the floc within a few minutes and suggested that an initial small mixing installation would suffice to produce a high floc loading. The presence or absence of aeration was not considered significant, and these researchers showed evidence that unaerated pump stations receiving raw sewage and wastewater provided effective bulking control.

Heide and Pasveer were perhaps the first investigators to define the biological mechanisms involved in the early successful selector work of Davidson (1949, 1957, 1959).

Wanner and Grau (1987) reported that it also may be necessary to control the environment by compartmentalization within the oxic zone. The rationale was that the hydrolysis of particulates to $SBOD_5$ can result in a low level of $SBOD_5$. This situation would favor the growth of high surface area/volume (weight) organisms, i.e., filamentous types.

Specific environmental factors have been linked to the presence of specific filamentous organisms. Jenkins et al. (1984) presented a listing of causative factors and the dominant filament organism found. This information is enclosed as Table 8.3.

Hinshelwood's (1946) observations of pure culture growth characteristics may explain some of the problems of filament identification, appearance, and disappearance. He reported that under specific growth conditions, bacterial cells could elongate up to 20 or 30 times their normal length and then sometimes return to normal length. He concluded that this was due to lack of adaption to the growth medium, and the result was a filamentous form of the organism. Organisms that showed these characteristics were *Bact. lactic aerogenes, Corynebacterium diptheriae, Bact. coli, Bacillus subtilis, Staphylococcus aureus,* and *Bact. typhosum.*

The filaments are formed as a result of slowed or inhibited division of the cells, which results in elongation. Hinshelwood reported a number of materials that cause organisms to become filamentous such as dye wastes, proflavine, and m-cresol.

NON-FILAMENTOUS BULKING

The occurrence of non-filamentous bulking is not commonly reported. In fact, only with the suppression of filamentous bulking will this mode be more apparent. This mode of bulking has been referred to as "hydrous" or "viscous" bulking due to the slime or jelly-like characteristics of the sludge solids.

TABLE 8.3. Dominant Filament Types as Indicators of Conditions Causing Activated Sludge Bulking.

Suggested Causative Factors	Indicative Filament Types
Low DO	type 1701, *S. natans, H. hydrossis*
Low *F/M*	*M. parvicella, H. hydrossis, Nocardia* sp. types 021N, 0041, 0675, 0092, 0581, 0961, 0803
Septic wastewater/sulfide	*Thiothrix* sp., *Beggiatoa* and type 021N
Nutrient deficiency	*Thiothrix* sp., *S. natans* type 021N, and possibly *H. hydrossis* and types 0041 and 0675
Low pH	fungi

Source: Jenkins et al., 1984.

Pipes (1969) identified sludge bulking with non-filamentous organisms, and this was considered as "zoogleal" bulking by Eikelboom and van Buijsen (1981). It is postulated that there is an excessive production of exocellular protoplasm that has a very high water retention; hence, the term "hydrous" bulking. Hale and Garver (1983) referred to this type of non-filamentous bulking as "viscous" bulking due to the viscous nature of the sludge and the poor settling and compaction of the solids.

The causes of non-filamentous bulking are not well defined. With selectors controlling most filamentous growths, the causative factors will be more visible. A number of facilities with selectors have reported periodic increases in SVI in the absence of filaments. Certainly, factors such as excessive floc loading and toxic agents can cause sheath or exocellular production. High floc loading and inadequate aeration capacity to exhaust stored food could be a cause. The presence of the exocellular material can be observed using the techniques set forth in Jenkins et al. (1984), and excellent photomicrographs accompany the text.

The earlier noted observations of Hinshelwood may be a factor in the rapid changes noted in non-filamentous bulking. Chlorine proved highly effective in eliminating this type of bulking at Phoenix, Arizona (Wendell, 1990). Chlorine may cause the extended cells to fracture along the division lines.

FOAMING

The reported observations of foaming fall into two distinct categories – a light-colored frothy foam, and a dense darker foam. The first category of foam is caused by non-degraded surfactants in the mixed liquor and the clarified effluent. These foams are generally unstable, especially when water sprayed. They may be more persistent in effluents with low TSS content, but are still considered to be only a minor nuisance.

The foam that may form on final clarifiers due to denitrification and flotation is controllable by the design and operation of the final clarifier and thus is not further mentioned in this section. Likewise, there may be incidents where industrial discharges have caused flotation of solids, but these are not general problems and are handled on a site-specific basis.

The foam resulting from a combination of floated biomass and surfactants can be quite serious and difficult to cope with. Often types are grouped together under the common heading *Nocardia* foam, which is not always accurate. Needless to say, these stable, persistent, often odorous and unsightly foams have frustrated operators and designers alike. The first reported presence of a heavy viscous layer of foam on the aeration and clarifier surfaces was at Milwaukee (Anon, 1969). However, it is unlikely that this is a recent problem and the problem may have increased in magnitude due to longer SRTs in AWT plants.

The analysis of biological foams shows the presence of many Gram-positive, branching filamentous bacteria. The most common organism has been identified as *Nocardia amarae* (Lechevalier, 1975). Other organisms that have been identified include *Rhodococcus rhodochrous* (Lemmer and Kroppenstedt, 1984), *Microthrix parvicella* (Jenkins et al., 1984), and a number of other nocardioform actinomycetes (Lechevalier and Lechevalier, 1974; Sezgin et al., 1988). A study of fifty plants by Blackall et al. (1991) in Australia revealed *Nocardia amarae* as the most common organism (eighteen plants), followed by *Nocardia pinensis* (three plants), both (six plants) for a total of 27/50 plants.

The causes of *Nocardia* foaming are not clear and are contradictory. The presence of significant quantities of the organism in the MLSS does not always cause foaming. An SBR may have *Nocardia* foaming in only the latter part of the aeration cycle and additional feed will suppress the foaming (Dana, 1988). Likewise, periodic cessation of aeration also will assist foam control in the SBR.

Nocardia organisms are widely dispersed in nature. Pipes (1978) suggested that the growth was associated with warmer weather ($t \geq 18°C$) and longer sludge ages (SRTs > nine days). Several investigators (Eikelboom, 1975; Pipes, 1978; Lewis, 1979; Matsche, 1982) found an association of *Nocardia* foam and emulsified fatty (oils and grease) materials. The foam is enriched *Nocardia*, activated sludge flocs, and air bubbles with fatty compounds that impart hydrophobicity to the foam, making it difficult to impossible to disperse with water sprays.

In the Queensland survey (Blackall et al., 1991), only eight of the fifty plants had never experienced a scum problem. The summary of the survey is shown in Table 8.4. No obvious cause for foaming was found.

There are types of industrial cleaners that caused excessive foaming (Dana, 1988) with *Nocardia* in the mixed liquor of a cheese plant SBR and also a plant receiving tank car wastes. When the type of detergent was changed, the foaming level was reduced but not eliminated.

PROCESS ALTERNATIVES FOR BULKING CONTROL

FIXED FILM

The use of fixed film reactors (FFR), trickling filters, or RBCs is a time-honored method for control of bulking organisms. The FFRs remove much of the soluble substrates required to support filaments. The presence of filaments in the FFRs has not proven detrimental to the directed coupled activated sludge process.

At this time, the FFR is unable to contribute to and is detrimental to phosphorus and nitrate removal and is not amenable to most BNR flow-

TABLE 8.4. Frequency of Scum and Bulking in Activated Sludge Plants in Queensland and Design and Operating Features Relevant to Scumming Plants.

	Frequency									
	Never		Occasionally		Frequently		Continuously		Total (50)*	
	No.	%	No.	%	No.	%	No.	%	No.	%
Scumming (50)	4	8	21	42	7	14	18	36	46	92
Plant design of scumming plants:										
Conventional	2	50	13	62	5	72	8	45	28	56
Package plant	0	0	4	19	1	14	6	33	11	22
Oxidation ditch	2	50	4	19	1	14	4	22	11	22
Plant parameters of scumming plants:										
Overloaded	0	0	9	43	4	57	3	17	16	32
Underloaded	4	100	9	43	3	43	12	67	28	56
Correctly loaded	0	0	2	9	0	0	1	5	3	6
Unknown	0	0	1	5	0	0	2	11	3	6
Surface aerators	3	75	13	62	5	72	14	78	35	70
Diffused aerators	1	25	8	38	2	28	4	22	15	30
MLSS:										
Above 2000 mg/L	2	50	15	71	3	43	13	72	33	66
Below 2000 mg/L	2	50	2	10	4	57	5	28	13	26
Unknown	0	0	4	19	0	0	0	0	4	8
Sludge age:										
Above 10 days	3	75	7	33	0	0	5	28	15	30
Below 10 days	0	0	7	33	3	57	6	33	16	32
Unknown	1	25	7	33	4	43	7	39	19	38
Bulking	20	40	23	46	6	12	1	2	30	60

*The total number of plants in the survey was 50.

Source: Blackall et al., 1991.

sheets. Studies are ongoing (Kelly, 1987) to expand the BNR capabilities of trickling filters beyond nitrification into denitrification and biological phosphorus removal.

SELECTORS

There are several types of selectors used to promote the selection of non-filamentous organisms. The types of selectors used are categorized as fully oxic, aerated low DO, anoxic, and anaerobic. Interrupted or intermittent feeding is a different type of approach employed for SBRs, complete-mix systems, and oxidation ditches, and is discussed separately.

The terminology employed for selector definitions is not necessarily that of the microbiologist, but is used in the BNR field to differentiate the design and operating conditions. The different types of selectors and operating conditions are set forth in Table 8.5.

Oxic

The oxic selector (Casey et al., 1975) design was based on maintaining a sufficiently high DO to discourage low DO filaments. A minimum DO of 2 mg/L occurs at 1.5 $SBOD_5$ (F_S)/active mass (M_a) ratio of $\leq$ 4 kg/kg·d. These conditions are not operable except with pure oxygen or with very low MLSS. Further, even with pure oxygen, many plants have experienced bulking. The use of large quantities of oxygen in the initial treatment stages is not compatible with BNR objectives.

Chudoba et al. (1973a, 1973b) conducted extensive testing with multi-compartmentalized aeration zones and demonstrated that bulking control was primarily accomplished by providing a *F/M* gradient. This is in

TABLE 8.5. Quantification of Criteria Used for Initial Contact Zones.

Environment	Criteria
Anaerobic (A_N)	: no added oxygen or air : total influent NO_XN < 0.5 mg/L
Anoxic (A_X)	: no added oxygen or air : total influent NO_XN ≥ 0.5 mg/L
Aerated (A)	: addition of oxygen or air : DO ≥0.0. mg/L
Oxic (O_X)	: addition of oxygen or air : DO ≥ 1.0 mg/L
Hi *F/M*	: ≥ 3 kg BOD_5/kg MLSS·d in ICZ
Lo DO (A_L)	: ≤ 0.3 mg/L DO in ICZ
Hi DO (A_H)	: ≥ 2.0 mg/L DO in ICZ

agreement with the earlier cited work of the British Ministry (1968), Koller (1966), and Heide and Pasveer (1974).

Aerated Low DO/High *F/M*

The unintentional use of low DO biological selection occurs in very long aeration basins with inadequate aeration capacity in the front end. Tomlinson (1976) reporting the bulking occurrences in sixty-five English plants, showed that initial zone *F/M* (calculated) to control bulking should be ≥4 kg/kg·d, as shown in Figure 8.6.

Many investigators have contributed to better understanding of selectors and their beneficial effects on the operation of activated sludge plants. The omission of works of some investigations is merely due to the lack of adequate space and opportunity. Some of the key investigations and the operating conditions are shown in Table 8.6.

Batch reactors, as selectors, have different operating conditions than a CMAS or a plug flow reactor. The SBR has a significantly different growth environment and thus has been addressed separately. Table 8.7 provides the comparative environments for batch reactors which could have all variations of selector modes.

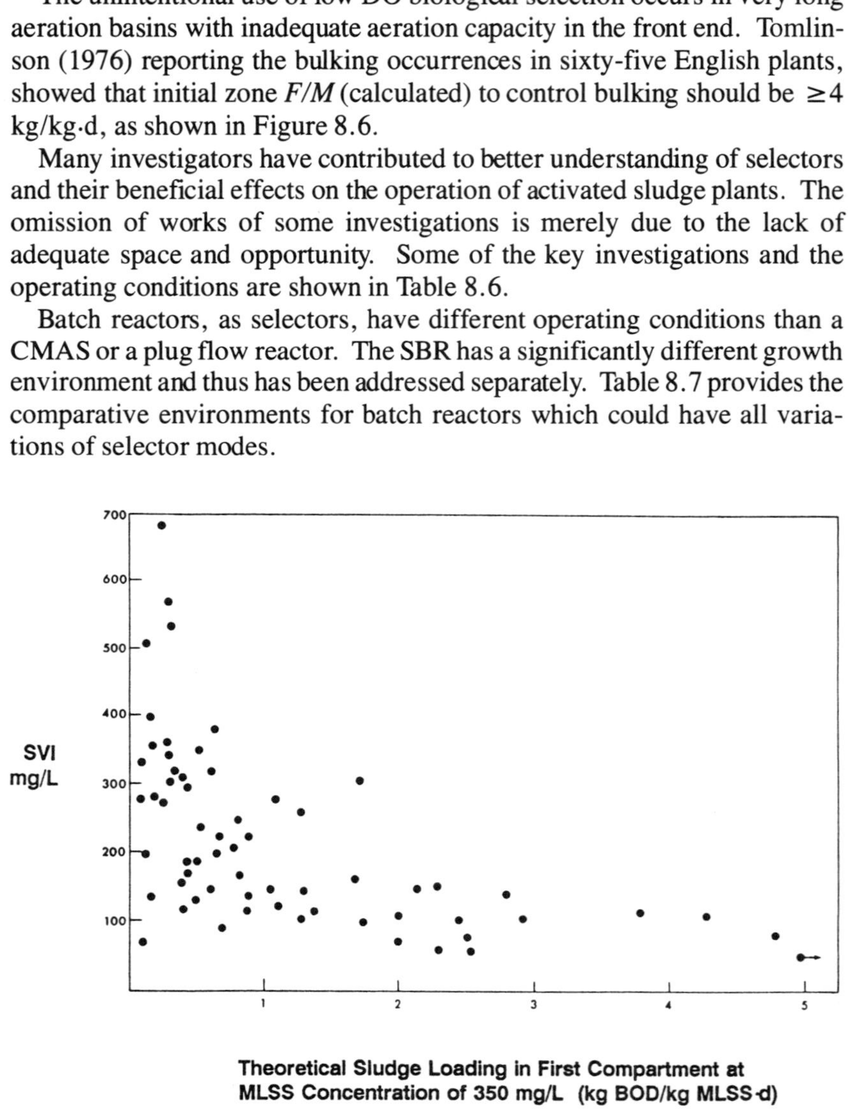

Figure 8.6 Relation between SVI and theoretical sludge loading in the first aeration compartment (Tomlinson, 1976).

TABLE 8.6. Operating Modes in Selector Studies.

Author	Selector Mode	DO, mg/L	ICZ *F/M*, lb/lb/d	SVI, mL/g
Davidson, 1959	A_N	0.0	1.0	34
Bhatla, 1967	A_L	0.0	>2.5	<120
British Ministry, 1969	A	Unknown	0.8	<75
Milbury, 1971	A_L	0.0	>2.0	<100
Chudoba, 1973	A_L	≤0.5	≥2.5	<100
Heide and	A_N/A_X	0.0	>5.0	<100
Pasveer, 1973	SBR	0.0	low	40
Rensink, 1974	A	Unknown	3.6	<100
Tomlinson, 1976	A	Unknown	>2.0	≤100
Spector, 1977	A_N	≤0.7	>3.0	<100
Chudoba and Wanner, 1988	A_H	≈1.0	12.0	<50

While the SBR has been successfully employed to model the SVI control of compartmentalized reactors in series, the actual growth conditions of the four systems are significantly different. Due to the design, the biomass in the SBR at the same SRT_{OX} will be 180% of the CMAS and PFR. Like the CMAS, the feed contacts the total mass and, thus, to maintain a loading gradient, the feed period should be short. Interestingly, while the growth environment of the PFR and SBR is totally different, bulking control is generally accomplished.

TABLE 8.7. Growth Environment in Typical Types of Biological Reactors.[1]

Mode	Floc Loading mg $SBOD_5$/g MLSS	Initial Zone *F/M* kg $CBOD_5$/kg·d	$SBOD_5$ Concentration 0 min[5] mg/L	$SBOD_5$ Concentration 10 min mg/L
CMAS	—	0.2	3	3
P.F. selector[1,2]	17.2	5.8	55	20
P.F. selector w/ IR[3]	7.6	5.8	25	10
SBR/selector[4]	2.9	0.41	2.0	6

[1] Q = 100 L/s, $CBOD_5$ = 160 mg/L, $SBOD_5$ = 80 mg/L RAS/Q = 0.5, MLSS = 3000 mg/L, Q + RAS = 4 hrs. O_X, RAS $SBOD_5$ = 5 mg/L.

[2] Selector detetion – 10 minutes (Q + RAS).

[3] IR = 3Q, $SBOD_5$ = 5 mg/L, MLSS = 3000 mg/L

[4] Decant/Fill – 20%, fill time = 20 minutes, O_X @ 55% of volume.

[5] Calculated $SBOD_5$ at initial and instantaneous mix (theoretical).

The use of aerated low DO/high *F/M* selectors has proven successful at Davenport, Iowa, Columbus, Ohio, Phoenix, Arizona, Lynnwood, Wisconsin, and other facilities. The recommended design (Albertson, 1987a, 1987b) for the aerated low DO selector is three zones with an initial zone *F/M* of 6 kg BOD_5/kg MLSS·d followed by a second zone of equal volume and the third zone with a volume equal to the first two zones. Coarse bubble diffusers in a cross-roll pattern, or jet aerators, are employed to provide aeration and mixing. The DO is maintained at <0.3 mg/L and generally approaches zero during the daytime loading periods.

The Columbus, Ohio Southerly (4990 L/s) and Jackson Pike (2630 L/s) WWTPs (Albertson et al., 1991) have low-DO anoxic and low-DO aerated selectors of the above design, respectively. Based on full-scale operation, there appears to be very little difference in performance between the unaerated, aerated, and anoxic selectors. The configuration and performance of the Southerly WWTP is shown in Figure 8.7.

While the Southerly WWTP has a high $SBOD_5$ due to brewery wastes, the Jackson Pike WWTP has a significantly lower content, and thus was provided with two selector zones without internal recycle and coarse bubble diffusers. The SVI has been less stable than Southerly, but this may not be associated with the selector design. The SVI ranged from 50-10 mL/g and has averaged 71 mL/g for 2.5 years since startup.

At Davenport, Iowa, the SVI is maintained at 80 mL/g by adjusting the percentage of time (hr/day) the first zone of four equal size zones is aerated. Without aeration in the first zone, the SVI decreased to 40-50 mL/g and a turbid effluent resulted. The selector and subsequent SVI control increased peak capacity of the plant from 1050 L/s to 1840 L/s (Resnick, 1988).

Chudoba and Wanner (1988) employed a four-stage aerated selector for the Leopoldsdorf sugar mill. The design was based on a *F/M* gradient of 12, 6, 4 and 3 kg BOD_5/kg MLSS·d at 4000 mg/L MLSS. Operating at an overall *F/M* of 0.3 kg/kg·d, the resulting SVI was 50 mL/g and the effluent BOD_5 was 10 mg/L. This four-stage selector design was equally successful at Middletown, Ohio (Albertson, 1991).

To ensure best control of bulking organisms, the $SBOD_5$ (SCOD) should be mostly removed in the selector zones. The air rate can be increased in the third zone (and second if necessary) to reduce selector effluent soluble substrates if the SVI is increasing.

The recommended design is a Σ *F/M* gradient of $\approx$ 6, 3, and 1.5 kg/kg·d in the initial three zones. The air rate should be at least 50% of the average air rate (m^3/m^3) in the basin.

Anoxic

The bulking control benefits derived from use of anoxic selectors may impact the efficiency of Bio-P removal. However, the anticipated adverse

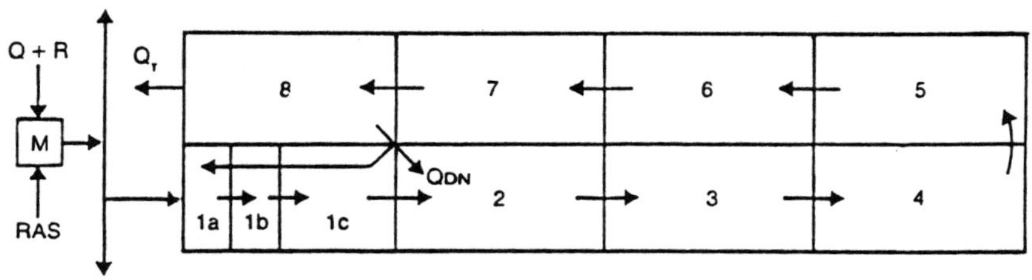

ZONE	MODE	ENVIRONMENT	EQUIPMENT
1a	MIX / AERATE	A_X	JET AERATORS
1b & 1c	MIX / AERATE	A_N / A_X	JET AERATORS
2	MIX / AERATE	A_N / A_X / O_X	JET AERATORS
3	AERATE	A_N / O_X	DIFFUSERS
4	AERATE	A_N / O_X	DIFFUSERS
5-8	AERATE	O_X	DIFFUSERS

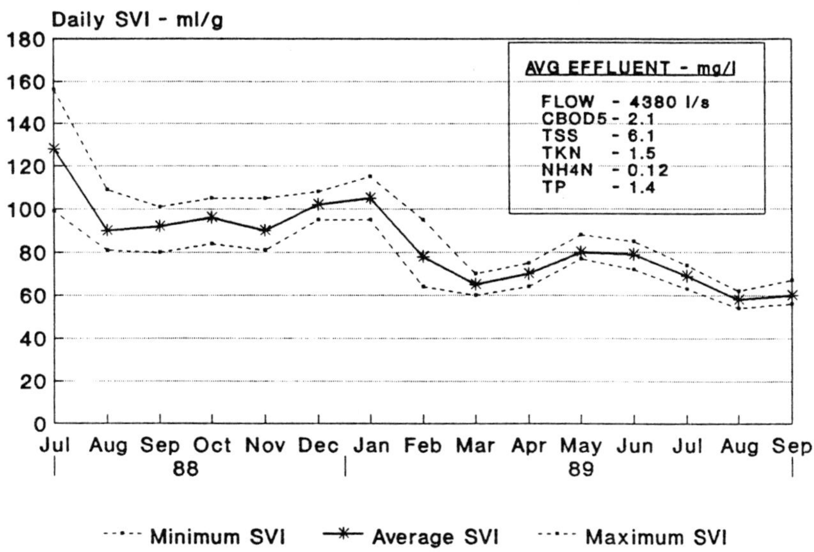

Figure 8.7 Columbus, OH semi-aerobic design and performance.

effect is not predictable and to an extent may be mitigated. The $SBOD_5$ (SCOD) removal in the Bio-P function is much more rapid than denitrification, and thus the small, high *F/M* zones will favor the Bio-P mode of soluble substrate removal. Further, the nitrate-containing recycle can be split between the selector zones and denitrification zones as was shown in Figure 8.7 and also employed for Phoenix 23rd Ave. WWTP and Baltimore Back River WWTP.

While the recycle of NO_3N has been reported to suppress the Bio-P mechanism (Barnard, 1976, 1982; Marsden and Marais, 1977; Simpkins and McLaren, 1978) the data is not consistent in this regard. Hascoet and Florentz (1985), using batch reactors, were able to maintain Bio-P functions even with the presence of high initial nitrate concentrations. An initial anoxic condition in the SBR would be normal for the feed cycle in an AWT mode. With aerated (low DO) anoxic high *F/M* selectors and an internal recycle to the influent, the Phoenix 23rd Ave. WWTP periodically produced high phosphorus removals (Albertson and Hendricks, 1992).

It is possible that the early South African BNR plants that employed single-stage anoxic and oxic zones had NO_3N interference with Bio-P mechanisms due to the reactor configuration. This comment is supported by the reports of bulking in many plants even though filamentous cultures are considered to be suppressed by anoxic conditions. The lack of a *F/M* gradient in both anoxic and oxic zones may have been the reason for the bulking and the Bio-P interference.

Kershaw and Finch (1936) observed that the recycle of nitrified trickling filter effluent to the activated sludge improved sludge density and settleability. Edmonson and Goodrich (1943) found that recycling an activated sludge effluent over a nitrifying filter and then returning to the aerator influent provided for bulking control. Since then, a number of investigators (Bailey and Thomas, 1974; Heide and Pasveer, 1974; Christepson and Harremöes, 1977; Tomlinson and Chambers, 1979; Price, 1982; Chambers, 1982; Wagner, 1982) have confirmed this.

Wanner et al. (1987a) reported that anoxic selectors were able to suppress the growth of type 021N and *Sphaerotilus natans*. Further, they established that filamentous cultures had NO_3N respiration (denitrification) rates that were 10% of non-filamentous cultures and concluded that some filamentous organisms cannot use NO_3N as an electron acceptor.

Shao and Jenkins (1989) concluded that the design of the anoxic selector should provide for a SCOD of less than 100 mg/L to ensure bulking control. Subsequent review by the authors reduced that value to 60 mg/L. When the SCOD entering the oxic period exceeded 60 mg/L, the SVI would increase. They found that after exhaustion of NO_3N, there was an appreciable leakage of soluble substrate into solution. Lack of an *F/M* gradient in the oxic zone could cause secondary bulking (Wanner and Grau, 1987).

The Phoenix 23rd Ave. WWTP employs anoxic selectors with coarse bubble aeration. The modified configuration is shown in Figure 8.8 (Albertson and Hendricks, 1992). The secondary plant was converted from a plug flow basin (with step feed alternatives) to the selector mode with internal recycle in an NDN mode. The process results are shown in Figure 8.9, wherein the SVI was reduced from a range of 200-300 mL/g to less than 100 mL/g. The flow capacity was also increased 35 to 45%.

The phosphorus removal at 23rd Ave. is very erratic with the West basin achieving significantly lower phosphorus content while both units receive the same RAS. The internal recycle is about 50% higher for the East unit, producing more competition for the soluble substrate. At a PE BOD_5 of 22,000 kg/d, the selector *F/M* gradient is 9.5, 3.5, and 1.6 kg $CBOD_5$/kg MLSS·d at 3000 mg/L MLSS. (Basin structure dictated baffle placement.) The plant expansion will provide for splitting the internal recycle between A_X-1 and A_X-3 which could improve Bio-P removal.

The Tri-Cities, OR WWTP was designed for NdeN in the dry weather and to be fully oxic during the high flow periods. To counteract solids washout in the winter, the plant was operated in a step feed mode. While the A_X-O_X mode provided excellent SVI control, step feeding and partially aerating the anoxic caused bulking, as shown in Figure 8.10. The *F/M* in the single (no stages) anoxic zone was 0.6-1.3 kg/kgd, possibly too low to be effective if aerated.

The design recommendations for the anoxic selector are the same as for the aerated selector. The *F/M* gradient is ≈ 6, 3, and 1.5 kg/kg·d for the three zones. The internal recycle should permit returning 20-50% of the flow to A_X-1 and the balance to A_X-3. Aeration can reduce Bio-P potential, but low air rates with coarse bubble diffusers in a side roll pattern will mix and minimize oxygen transfer. Submerged mixers will limit oxygen introduction if phosphorus removal is required.

Anaerobic

The first reported use of an anaerobic selector was by Davidson (1957, 1959) who conducted studies starting in the late 1940s and subsequently received patents on the anaerobic oxic flowsheet. His lab tests proved successful in controlling bulking in the activated sludge treatment of distillery wastewater, which is notorious for the proliferation of filamentous organisms. Table 8.8 provides a brief chronological review of several investigators' contributions to this type of selector.

Barnard (1974) noted that there was a need to provide an anaerobic zone (as distinct from anoxic) in the Bio-P process and incorporated an initial anaerobic zone in front of the first anaerobic zone (Nicholls, 1975). However, while the anaerobic zone improved the phosphorus removal, many of

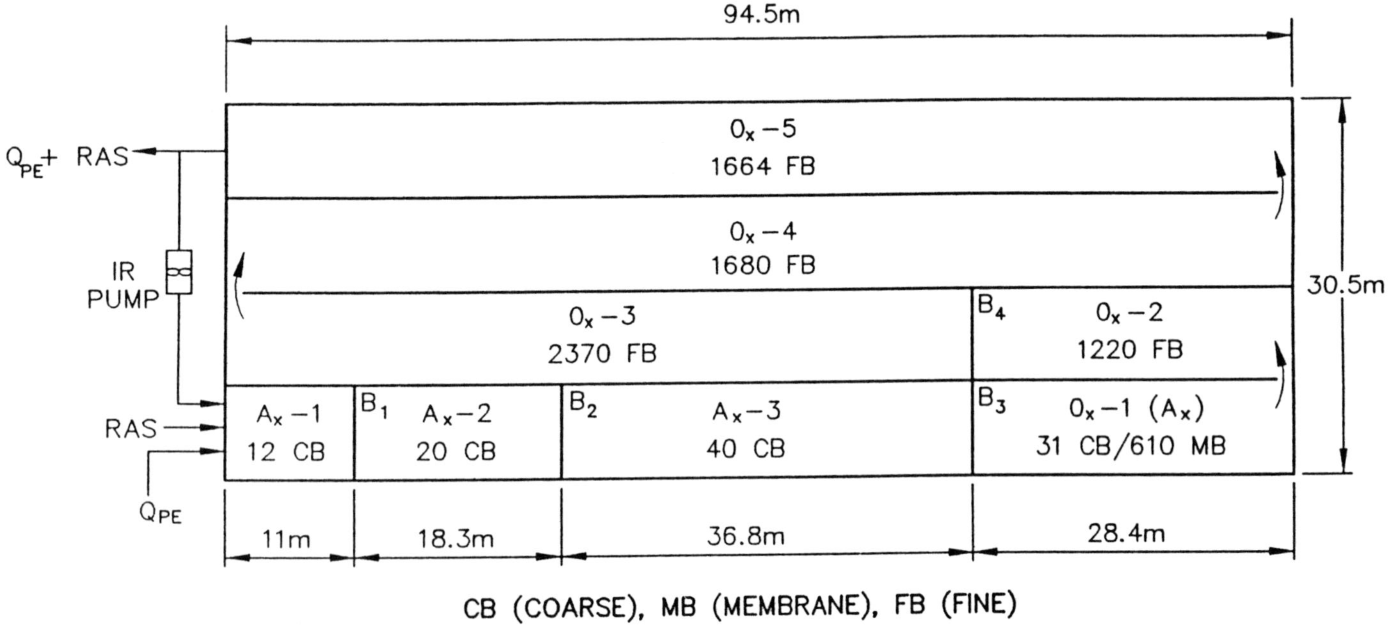

Figure 8.8 No. 1 (east) aeration basin zones, no. and type diffuser and baffle (B) arrangement at Phoenix 23rd Avenue WWTP.

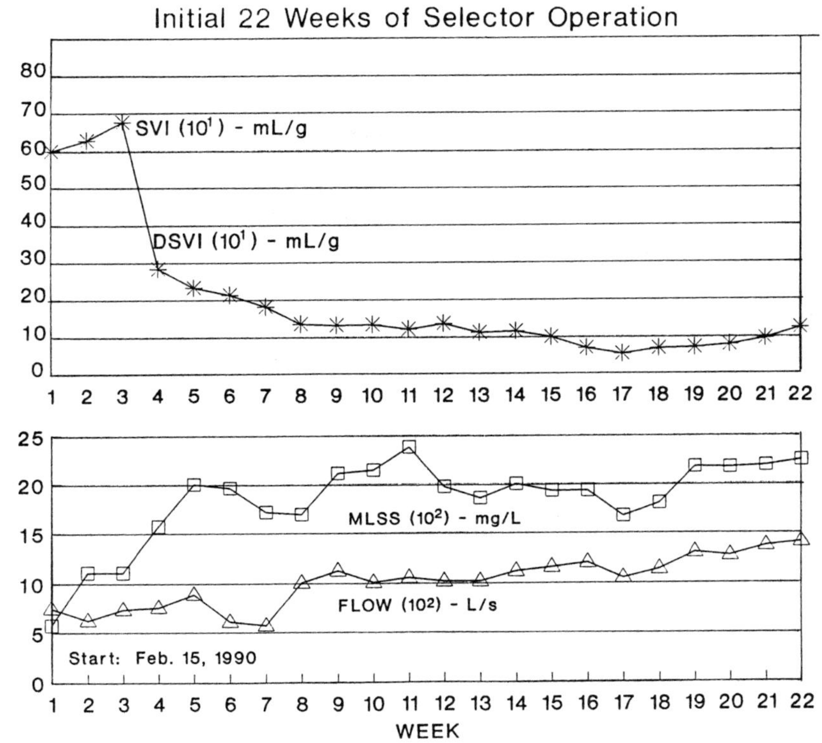

Figure 8.9 The benefits of anoxic selector at 23rd Avenue, Phoenix, Arizona.

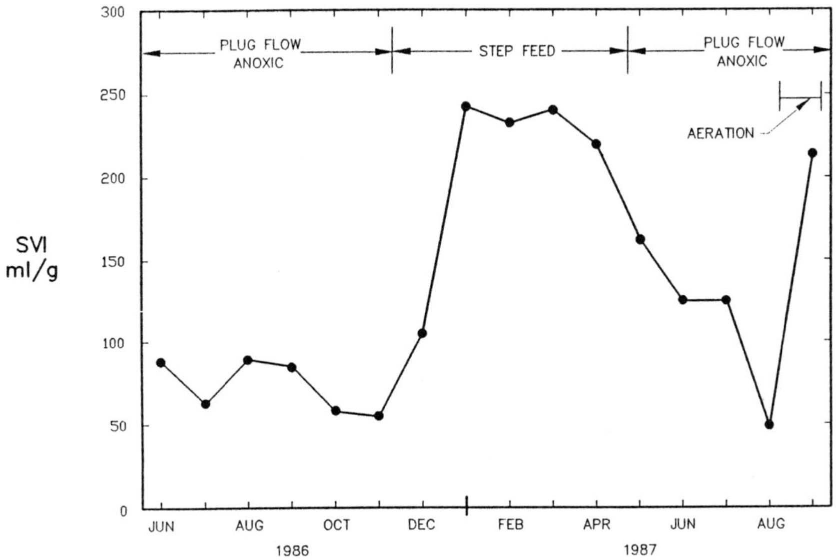

Figure 8.10 The effect of anoxic zone on SVI at Tri Cities, OR.

TABLE 8.8. Comparison of Elements to the Anaerobic/Oxic Process.

Component or Mode	Davidson, 1949	Garber, 1972	Chudoba et al., 1973a	Heide and Pasveer, 1974	Barnard, 1976	Spector, 1977
Mix initial zone w/o air	Yes	No	No	Yes	Yes	Yes
Oxic zone	Yes	Yes	Yes	Yes	Yes	Yes
Clarifier	Yes	Yes	Yes	Yes	Yes	Yes
Oxic sludge to initial zone	Yes	Yes	Yes	Yes	Yes	Yes
Mix influent with RAS	Yes	Yes	Yes	Yes	Yes	Yes
Claim bulking sludge control	Yes	Yes	Yes	Yes	No	Yes

the plants experienced moderate to severe bulking (Blackbeard and Ekama, 1984; Pitman, et al., 1983; Gabb et al., 1988) in both South Africa and the U.S.A. Albertson (1987b) conjectured that the bulking sludge problems in these systems may be due to the early practice of single-stage A_N, A_X and O_X zones that lack an organic loading gradient considered necessary for control of bulking organisms as set forth earlier. Further, the high internal recycle rates common to these systems would produce very low floc (contact) loadings and soluble substrate concentrations that would also favor filaments.

The initial anaerobic zone is a salient part of the Bio-P systems, such as the five-stage Bardenpho, three-stage Phoredox, A^2O, UCT, and VIP processes, but many publications do not define the SVI characteristics. Further, the standard SVI test is not representative in the 3500-4500 mg/L MLSS common to many of these systems.

Only the A/O system is also directed to bulking sludge control. The A/O system contains the same elements in the same orientation as the earlier cited Davidson process and was also patented (Spector, 1977). The anaerobic contact zone in the Spector reference was defined as a treatment zone that receives influent wastewater and return sludge, and is not aerated. The NO_3N input is limited to <0.5 mg/L of the total influent flow to the initial zone to prevent it from becoming anoxic. The process is claimed to be operable up to 0.7 mg/L, but this level of DO would require a very intense level of aeration, which is excluded from the zone.

Tracy et al. (1986) conducted SBR tests with an initial anaerobic contact period of 1 hour, followed by 5.5 hours at DO = 4.0 mg/L vs. a low DO (<0.5 mg/L) for one hour, which was increased eventually to 3.5 hours, followed by an oxic 4 mg/L DO period of 6.5 hr less the low DO time. The SVI were roughly comparable at 1 hr selector time, but the low DO SBR bulked when the low DO period was extended up to 3.5 hrs. However, it was not a comparable test since the anaerobic period remained one hour and the conclusions could have been biased by that fact.

Deakyne et al. (1983) reported excellent sludge properties (SVIs 60-75 mL/g) from A/O pilot plant treating the Baltimore Patapsco WWTP wastewater. Watanabe et al. (1984), using a synthetic wastewater in a lab pilot unit, produced SVIs of 40 mL/g in an anaerobic-oxic mode. Inamore et al. (1986) found the anaerobic selector to be superior to the oxic selector.

The 347 L/s (Newark, Ohio) treatment plant utilizes an A_N-A_X-O_X flow sheet for SVI control, nitrification, and denitrification. The original plant has had a long history of poor treatment performance due to bulking sludge and foaming episodes. Significant industrial discharges were at least part of the cause. The new plant was brought on line in June 1988. The SVI has decreased from 100-400 mL/g to 80-200 mL/g, and averaged 124 mL/g. No adverse bulking in the final clarifiers has occurred in the new plant. The SVI first increased in the colder months, but decreased below 100 mL/g when waste activated sludge (WAS) was not returned to the primary clarifiers.

The Newark plant has a three-stage A_N-A_X selector currently operating with a *F/M* of 1.59, 0.79, and 0.53 kg BOD_5/kg MLSS·d, respectively. The fourteen-month average clarified effluent quality of the plant is as follows: 2.7 mg/L $CBOD_5$, 3.8 mg/L TSS, 1.0 mg/L NH_4N, 2.4 mg/L TKN, 8.3 mg/L NO_XN, and 4.1 mg/L TN. Longer-term data are needed to determine whether or not the higher winter SVI values were an aberration.

Wanner et al. (1987b) concluded that anaerobic conditioning could suppress the growth of type 021N and *Sphaerotilus natans*, but that the positive results could be offset by simultaneous sulfate reduction by *Thiothrix* and other similar filamentous organisms. However, *Thiothrix* was eliminated when three anaerobic zones were used in place of one unit, furthering support for an anaerobic gradient. Yamamoto et al. (1991) conducted studies in SBRs with synthetic media and filamentous conditions developed from sulfur-reducing bacteria. The organisms were type 021N and *Beggiatoa* sp., and SVIs exceeded 1000 mL/g.

Gerber and Winter (1984) evaluated the effects of the length of the anaerobic period on the viability of the process. The anaerobic retention period varied from six to twenty-four hours, and up to 78% of the total time. The denitrification rates and total P removal increased appreciably with the longer retention times and the SVIs were 50-60 mL/g.

INTERMITTENT FEED/SBR

The initial use of intermittent feeding originated with the Aldern and Lockett (1914) experiments leading to the development of the activated sludge process. Unfortunately, this mode of batch feeding pilot units was also common practice for industrial waste treatability studies in the 1950-1960s for scaling to continuous CMAS operation. Predictably, bulking was avoided in the pilot tests, but was a serious problem in many of the full-scale units.

As was shown in Table 8.7, the use of intermittent rapid feeding of the total reactor still results in a very low floc (contact) loading and *F/M* during the feed cycle. While the SBR is a perfect plug flow reactor in organic removal, it differs in process because all of the biomass is in contact with the feed in the manner of the CMAS unit. However, like the PFR (and unlike CMAS), a substrate gradient can be developed and the environment (A_N, A_X, etc) can be controlled by the operating mode. Due to the large biomass/food ratio, biosorption can be quite complete prior to the oxic cycle.

Eikelboom (1982), in agreement with Pasveer (1969), Heide and Pasveer (1974), Chudoba et al. (1973a, 1973b, 1974), and Rensink (1974), considered that the proper floc loading (mg COD/g MLSS) and subsequent biosorption capacity was a means of providing bulking control. He found when the biosorption capacity was low, the sludge settling properties would deteriorate. Both high and low floc loadings could cause bulking, and he considered a floc loading between 50-150 mg COD/g MLSS to be the proper range. The COD level was based on lab-centrifuged supernatant.

Yamamoto and Matsui (1989), employing synthetic feed, found the optimum contact loading to be in the range of 300-700 mg COD/g MLSS or much higher than other researchers. Rensink et al. (1982) produced low SVIs with a contact loading of 60-120 mg COD/g MLSS. Goronszy (1990), treating a strong biodegradable industrial wastewater, produced SVIs < 100 mL/g with a contact loading range of 120-150 mg SCOD/g MLVSS.

The contact or floc loading concept has an understandable formulation. Currently, inadequate data is available to project design methodology. Also, it would appear that the design values for PFR with selectors and PFR reactors with internal recycle are different than for SBRs.

TOXICANTS

The control of bulking sludges with toxicants has both been purposeful and inadvertent. Chlorine and hydrogen peroxide are commonly employed for bulking control. Many industrial discharges contain components that also inhibit filamentous growth, but not necessarily BOD_5 removal. Until recent years, the toxicants and fixed film pretreatment were the primary means of bulking sludge control, and chlorine still has a place for *Nocardia* control.

While chlorine was the SVI control means of choice (Jenkins et al., 1984), the success of the selectors for biological control of bulking has moved chlorine's position to a backup for newer and modified plants employing selectors.

Chlorine

Chlorine has been used in the U.S. for many years and became increasingly popular as the need to produce a better effluent quality was required in the presence of bulking sludge. The cost of chlorine is less than polymers (Jenkins et al., 1984; Chambers and Tomlinson, 1982) for control of bulking sludges in clarifiers.

Chlorination for bulking control is also practiced in Germany, Austria, Great Britain, and South Africa. Chlorine is particularly popular in the U.S. since chlorine is already in use at nearly all of the facilities.

Jenkins et al. (1984) suggests that the chlorine be employed at a point of lowest oxygen demand and where there will be rapid dispersion of the chlorine. Good locations are found in the RAS line, particularly where there are elbows of crossovers. Chlorine addition to wet wells or clarifier feedwells will result in over- and under-dosing and lower effectiveness, as well as greater effluent turbidity.

Other locations for chlorine addition include weir discharges, fast-flowing channels (use multi-point diffuser or mixer), and pump recycle flows, such as an internal recycle pump. The dosage point should be located such that the total MLSS is exposed about every eight to twelve hours.

Chlorine is dosed on a kg/kg TSS basis. Jenkins (1980), Beebe and Jenkins (1981), Beebe et al. (1982), and Jenkins et al. (1984) provide guidelines for chlorine dosing. The initial dosage range for severely bulking sludge will be 8-12 Cl_2/1000 kg TSS. Low return sludge solids concentrations will require a higher dosage due to liquid demands for chlorine. If the SVI level is not critical, it is suggested that the dosage be started at 2-3 kg/1000 kg and increased if not effective within two to three days. The maintenance dosage should not exceed 5 kg/1000 kg.

Best practice is to pre-select a target DSVI that is not to be exceeded, as well as a desired and lower operating range; for example, an operating range of 90-110 mL/g and maximum of 125 mL/g. As the DSVI (or SSVI) increases above 110 mL/g, the Cl_2 dosage is initiated (or increased) to return the settling characteristics to the lower end of the operating range. At that point, the chlorine feed ceases (or is reduced) until the DSVI again exceeds the operating range.

While biological selectors have provided excellent control of bulking, the availability of chlorine as a backup is warranted where the wastewater tends to promote excessive growths of filament. Chlorine was reported by Wendell (1990) to be very effective against non-filamentous bulking, but this could be an isolated reaction. Chlorine will provide a more rapid control of filamentous bulking than will biological selection.

Hydrogen Peroxide

Hydrogen peroxide has been employed for control of bulking. Jenkins et al. (1984) reported on the results of several investigators using hydrogen peroxide. Hydrogen peroxide has an advantage in that there would not be a production of chlorinated hydrocarbons. Further, the decomposition of 2 moles of H_2O_2 yields 1 mole of oxygen in solution.

The competing reactions for H_2O_2 are not reported and the dosages vary widely. Caropresso et al. (1974) reported that H_2O_2 was used to cure bulking in the Petaluma, California plant, which treated a mixture of domestic and industrial wastewater. The SVI of 550 mL/g was reduced to 300 mL/g in four days and the SVI thereafter dropped to 200 mL/g even though the H_2O_2 addition was stopped. The H_2O_2 (as 50%) dosage was 18-136 mg/L and 2100 kg was used over the four-day period.

Hydrogen peroxide appears more costly than Cl_2 for this service, but information is limited.

FOAMING ORGANISM CONTROL

The most successful means of controlling *Nocardia* foams has been the reduction in SRT and thus washing (wasting) out the slower-growing organisms. This is not usually an alternative for BNR plants, which operate at longer SRTs. Wheeler and Rule (1980) reported that the addition of 1-5% (V/V) digester supernatant did not reduce *Nocardia* foam nor were predatory organisms effective.

Antifoams and typical water sprays in aeration basins are generally ineffective on these stable foams. *Nocardia* foam is controlled in San Jose/Santa Clara plant (Beebe, 1983) by combined sludge wastage to an *F/M* of 0.5 kg/kg·d and chlorination at 4 kg/kg MLSS·d for four to five consecutive days.

A severe *Nocardia* foaming problem at the North San Mateo Sanitation District plant, Daly City, California was made manageable by flushing the scum into flotation thickener using high volumes of water. The addition of the diluted scum to the DAF unit has improved the performance while ensuring that the foam doesn't recycle to the plant.

Richards et al. (1990), reported on successful procedures to trap and remove *Nocardia* foam and transport the scum directly to dewatering. The sludge scum mixture does not impair the dewatering of the sludge. Trapping and continuously removing the foam will reduce the amount of *Nocar-*

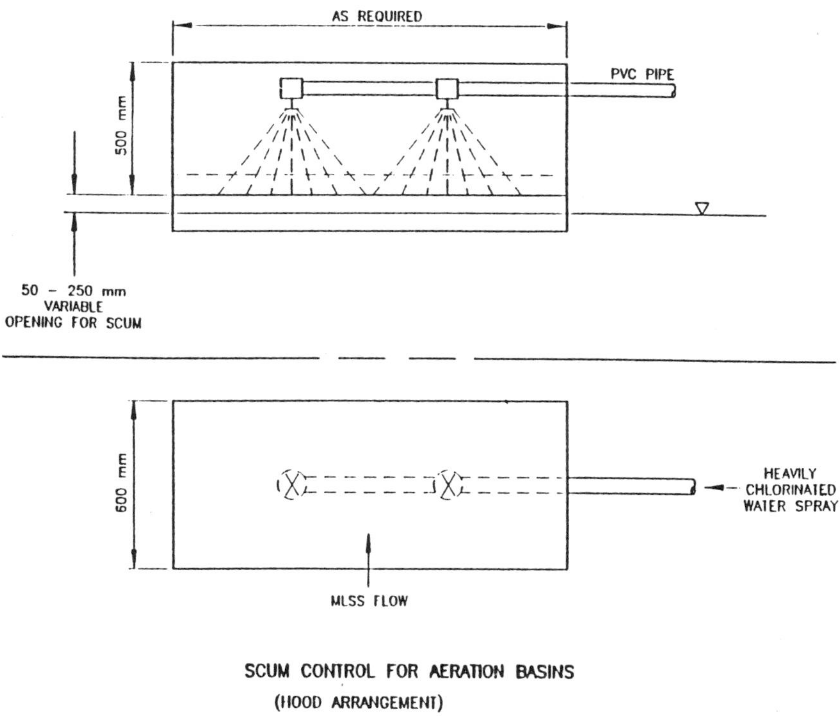

Figure 8.11 *Nocardia* control by foam chlorination at Phoenix.

dia produced by eliminating the reseeding effects of trapped *Nocardia*. The Atlanta plant has a "window" of nine to twenty days SRT in which the growth of *Nocardia* is enhanced, and above twenty days SRT *Nocardia* is not a serious problem.

The Phoenix 23rd Ave. WWTP has had a history of *Nocardia* infestation when the SRT exceeds two days. When the plant SRT was increased for AWT, *Nocardia* foam developed (Albertson and Hendricks, 1992). The return sludge was dosed with 1800 kg/d but was not successful in controlling the growth. Spray hoods, as shown in Figure 8.11, were placed across the aeration basins. Foam that entered the hood was lightly sprayed at about 10 L/min of 2000-3000 mg/L Cl_2 solution. *Nocardia* foam was eliminated in two days using 135 kg/d of Cl_2. Three subsequent foaming events were controlled in one to one and a half days. The low chlorine quantity applied directly onto the foam surface did not impair the effluent quality nor did it increase turbidity.

The anoxic selector flowsheet with an SRT of five to twelve days provided control of the *Nocardia* over 95% of the time. Except for natural traps in the biological system, the 23rd Ave. flowsheet, shown in Figure 8.8, may provide for complete foam control. However, it is not a universal solution.

Stamford, Connecticut, Ocean Co., New Jersey, and Trenton, New Jersey have also controlled *Nocardia* by spraying concentrated chlorine solution on the *Nocardia* foam. However, this procedure can only be fully effective when natural traps in the system are eliminated.

The basic strategies for control of *Nocardia* foam are as follows:

(1) Remove all natural traps within the aeration basins. Baffles should be overflow type with 2-3 cm of headloss to ensure forward flow of scum.
(2) Provide an engineered *Nocardia* trap at the end of the aeration basin or in the effluent channel where the scum can be either skimmed for disposal, or chlorinated in-situ.
(3) Do not permit the *Nocardia* scum to recycle to the wet well or to other points where it will ultimately reach the aeration basins.

Low-intensity fine bubble aeration can be used to encourage the *Nocardia* foam to float before or within the trap. When chlorine sprays are used, the design must ensure containment of the chlorine vapors. Spray nozzles or other distribution devices that produce larger droplets would be best.

REFERENCES

Albertson, O. 1987a. "The Control of Bulking Sludges: From the Early Innovators to Current Practice," *J. Wat. Poll. Con. Fed.*, 59(4):172-180, errata, *J. Wat. Poll. Con. Fed.*, 59(7):656.

Albertson, O. 1987b. 1987a errata and response to discussion by A. Busch and R. S. Sykes of Albertson 1987a, *J. Wat. Poll. Con. Fed.*, 59(12):1078-1082.

Albertson, O. 1991. ''Bulking Sludge Control–Progress, Practice and Problems,'' *Wat. Sci. Tech.*, 23(4-6):835-846.

Albertson, O., R. Scott, H. D. Stensel and R. W. Okey. 1991. ''Expansion and Upgrading of Columbus, OH WWTPs to Advanced Wastewater Treatment,'' *Proc. of 6th IAWPRC Conference on Design and Operation of Large Wastewater Treatment Plants, Prague, Czechoslovakia.*

Albertson, O. and P. Hendricks. 1992. ''Bulking and Foaming Organism Control at Phoenix, AZ WWTP,'' To be presented at the *1992 IAWPRC Biannual Conference, Washington, D.C.*

Aldern, E. and W. T. Lockett. 1914. ''Experiments on the Oxidation of Sewage without the Aid of Filters,'' *J. Wat. Poll. Con. Fed.*, 59(4):172-182.

Anon. 1969. ''Milwaukee Mystery: Unusual Operating Problem Develops,'' *Wat. and Sew. Works*, 116:213.

Bailey, D. A. and E. V. Thomas. 1975. *Journal of the Institute of Water Pollution Control*, No. 5:497-515.

Barnard, J. L. 1974. ''Cut N and P without Chemicals,'' *Water Waste Eng.*, 2(7):33.

Barnard, J. L. 1976. ''A Review of Biological Phosphorus Removal in the Activated Sludge Process,'' *Water SA*, 2(3):136-144.

Barnard, J. L. 1982. ''The Influent of Nitrogen on Phosphorus Removal in Activated Sludge Plants,'' *Water Sci. Tech.*, 14(1/2):31-45.

Baumann, M., H. Lemmer and H. Opp. 1986. ''Microbial Populations in Activated Sludge in Relation to Sewage Treatment Plant Operation,'' Abstract of the *Fourth International Symposium on Microbial Ecology, Ljubojana, Yugoslavia, September.*

Beebe, R. D. 1983. Privately Communicated. San Jose/Santa Clara Water Pollution Control Plant, San Jose, CA (In Jenkins, Richards and Daigger, 1984).

Beebe, R. D. and D. Jenkins. 1981. ''Control of Filamentous Bulking at the San Jose/Santa Clara Water Pollution Control Plant,'' Presented at the *53rd Annual Conference, Calif. WPCF, Long Beach, CA.*

Beebe, R. D., D. Jenkins and G. T. Daigger. 1982. ''Activated Sludge Bulking Control at the San Jose/Santa Clara, California Water Pollution Control Plant,'' Presented at the *55th Annual Conference of the Water Pollut. Control Fed., St. Louis, MO.*

Bhatla, M. 1967. ''Relationship of Activated Sludge Bulking to Oxygen Tension,'' *J. Wat. Poll. Con. Fed.*, 39(12):1978.

Blackall, L., A. Harbers, P. Greenfield and A. Hayward. 1991. ''Foaming in Activated Sludge Plants in Queensland, Australia and an Evaluation of Some Control Methods,'' *Wat. Res.*, 25(3):313-317.

Blackbeard, J. R. and G. A. Ekama. 1984. ''Preliminary Report on Filamentous Micro-organisms Responsible for Bulking and Foaming in Activated Sludge Plants in Southern Africa,'' Dept. of Civil Eng. Univ. of Cape Town, RSA.

Blackbeard, J. R., G. A. Ekama and G. v. R. Marais. 1986. ''Identification of Filamentous Organisms in Nutrient Removal Activated Sludge Plants in South Africa,'' *Wat. Poll. Control*, 85(1):90-100.

Blackbeard, J., D. Gabb, G. Ekama and G. Marais. 1988. ''Identification of Filamentous Organisms in Nutrient Removal Activated Sludge Plants in South Africa,'' *Water SA*, 14:29-34.

British Water Pollution Research Laboratory. 1969. Ministry of Technology, H.M.S.O., London, England.

CDM. 1989. Orange County Re-rating Study – Eastern Service Area WWTF, a report to Orange County Public Utilities Division, Orlando, FL.

Caropreso, F. E., C. W. Raleigh and J. C. Brown. 1974. ''Attack Bulking Sludge with H_2O_2 and a Microscope,'' Bull., Calif., WPCF, p. 44.

Casey, J., C. S. McDowell, M. Spector and A. J. Zupko. 1975. ''Non-bulking Activated Sludge Process,'' US Patent No. 3,864,246 (Feb. 4, 1975).

Chambers, B. 1982. ''Effect of Longitudinal Mixing and Anoxic Zones on Settleability of Activated Sludge,'' in *Bulking of Activated Sludge: Preventative and Remedial Methods*, B. Chambers and E.J. Tomlinson, eds. Chichester, England: Ellis Horwood Ltd.

Chambers, B. and E. J. Tomlinson. 1982. ''The Cost of Chemical Treatment to Control Bulking of Activated Sludge,'' in *Bulking of Activated Sludge: Preventative and Remedial Methods*. B. Chambers and E. J. Tomlinson, eds. Chichester, England: Ellis Horwood Ltd., pp. 264-270.

Christepsen, M. H. and P. Harremöes. 1977. *P. Prog. Wat. Tech.* 8(4/5):509-555, Pergamon Press.

Chudoba, J., J. Blaha and V. Madera. 1974. ''Control of Activated Sludge Filamentous Bulking – III. Effect of Sludge Loading,'' *Wat. Res.*, 8:231-237.

Chudoba, J., P. Grau and V. Ottova. 1973a. ''Control of Activated Sludge Filamentous Bulking – II. Selection of Micro-organisms by Means of a Selector,'' *Wat. Res.*, 7:1389-1406.

Chudoba, J, V. Ottova and V. Madera. 1973b. ''Control of Activated Sludge Filamentous Bulking – I. Effect of Hydraulic Regime or Degree of Mixing in an Aeration Tank,'' *Wat. Res.*, 7:1163-1182.

Chudoba, J. and J. Wanner. 1988. Discussion of Albertson (1987), submitted to JWPCF (Unpublished).

Dana, F. 1988. Plant Manager, Star Valley Cheese Coop, Thayne, WY. Private communications.

Davidson, A. 1949. Internal test reports and files provided by Schenley Distilleries of the anaerobic-oxic studies.

Davidson, A. 1959. US Patent No. 2,788,127, applied Apr. 22, 1952, issued Apr. 9, 1959.

Deakyne, C. H., M. Patel and D. Kritchen. 1983. ''Demonstration of Biological Phosphorus Removal by A/O Process at 70 MGD Patapsco Wastewater Treatment Plant,'' A report by Whitman, Requardt and Associates, Baltimore, MD.

Dick, R. I. and P. A. Vesilind. 1969. "The Sludge Volume Index – What Is It?" *J. Wat. Poll. Con. Fed.*, 41:1285-1291.

Donaldson, W. 1932a. ''Some Notes of the Operation of Sewage Treatment Works,'' *Sew. Works J.*, 4:48-59.

Donaldson, W. 1932b. ''Use of Activated Sludge Increasing,'' *Civil Eng.*, 2(3):167-169.

Edmonson, J. H. and S. R. Goodrich. 1943. *Journal of the Institute of Sewage Purification*, pp. 57-92.

Eikelboom, D. H. 1975. ''Filamentous Organisms Observed in Bulking Activated Sludge,'' *Wat. Res.*, 9:365.

Eikelboom, D. H. 1977. ''Identification of Filamentous Organisms in Bulking Activated Sludge,'' *Prog. Water Technol*, 8, 153.

Eikelboom, D. H. 1982. ''Biosorption and Prevention of Bulking Sludge by Means of a High Floc Loading,'' in *Bulking of Activated Sludge: Preventative and Remedial*

Methods. B. Chambers and E. J. Tomlinson, eds. Chichester, England: Ellis Horwood Ltd., pp. 90-110.

Eikelboom, D. H. and H. J. J. van Buijsen. 1981. *Microscopic Sludge Investigation Manual*. TNO Research Institute for Environmental Hygiene. The Netherlands.

Freese, S. W. 1938. "Aeration Tanks for Activated Sludge Plants," American Society of Civil Engineers, *J. of Environ. Eng. Div.*, 103:1620.

Gabb, D., G. A. Ekama, D. Jenkins and G. v. R. Marais. 1988. "Evaluation and Development of Specific Control Methods for Ameliorating Low F/M Bulking," Presented at the Wat. Inst. of S.A. monthly meeting, Halfway House (May).

Garber, W. F. 1972. "Phosphorus Removal by Chemical and Biological Mechanisms," *Appl. New Concepts Phys. Chem. Treat.*, 243.

Gerger, A. and C. T. Winter. 1984. "The Influence of Extended Anaerobic Retention Time on the Performance of Phoredox Nutrient Removal Plants," *Wat. Sci. Tech.*, 17(Amsterdam):81-92.

Goronszy, M. 1990. "Selector Design for High COD Wastewaters – A Case History," Presented at the *62nd Annual Conference of the WPCF, San Francisco, CA.*

Grau, P. and M. Dohanyos. 1971. "Bulking of Activated Sludge," *Proceedings of the 19th Seminar on Water Management, Prague, Czechoslovakia*, 240.

Hale, F. D. and S. R. Garver. 1983. "Viscous Bulking of Activated Sludge," Presented at the *56th Annual Conference Water Polln. Control Fedn.*, *Atlanta, GA*, Oct. 2-7.

Hascoet, M. C. and M. Florentz. 1985. "Influent of Nitrates on Biological Phosphorus Removal from Wastewater," *Water SA*, 11(1):1-8.

Heide, B. A. and A. Pasveer. 1974. *H2O*, 18:373 (August).

Hinshelwood, C. N. 1946. *The Chemical Kinetics of the Bacterial Cell*, pp. 230-250, Oxford Univ. Press, Amen House, London, England.

Inamori, U., T. Takahashi and R. Sudo. 1986. "Effects of Anaerobic Conditions on Activated Sludge Process," *J. Jap. Sew. Works Assoc.* 23(5):61-69 (In Japanese).

Jenkins, D. 1980. "The Control of Activated Sludge Bulking," Presented at the *52nd Annual California Water Polln. Control Assoc. Conference, Monterey, CA.*

Jenkins, D., M. G. Richard and J. B. Neethling. 1983. "Causes and Control of Activated Sludge Bulking," Presented at the *1983 Annual Conference of the Institute of Water Polln. Control, Southern African Branch, East London, RSA.*

Jenkins, D., M. G. Richard and G. T. Daigger. 1984. "Manual on the Causes and Control of Activated Sludge Bulking and Foaming." Water Research Commission, Pretoria, S.A.

Keinath, T. M. and E. J. Wahlberg. 1990. Private communication providing test data to O. Albertson for plots.

Kelly, H. "Bio-P Treatment Using a Fixed-film and Suspended Growth Combination: A Demonstration Project," *Proc. EED/ASCE meeting*, pp. 79-90, Orlando, FL.

Kershaw, J. H. and J. Finch. 1936. *Journal of the Institute of Sewage Purification*, pp. 397-405.

Koller, J. 1966. "Comparison of Some Activated Sludge Modifications." M.S. Thesis, Department of Water Technology, Institute of Chemical Technology, Prague, Czechoslovakia.

Koopman, B. and K. Cadee. 1983. "Prediction of Thickening Capacity Using the Dilution Sludge Volume Index," *Wat. Res.*, 17:1427.

Lechevalier, H. A. 1975. "Actinomycetes of Sewage Treatment Plants." U.S. Department of Commerce, NTIS Report No. PB 245 914.

Lechevalier, M. P. and H. A. Lechevalier. 1974. "*Nocardia amarae* sp. nov., An Actinomycete Common in Foaming Activated Sludge," *Int. J. Sys. Bact.*, 24:278-288.

Lee, S. E., B. Koopman, D. Jenkins and R. Lewis. 1982. "The Effect of Aeration Basin Configuration on Activated Sludge Bulking at Low Organic Loading," *Wat. Sci. Technol.*, 14:407-427.

Lee, S. E., B. Koopman, H. Bode and D. Jenkins. 1983. "Evaluation of Alternative Sludge Settleability Indices," *Wat. Res.*, 17:1421.

Lemmer, H. and R. M. Kroppenstedt. 1984. "Chemotaxonomy and Physiology of Some Actinomycetes Isolated from Scumming Activated Sludge," *Sys. Appl. Microbiol*, 5:124-135.

Lewis, R. 1979. Privately communicated. U.S.E.P.A., AW Briedenbach Environmental Research Center, Cincinnati, OH (In Jenkins, et al., 1984).

Marsden, M. G. and G. v. R. Marais. 1977. "The Role of the Primary Anoxic Reactor in Denitrification and Biological Phosphorus Removal." Research Report W 19, Dept. of Civil Eng., University of Cape Town.

Matsche, N. F. 1982. "Control of Bulking Sludge–Practical Experience in Austria," *Water Sci. and Tech.*, 14:311.

Matsche, J. P., J. P. Rideau and J. Serns. 1975. "La Traitement Par Boues Activees des Eaux Residuaires de Fabrication de Marrons Glaces," *Ind. Aliment. Agric.* 92:331-342.

Milbury, W. F. et al., 1965. "Compartmentalization of Aeration Tanks," American Society of Civil Engineers, *J. Environ. Eng. Div.*, SA3:45.

Nicolls, H. 1975. "Full Scale Experimentation of the New Johannesburg Extended Aeration Plants," *Water SA*, 1(3):121-132.

Palm, J. C., D. Jenkins and D. S. Parker. 1980. "Relationship Between Organic Loading, Dissolved Oxygen Concentration and Sludge Settleability in the Completely-mixed Activated Sludge Process," *J. Wat. Poll. Con. Fed.*, 52:2484-2506.

Pasveer, A. 1969. "A Case of Filamentous Activated Sludge," *J. Wat. Poll. Con. Fed.*, 41(7):1340.

Pipes, W. O. 1969. "Types of Activated Sludge Which Separate Poorly," *J. Wat. Poll. Con. Fed.*, 41:714-724.

Pipes, W. O. 1978. "Actinomycete Scum Formation in Activated Sludge Processes," *J. Wat. Poll. Con. Fed.*, 50:628-634.

Pitman, A. 1984. "Settling of Nutrient Removal Sludges," Presented at the *12th Biennial Conf. IAWPRC, Amsterdam, Holland.*

Pitman, A., S. Venter and H. Nicholls. 1983. "Practical Experience with Biological Phosphorus Removal Plants in Johannesburg," *Wat. Sci. Tech.*, 15:233-259.

Price, G. J. 1982. "Use of an Anoxic Zone to Improve Activated Sludge Settleability," in *Bulking of Activated Sludge: Preventive and Remedial Methods*, B. Chambers and E. J. Tomlinson, eds. Chichester, England: Ellis Horwood Ltd., pp. 259-260.

Rachwal, A., D. Johnstone, M. Hanbury and D. Critchard. 1982. "The Application of Settleability Tests for the Control of Activated Sludge Plants," in *Bulking of Activated Sludge: Preventive and Remedial Methods*, B. Chambers and E. J. Tomlinson, eds., Chichester, England: Ellis Horwood Ltd., pp. 224-244.

Rensink, J. H. 1974. "New Approach to Preventing Bulking Sludge," *J. Wat. Poll. Con. Fed.*, 46(8):1888-1901.

Rensink, J. H., H. J. G. W. Donker and T. S. J. Ijwema. 1982. "The Influence of Feed Pattern on Sludge Bulking," Chapter 9 in *Bulking of Activated Sludge: Preventative and Remedial Methods*, B. Chambers and E. J. Tomlinson, eds. Chichester, England: Ellis Horwood Ltd., pp. 147-165.

Resnick, J. 1988. Supt., Davenport WWTP, Annual reports for 1987, 1988 and private communications.

Richard, M. G., D. Jenkins, O. Hao and G. Shimizu. 1982a. *The Isolation and Characterization of Filamentous Micro-organisms from Activated Sludge Bulking*. Report No. 81-2, Sanitary Engineering and Environmental Health Research Laboratory, Univ. of Calif, Berkeley, CA.

Richard, M. G., G. Shimizu, D. Jenkins, T. Williams and R. F. Unz. 1983. "Isolation and Characterization of *Thiothrix* and *Thiothrix*-Like Filamentous Organisms from Bulking Activated Sludge," Presented at the *1983 Annual Meeting of the Amer. Soc. for Microbiology, New Orleans, LA* (*Abs. Ann. Meet.*, *Q60*, 270).

Richard, M. G., G. P. Shimizu and D. Jenkins. 1984. "The Growth Physiology of the Filamentous Organism Type 021N and Its Significance to Activated Sludge Bulking," Presented at the *57th Annual Conference, Water Polln. Control Fedn, New Orleans, LA*.

Richards, T., P. Nungesser and C. Jones. 1989. "Solution of *Nocardia* Foaming Problems," *JWPCF*, 62(7):915-919

Richards, T., P. Nungesser and C. Jones. 1990. "Solution of *Nocardia* Foaming Problems," *JWPCF*, 62(7):915-919.

Ryder, R. A. 1964. "Dissolved Oxygen Control in Activated Sludge," *Proceedings of the 24th International Waste Conference*, Purdue University, Lafayette, Indiana, Engineering Extension Series, 135.

Scalf, M. R., et al., 1969. "Phosphate Removal at Baltimore, Maryland," American Society of Civil Engineers, *J. Environ. Eng. Div.*, SA5:817.

Sezgin, M., D. Jenkins and D. Parker. 1978. "A Unified Theory of Filamentous Sludge Settling," *Prog. Water Techn.*, 12:97.

Sezgin, M., M. P. Lechevalier and P. R. Karr. 1988. "Isolation and Identification of Actinomycetes Present in Activated Sludge Scum," *Proc. of IAWPRC Conf. on Wat. and Wastewat. Microbiol.* pp. 41-1-41-7. Newport Beach, Calif.

Shao, Y., and D. Jenkins. 1989. "The Use of Anoxic Selectors for the Control of Low F/M Activated Sludge Bulking," *Wat. Sci. Tech.*, 21(Brighton):609-619.

Simpkins, M.J. and A. R. McLaren. 1978. "Consistent Biological Phosphate and Nitrate Removal in an Activated Sludge Plant," *Prog. Wat. Tech.*, 10(5/6):433-442.

Skerman, V. B. D. 1967. *A Guide to the Identification of Genera of Bacteria*, 2nd edition. Baltimore: Williams & Wilkins.

Spector, M. L. 1977. "Production of Non-bulking Activated Sludge." U.S. Patent No. 4,056,465 and the supporting documents in the PTO file wrapper.

Strom, P. F. and D. Jenkins. 1984. "Identification and Significance of Filamentous Micro-organisms in Activated Sludge," *J. Water Pollut. Control Fed.*, 56:52.

Thomas, H., Jr., and J. E. McKee. 1944. "Longitudinal Mixing in Aeration Tanks," *Sew. Works J.*, 16(1):42.

Tomlinson, E. 1976. "Bulking—A Survey of Activated-Sludge Plants." Technical Report TR35, Water Res. Center, pp. 1-32.

Tomlinson, E. J. and B. Chambers. 1979. "The Use of Anoxic Mixing Zones to Control the Settleability of Activated Sludge," Water Research Centre, Technical Report, TR116, April, pp. 1-24.

Tracy, K. D., M. E. Adams and A. Flammino. 1986. "Control of Activated Sludge Settling Characteristics with Anaerobic Selectors," Presented at the *59th Annual*

Conference Water Pollution Control Federation, Los Angeles, California, October.

Tsukamura, M. 1981a. "Differentiation Between the Genera *Mycobacterium*, *Rhodococcus* and *Nocardia* by Susceptibility to 5-fluorouracil," *J. Gen. Microbiol.* 125:205-208.

Tsukamura, M. 1981b. "Test for Susceptibility to Mitomycin C as Aids for Differentiating the Genus *Rhodococcus* from the Genus *Nocardia*, and for Differentiating *Mycobacterium fortuitum* and *Mycobacterium chelonei* from Other Rapidly Growing Mycobacteria," *Microbiol. Immun.*, 25:1197-1199.

Vacker, D. C., et al. 1967. "Phosphate Removal Through Municipal Wastewater Treatment at San Antonio, Texas," *J. Wat. Poll. Con. Fed.*, 39:750.

Wagner, F. 1982. "Study of the Causes and Prevention of Sludge Bulking in Germany." Chapter 2 in *Bulking of Activated Sludge: Preventative and Remedial Methods*, B. Chambers and E.J. Tomlinson, eds. Chichester, England: Ellis Horwood Ltd.

Wanner, J., J. Chudoba, K. Kucian and L. Proske. 1987a. "Control of Activated Sludge Filamentous Bulking VII. Effect of Anoxic Conditions," *Wat. Res.*, 21:1447-1551.

Wanner, J. and P. Grau. 1987. "Filamentous Bulking in Nutrient Removal Activated Sludge Systems," Proc. IAWPRC Workshop *The Design and Operation of Large Wastewater Treatment Plants*, p. 1, Budapest, Hungary.

Wanner, J., V. Ottava and P. Grau. 1987b. "Effect of an Anaerobic Zone on Settleability of Activated Sludge," *Proc. IAWPRC Conf., Rome, Italy*, pp. 155-164.

Wanner, J. and P. Grau. 1989. "Identification of Filamentous Microorganisms from Activated Sludge: A Compromise Between Wishes, Needs and Possibilities," *Wat. Res.* 23(7):883-889.

Watanabe, A., A. Miya and Y. Matsuo. 1984. "Laboratory Scale Study on Biological Phosphorus Removal Using Synthetic Wastewater." Newsletter of the IAWPRC Study Group on Phosphate Removal in Biological Systems, 2(1):40-43.

Wells, W. F. 1969. "Differences in Phosphorus Uptake Rates Exhibited by Activated Sludge," *J. Wat. Poll. Con. Fed.*, 41(5):765.

Wendell, D. 1990. Process supervisor, private communication regarding non-filamentous bulking and chlorination, 23rd Avenue WWTP, Phoenix, AZ.

Westgarth, W. C., D. A. Sulzer and D. A. Okun. 1964. "Anaerobiosis in the Activated Sludge Process," paper presented at the *2nd IAWPRC, Tokyo.*

Wheeler, D. R. and A. M. Rule. 1980. "The Role of *Nocardia* in the Foaming of Activated Sludge: Laboratory Studies," Presented at *Georgia Water and Polln. Control Assoc. Annual Meeting, Savannah, GA.*

Witherow, J. L. 1969. "Phosphate Removal by Activated Sludge," *Proceedings of the 24th International Waste Conference*, Purdue University, Engineering Extension Series, Lafayette, Indiana, 135, 1169.

Yamamoto, R. and S. Matsui. 1988. "Measuring the Growth Rate of Filamentous Bacteria and Influence of the Bacteria on Settleability, Thickening Properties and Floc Configuration of the Activated Sludge." *Adv. in Wat. Poll. Control, Wat. Poll. Control in Asia IAWPRC*, pp. 329-339.

Yamamoto, R. and S. Matsui. 1989. "Influence of Contact Loading on Polysacharide Storage and Settleability of Activated Sludge," *Wat. Sci. Tech.*, 21:1639-1642.

Yamamoto, R.I., T. Komori and S. Matsui. 1991. "Filamentous Bulking and Hindrance of Phosphorus Removal Due to Sulfate Reduction in Activated Sludge," *Wat. Sci. Tech.*, 23:927-935.

CHAPTER 9

Sludge Handling and Processing

DEWATERING AND DISPOSAL

IN BNR processes the biologically removed phosphates are taken up in the bacterial cells. Under anaerobic conditions, the phosphates are released to the liquid phase and after a subsequent dewatering stage the liquid may have to be treated for the removal of phosphates before recycling it to the plant. It is more advantageous to dewater the waste BNR sludge as fast as possible before any release of phosphates can take place. Gravity thickening is not advisable due to the release of phosphates.

SLUDGE DEWATERING

A diagram of possible sludge thickening and dewatering processes is shown in Figure 9.1. Sludge thickening, and in some cases dewatering, would have advantages for most subsequent sludge treatment processes. This also applies to sludge digestion, which will be discussed separately.

The advantages of thickening primary sludge have been discussed before in terms of advantages for the BNR processes. The thickener can be used as a fermenter and the fatty acids elutriated for use as feedstock for the organisms responsible for removing the phosphates as well as for denitrification. For this reason it is recommended that the thickening of primary sludge and waste activated sludge be separated as far as possible. Even when the sludge is being dewatered for heat treatment or composting, dewatering of the primary sludge will yield a liquid that would be very beneficial to the BNR process, provided the ammonia return load is balanced and the polymer is not toxic.

Belt pressing or centrifugation will produce a filtrate or centrate that can be returned to the anaerobic contact zone. However, there is a note of caution. In the first place, some polymers inhibit nitrification. In the second place, after fermentation the ammonia content of the centrate may be high and can cause shock loading conditions if not balanced. From both points

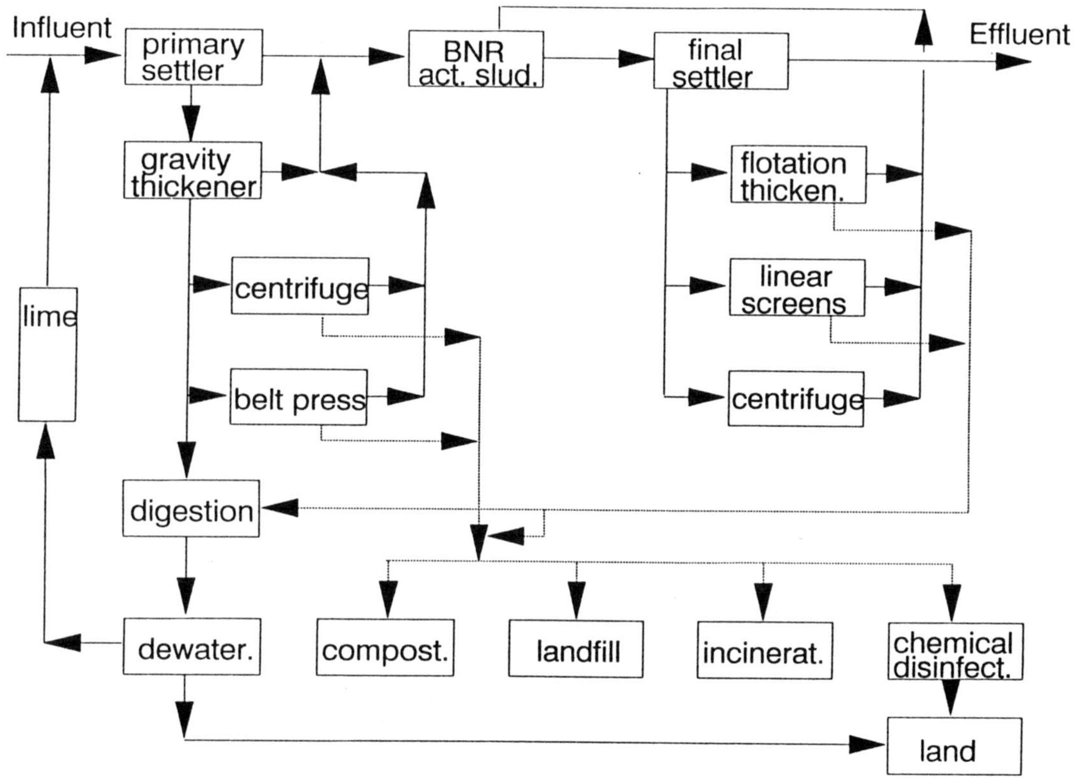

Figure 9.1 Sludge treatment alternatives for BNR WWTPs.

of view, the centrate or filtrate should be discharged back to the plant over an extended period, preferably during the low load periods.

The waste activated sludge should be thickened and dewatered in such a way that the phosphates will be concentrated in the smallest possible volume of liquid. Flotation thickening is ideal for keeping the sludge fresh and for reducing the volume without releasing phosphates to the liquid. The underflow from the thickener can be returned to the aeration basin.

Flotation thickening can be used to thicken the sludge up to 5% without the use of polymers and without the release of phosphates, which allows the underflow to be returned to the aeration basin. When using flotation thickening, the mixed liquor can be wasted directly from the aeration basin, allowing for volumetric control of the MCRT. An air to solids ratio of at least 0.03 must be provided to achieve 4 to 5% solids without the addition of polymers.

At 5% solids, the sludge scraped off the tanks is like a thick mousse and does not flow readily. Positive displacement pumps are required for moving the sludge. After thickening, the sludge may be further dewatered by belt press or centrifuge with the addition of polymers. If done immediately, most of the phosphates will be retained in the sludge.

An alternative to flotation thickening is the linear screen normally used as the first section of a belt press. This requires the use of polymers but the sludge can be thickened to about 6-7% in a very short time, which will prevent release of phosphates to the liquid stage.

After thickening or dewatering the sludge may be treated by any of the following methods:

- composting
- digestion
- landfill
- incineration
- heat treatment
- disinfection by high alkaline treatment

COMPOSTING

Composting normally requires undigested sludge dewatered to the highest economical density. This may be achieved with either centrifuges or belt presses, and separate processing of primary sludge and waste activated sludge for the reasons mentioned above. Primary sludge can be dewatered to about 22% solids and waste activated sludge to about 16%. The sludge can then be composted with the use of filler material such as sawdust or even prairie grass. No phosphates will be released. The nutrients will enhance the final product.

DIGESTION

Anaerobic or aerobic digestion of the sludge will lead to the release of phosphates from the microbial cells. In some instances, as will be discussed later in this chapter, phosphates may be precipitated during anaerobic digestion. If not, the liquid from the sludge dewatering process will contain a high concentration of phosphates, which could most economically be removed by lime treatment if the liquid is to be returned to the main stream. Alternatively, the liquid could be irrigated on land.

After anaerobic digestion, which will be discussed below, the effluent contains a high concentration of ammonia, which adds alkalinity. This will require a high lime dosage to first neutralize the alkalinity before the lime will be available for phosphate precipitation. The lime requirement can be reduced substantially by aeration, which will also make the final effluent more amenable for land treatment or for recycling to the aeration basin.

LANDFILL

Phosphate-rich secondary sludge can be landfilled after dewatering. It is generally expected that the phosphates will leach out under the anaerobic conditions that develop in landfills, but more likely the phosphates will be bound by the heavy metal content of the leachate.

When treating the dewatered sludge with lime or lime kiln ash, the sludge can be turned into a friable material for dumping but will produce odors if not covered properly. Generally, there should be few problems from landfilling of the dewatered BNR sludge.

INCINERATION

When incineration of the sludge is considered, separation of the primary and secondary sludges will again have merit. Plate and frame dewatering of the primary sludge will produce a dry sludge, which would be advantageous for incineration, while the filtrate can be fed to the anaerobic zone to enhance phosphate removal. The secondary sludge is best belt pressed to about 18% and the filtrate returned to the aeration basin.

HEAT TREATMENT

Heat treatment may return unwanted compounds and phosphates to the liquid stream. BNR processes are usually used to obtain high-quality secondary effluents. The return of highly contaminated return streams containing nondegradable compounds would tend to degrade the high-quality effluents and make it very difficult to achieve strict standards. The high nondegradable TKN would in most cases increase the effluent total N to above the required standards.

The toxic and other effects of return streams from heat treatment have not been fully investigated and a thorough study of such effects should be made before any application.

HIGH ALKALINE TREATMENT

There are several patented processes for disinfecting sludge by the application of either lime, cement, or ammonia. The addition of either lime or cement tends to turn the sludge into a friable soil that can be handled as a solid. This also has the effect of binding up the phosphates for slow release to plants when disposed of on land.

A more recent process called the Active Sludge Pasteurization process (ASP) consists of dewatering the sludge to about 15% solids, adding ammonia which raises the pH to 11.5 and the temperature by about 20°C, then adding phosphoric acid to neutralize the pH and push the temperature up further to about 75°C. The heat is recovered through heat exchangers. The sludge can then be applied as a fertilizer in wet form. In this way, all the phosphates removed biologically will be available to the plants.

IMPACT OF SOLIDS HANDLING UNITS ON BNR PERFORMANCE

Biological phosphorus and nitrogen removal have been implemented at several facilities in the Chesapeake Bay drainage basin to reverse the observed trend of eutrophication in the Bay. The quantities of phosphorus and nitrogen recycle from solids handling units were examined at several WWTPs after they reached steady state following the modifications. It was observed that the recycles exerted additional loadings on the secondary treatment system. The magnitudes of the nutrient recycles were a function of the type of solids handling unit. However, if adequate biological removal capacity was available at the treatment facility, the secondary effluent nitrogen and phosphorus levels were unaffected. The impact on the performance of the system may be categorized in terms of the type of the solids handling unit and the capacity of the secondary treatment system.

This section presents data from the modified plants and discusses the magnitudes of phosphorus and nitrogen recycles from gravity thickeners, dissolved air flotation thickeners, aerobic and anaerobic digesters, belt filter presses, and primary clarifiers processing waste activated sludge from BNR processes. Sludge production, nutrient release mechanisms, and forms in which nutrients are removed with processed sludge are also discussed.

DESCRIPTION OF WWTPS: PROCESSES AND SOLIDS HANDLING UNITS

Bowie, Maryland WWTP

The Bowie WWTP was modified to operate with two oxidation ditches in series preceded by an anaerobic cell, as shown in Figure 9.2. The anaerobic cell received a variable fraction of the influent flow during the experimental phase, usually from 30 to 40 percent of the total influent flow. The plant was rated at a capacity of 11.36 ML/d, and was receiving 7.57 ML/d of flow during the period of modification and performance evaluation. Since modification, the plant has achieved excellent biological nitrogen and phosphorus removal without the use of supplemental chemicals. The effluent concentrations have monthly average values of 0.6 mg/L total phosphorus and 4 mg/L total nitrogen on a year-round basis.

The mixed liquor recycle flow from oxidation ditch 1 to the anaerobic cell was paced with the influent flow. This flexibility allowed the plant to increase the actual anaerobic cell detention time in winter to enhance conversion of COD to short-chain fatty acids (SCFAs) and maintain BPR (Sen et al., 1991a and b). The raw influent that entered oxidation ditch

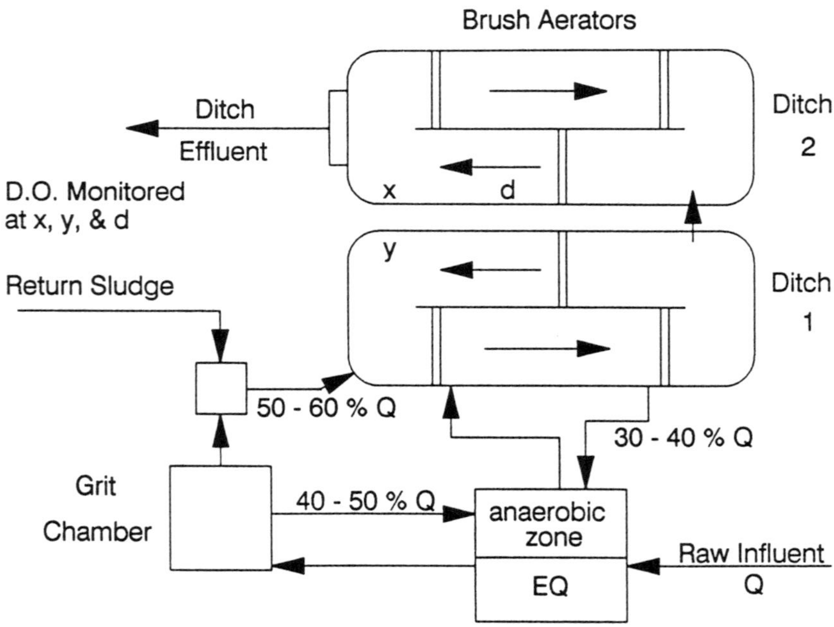

Figure 9.2 Schematic of the VT2 BNR process, Bowie, Maryland WWTP.

TABLE 9.1. Phosphorus Distribution at Bowie BNR WWTP.

Location	Flow ML/d	Total Phosphorus Concen. mg/L	Total Phosphorus Loading kg/d	Total Phosphorus Removal kg/d	% of P Removal Recycled
Raw Influent	8.71	6.0	220		
Primary Effluent	8.71	6.9	290		
Final Effluent	7.57	0.6	22		
Waste Act. Sludge	0.19	293	268	268	
Grav. Thick Recycle	0.11	20	11		4
Belt Press Recycle	1.02	12	59		22

1 directly helped increase denitrification rates and encouraged growth of non-polyP heterotrophs (Dold, 1991; Sen and Randall, 1991).

Phosphorus uptake, nitrification, and denitrification were controlled by monitoring secondary effluent alkalinity and turbidity and then cycling four or five brush aerators to control the oxygen inputs (Sen et al., 1990a and b). When the correct amount of oxygen was transferred into the mixed liquor, the secondary effluent turbidity was low and the alkalinity was within the desired operating range of 70 to 75 mg/L, which maintained the effluent pH above 6.6. Excellent phosphorus removal was observed when the correct amount of oxygen was transferred and nitrates were excluded from the anaerobic cell. However, when insufficient oxygen was supplied, the resulting conditions resulted in both reduced phosphorus uptake and floc fragmentation, which increased the effluent turbidity.

Plant performance data are shown in Figure 9.3. The phosphorus loadings and concentrations in the influent and recycles are listed in Table 9.1. The phosphorus content of the waste sludge averaged 4.5% on a total suspended solids (TSS) basis and 6.0% on a volatile suspended solids (VSS) basis at an operating MCRT of twenty days and average temperature conditions. Higher values were observed when phosphorus recycle from the solids handling units were higher, or the sludge production was lower. The nitrogen content of the waste sludge averaged 7% on a TSS basis and 9.5% on a VSS basis during the course of the study.

Solids handling systems at the Bowie WWTP were comprised of two gravity thickeners and a pair of belt filter presses. The thickeners were operated in series. The first thickener was utilized to thicken the waste sludge from 0.5% to 0.9% over a period of twenty-four hours. Dilution water was not added to the thickener to avoid excessive recycle of phosphorus with the supernatant. There was some phosphorus release under anaerobic conditions in the thickener. However, the release was slow (kinetically) because of the absence of an external source of substrate. The

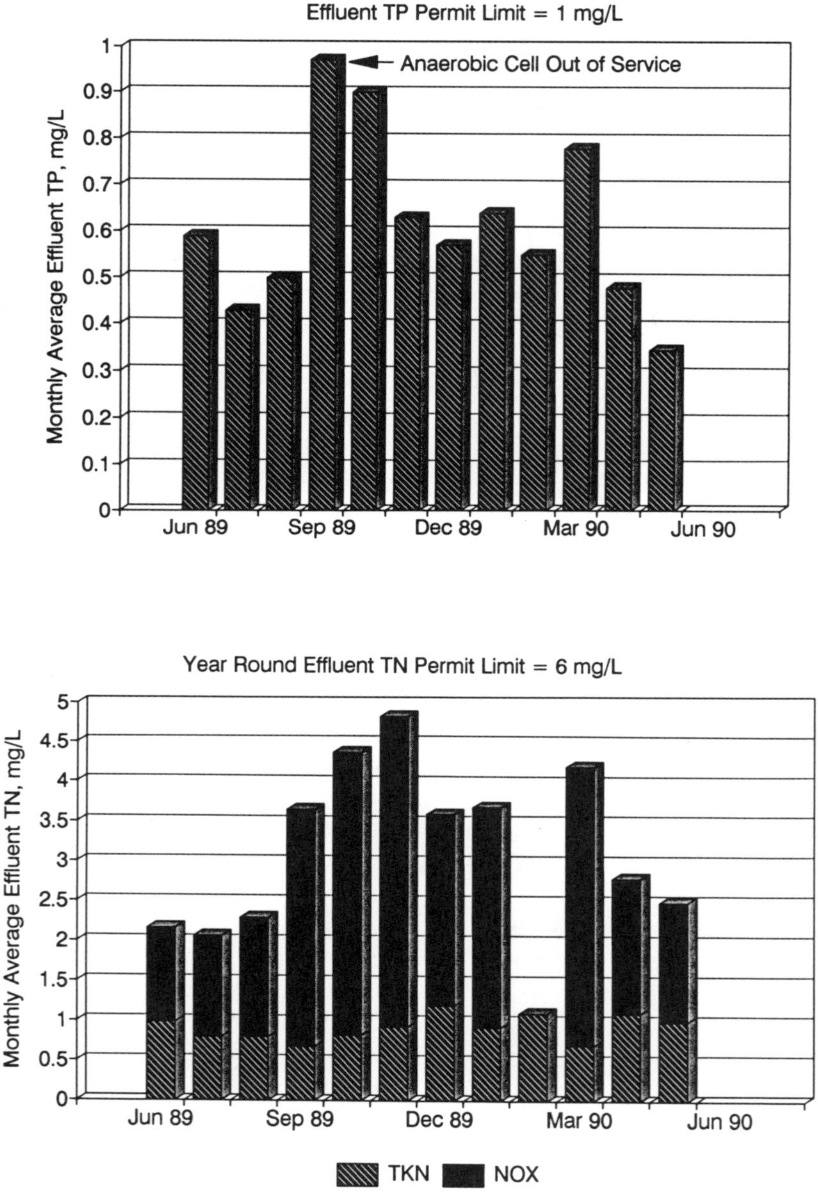

Figure 9.3 Bowie, Maryland WWTP biological nutrient removal performance.

quantity of phosphorus released into solution at the bottom of the thickener was 10 to 15% of the phosphorus content of the sludge but only a fraction of it escaped with the supernatant. Precautions were taken to maintain the sludge blanket levels at least one meter below the overflow weir of the first gravity thickener.

The second gravity thickener was operated as a holding tank and sludge from this gravity thickener was dewatered by the belt filter presses. The dewatered sludge contained 15 to 20% total solids. Mass balance on the washings from the filter presses indicated that as much as 22% of the phosphorus removed in the secondary treatment process was recycled to the headworks (Table 9.1). This could have been reduced slightly by installing newer filter presses to improve the solids removal efficiencies. However, at least 15% of the phosphorus removed in the biomass would have been recycled because it was released into solution in the thickeners.

The sum total of phosphorus recycle from the solids handling units averaged 26% of the phosphorus removed across the secondary treatment system (Table 9.1).

Old Maryland City WWTP

Modifications were implemented for high rate biological nitrogen and phosphorus removal at the Old Maryland City, Maryland WWTP, which was treating a flow of 2.65 ML/d. The activated sludge basin had a nominal hydraulic retention time (HRT) of four hours at average daily flow. A schematic of the process is shown in Figure 9.4.

Waste activated sludge was pumped into the raw influent stream to settle with the primary sludge, which gave the plant a single point of primary and waste activated sludge withdrawal. However, influent flow preaeration was minimized to promote formation of SCFAs by fermenters, and this resulted in anaerobic conditions in the primary clarifier. The high-phosphate activated sludge entering the clarifier utilized the readily available COD and released phosphate to solution. The net effect was that 43% of the phosphorus removed in the process by the waste activated sludge was released into solution in the primary clarifier (Table 9.2). This had a negative impact on the performance of the secondary treatment system.

Phosphorus levels increased from 5.3 mg/L in the raw influent to 6.5 mg/L in the primary clarifier effluent (Table 9.2). The increase occurred in spite of the removal of 2.64 kg/d of influent total phosphorus during primary sludge settling. The removal by settling was measured during operation of the plant as a conventional activated sludge process, i.e., before modification for BPR. The net result of the releases and removals was an average effluent TP concentration of 1.5 mg/L. The mass balances for the plant are given in Table 9.2 and illustrated in Figure 9.4.

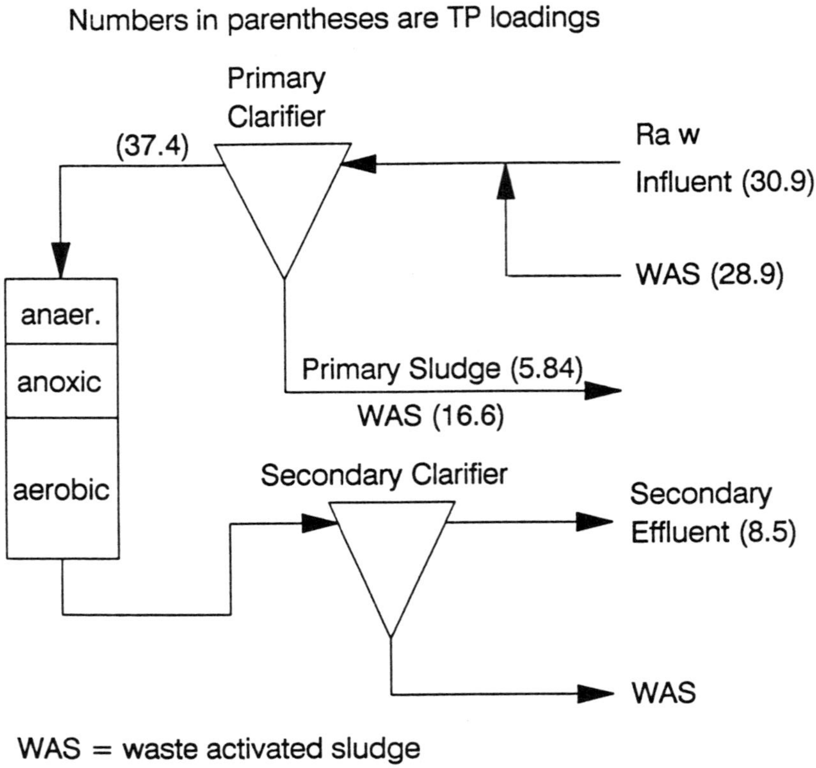

Figure 9.4 Process schematic of the Maryland City, Maryland BNR WWTP.

The phosphorus content of the waste sludge averaged 5% on a TSS basis and 6.5% on a VSS basis. The effluent soluble phosphorus averaged 1 mg/L but was periodically disturbed by high loadings. Lack of adequate separation of the aerobic, anoxic, and anaerobic zones was also detrimental to the performance of the BPR process.

Nitrogen recycle from waste activated sludge settling in the primary clarifier was less than 10%. The levels were this low because nitrogen is not released from waste sludge except through decay and hydrolysis of particulate forms (Wentzel et al., 1991), which did not occur extensively in the clarifier.

York River, Virginia WWTP

The York River WWTP was initially modified for BPR using the two-stage anaerobic-aerobic process, and was then modified to include nitrogen removal, first as the A2/O process and then as the VIP process. The plant was operated for BPR for more than three years, including periods both before and after implementation of a phosphate detergent ban. A plant schematic that includes waste sludge processing is shown in Figure 9.5. The plant had dissolved air flotation thickeners that were used to thicken the waste activated sludge. The thickened activated sludge was then mixed with primary sludge that had been thickened by gravity, and the mixture was pumped to the first digester of a two-stage anaerobic digestion system. The chemistry of phosphorus release and precipitation in the anaerobic digesters, and the magnitude of the phosphorus recycles to the process, were studied and modeled mathematically (Sen et al., 1987).

Prior to the detergent ban, the COD/TP ratio in the raw influent at the York River WWTP averaged 27/1. The corresponding ratio in the primary effluent, which included phosphorus recycle from the solids handling units, was 20/1. The primary effluent phosphorus concentration increased to 12 mg/L as a result of the phosphorus-rich recycle stream.

A mass balance conducted on the treatment system indicated that 30% of the phosphorus removed in the secondary treatment system was recycled to the headworks from the solids handling units. Because of the additional loading contributed by the recycle, biological phosphorus removal was COD limited prior to the phosphate detergent ban, i.e., phosphorus could not be removed to less than 1.0 mg/L because of insufficient COD. The system was not COD limited during steady state periods after the phosphate detergent ban because of the 35% decrease in raw influent phosphorus loadings.

TABLE 9.2. Phosphorus Distribution at the Maryland City BNR WWTP.

Location	Flow ML/d	Total Phosphorus: Concen. mg/L	Total Phosphorus: Loading kg/d	Total Phosphorus: Removal kg/d	% of P Removal Recycled
Raw Influent	2.65	5.3	68.0		
Primary Influent			131.6[1]		
Primary Settled Solids			12.8[2]		
Primary Effluent	2.61	6.5	82.3		
Final Effluent	2.57	1.5	18.7		
Waste Act. Sludge	0.04		63.6	63.6	
Primary Settled WAS			36.4[3]		
WAS Release in Pr.			27.2[4]		42.5

[1]Raw Influent TP + WAS TP.
[2]Amount of Raw Influent TP settled in primary clarifier was determined from data collected during conventional operation, i.e., without BPR.
[3]Raw Inf. TP + WAS TP − Primary Settled Solids TP − Primary Eff. TP.
[4]WAS TP − Primary Settled WAS TP.

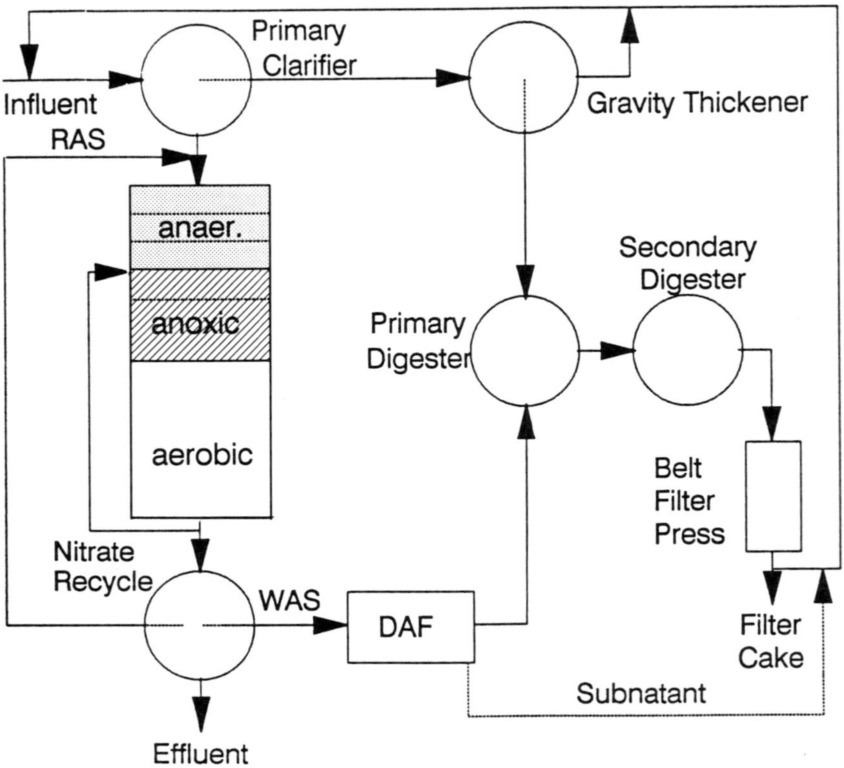

Figure 9.5 Schematic of York River, Virginia BNR WWTP.

The quantities of phosphorus removed and released in different units of the treatment system prior to the phosphate detergent ban are presented in Figure 9.6. The thickened sludge mixture entering the primary digester contained 1100 mg/L of total phosphorus. Of this amount, 750 mg/L was phosphorus removed in the secondary treatment system and stored as polyP in the bacterial cells.

The actual quantity of phosphorus solubilized in a digester processing BPR sludge can be calculated by measuring the quantity of potassium released simultaneously with phosphorus during the solubilization of polyP. The ratios of potassium and magnesium released compared to phosphorus tend to be wastewater specific (Comeau, 1985; Brannan, 1986), but can be determined for a specific wastewater and used for mass balance analysis. Because potassium is not precipitated at the ion concentrations present in the digesters, the observed increase in soluble potassium concentration during digestion is representative of increased ion releases resulting from

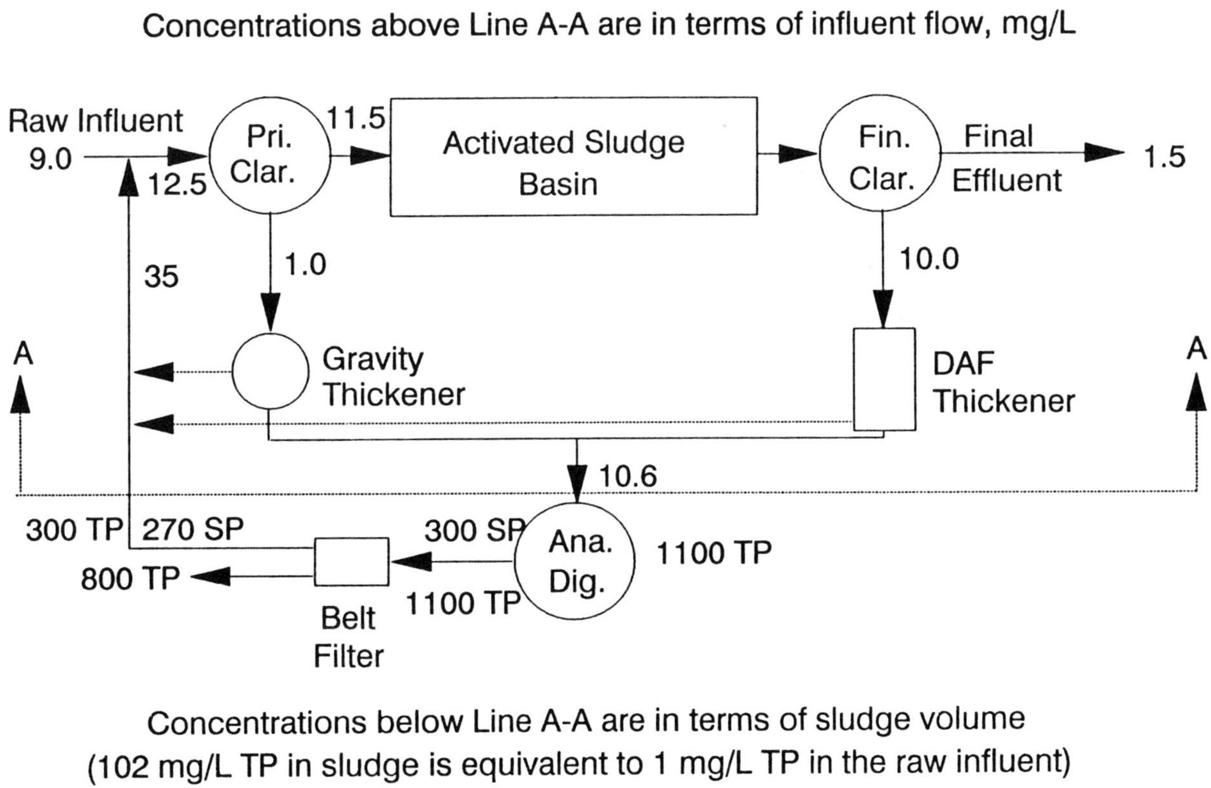

Figure 9.6 Phosphorus distribution throughout the York River, Virginia WWTP.

BPR (see Table 9.3). The average molar ratio of potassium to phosphorus release observed under anaerobic conditions at the York River WWTP was 0.24. Therefore, the quantity of P released into solution because of the breakdown of polyP can be calculated as follows:

$$\text{P released} = (230 - 55) * 31.1/(39.9 * 0.24) = 575 \text{ mg/L}$$

It was observed that a substantial quantity of the phosphorus that was solubilized was reprecipitated. The precipitation reactions were essentially completed in the primary digester with no measurable precipitate formation in the secondary digester. The soluble P levels in the secondary digester underflow during steady state (nine months after startup) increased from 50 mg/L to 300 mg/L (net increase of 250 mg/L, Table 9.3). Therefore, the quantity of phosphorus reprecipitated was (575 − 250) = 325 mg/L.

The three principal forms of phosphorus precipitates were struvite, i.e., magnesium-ammonium-phosphate ($MgNH_4PO_4$), brushite ($CaHPO_4{\cdot}2H_2$), and vivianite ($Fe_2(PO_4)_3{\cdot}8H_2O$). Batch studies were conducted in the laboratory with solutions in which concentrations similar to those observed in the digester were maintained (Sen, 1987). These studies confirmed the rate of precipitate formation and the time required to approach equilibrium. The precipitation of 140 mg/L of phosphorus with magnesium as struvite was followed by the precipitation of 60 mg/L of phosphorus as brushite. The precipitates were confirmed by measurements of phosphorus and cations both in solution and in the solid phase across the digesters, prior to and after modification of the plant for BPR (see Table 9.3), and by X-ray diffraction analysis identification of struvite in the digested sludge (Sen, 1987). The calculations for magnesium and calcium are shown as follows:

TABLE 9.3. Average Concentrations in the Primary Anaerobic Digester Sludge at the York River BNR WWTP.

	Sludge Concentrations, mg/L	
Parameter	Before BPR Operation	After 9 Months of BPR Operation
Total Phosphorus	350	1100
Soluble Phosphorus	50	300
Total Potassium	60	300
Soluble Potassium	55	230
Soluble Magnesium	10	10
Soluble Calcium	100	30
Ammonium	316	390

Mg/P molar ratio for polyP phosphorus release
= 0.26

Mg solubilized with polyP release in primary digester
= 0.26(575/31.1) * 24 = 115 mg/L

Net increase in soluble magnesium as a result of BPR (primary digester effluent from measurement at the WWTP)
= 0 mg/L (Table 9.3)

Quantity of magnesium precipitated
= 115 − 0 = 115 mg/L

Quantity of magnesium incorporated in calcium precipitates (determined by batch studies, Sen, 1987)
= 5 mg/L

Quantity of magnesium incorporated in struvite
= (115 − 5)
= 110 mg/L

Quantity of phosphorus incorporated in struvite
= 110 * 31.1/24
= 140 mg/L

Ca/P molar ratio for polyP phosphorus release at York River
= 0.0 mg/L

Calcium solubilized with polyP phosphorus in primary digester
= 0.0 mg/L

Net increase in soluble calcium as a result of BPR (primary digester effluent measurements at the WWTP)
= −70 mg/L (Table 9.3)

Quantity of calcium precipitated
= 0 − (−70)
= 70 mg/L

Substitution of magnesium for calcium in calcium precipitates (determined by batch studies, Sen, 1987)
= 5 mg/L

Moles of calcium + magnesium in brushite
= (70/40 + 5/24)
= 1.96

Quantity of phosphorus incorporated in brushite
= 1.96 * 31.1/1
= 60 mg/L

Although direct confirmation was not obtained, the measured concentrations of iron and phosphorus and the saturation constants for different iron and phosphorus compounds indicated that the most logical compound that could have precipitated another 125 mg/L of phosphorus was vivianite.

The phosphorus that remained in solution in the digesters was only 27% of the total phosphorus entering with the sludge mixture. The soluble phosphorus was then recycled to the primary train with the filtrate from the belt filter press. Another 5% of the removed phosphorus was recycled with the subnatant from the DAF unit processing waste activated sludge.

After implementation of the phosphate detergent ban, the quantity of phosphorus in the influent decreased from 9 mg/L to 6 mg/L. Because the concentration of COD in the wastewater was unchanged while the quantity of phosphorus that had to be removed decreased, the phosphorus content of the waste sludge decreased from 12% to 9% on a VSS basis. The decrease in polyP content of the sludge reduced the magnitudes of phosphorus release and reprecipitation in the digester. Soluble phosphorus levels in the secondary digester underflow decreased from 300 mg/L to 175 mg/L. The overall impact was a 35% reduction in total phosphorus loading in the recycle stream from the solids handling units. The effluent soluble phosphorus levels decreased below 1 mg/L because of an increase in the relative amounts of SCFAs available for the removal of phosphorus through excess P uptake.

Nitrogen was recycled in soluble form, principally as ammonium, in the filtrate from the filter press. Ammonium levels in the digester underflow averaged 700 mg/L and were unchanged after the implementation of biological nitrogen removal. The recycled unoxidized nitrogen contributed an additional loading of 25% to the secondary treatment system (see Table 9.3). The ammonium that was recycled with the filtrate had to be removed through nitrification and denitrification.

Little Patuxent, Maryland WWTP

A full-scale pilot BNR study was conducted at the Little Patuxent WWTP, located in Howard County, Maryland. Formerly a PhoStrip plant, the A2/O and PhoStrip processes were operated side by side with both processes, achieving effluent TP concentrations less than 1.0 mg/L. Batch studies conducted on the sludge from the A2/O process indicated that aerobic digestion would release 20% of the phosphorus in the MLVSS (Metcalf and Eddy, 1991). As a result of a phosphate detergent ban implemented in Maryland, the plant wastewater had an adequate concentration of readily degradable COD to take up the phosphorus contained in the recycle stream from the aerobic digester, along with the influent phosphorus.

Nitrates recycled with the supernatant from aerobic digesters can add 2 to 3 mg/L of nitrates to the primary effluent (plant performance data, Little Patuxent and Thurmont WWTPs, Maryland, 1990-91). The recycle increases the quantity of nitrates which must be denitrified by 15%. As a result, the sum total of the anaerobic and anoxic volumes required for BPR and denitrification can increase by 10 to 15%.

DISCUSSION

MANAGEMENT OF NUTRIENT BALANCE

Recycle stream loadings from solids handling units can have a substantial impact on the performance of BNR systems. Phosphorus is released during the breakdown of polyP granules when phosphorus storing bacteria are subjected to anaerobic conditions. Therefore, anaerobic digestion of BPR sludges will result in considerable phosphorus release. The percent breakdown of polyP was shown to be as high as 75% during anaerobic digestion of BPR waste activated sludge at the York River WWTP. However, the resulting elevated phosphorus and cation concentrations in the primary digester caused the formation of chemical precipitates that reduced the amount of phosphorus recycled with belt filter press filtrate to only 30% of the total quantity of phosphorus that entered the primary digester.

The precipitation of magnesium as struvite, calcium as brushite, and iron as vivianite was essentially complete in the York River WWTP primary anaerobic digester, which had a hydraulic detention time of ninety days. The precipitated crystals formed became attached to the sludge particles and were removed with the digested sludge. There was insignificant coating of the effluent and external heater piping by struvite during the three and a half years of BNR operation.

Parallel batch studies indicated that the kinetics of struvite precipitation are rapid and the bulk of the precipitation takes place within five days at the ion concentrations and conditions observed in the primary digester. The presence of seed crystals in the digester sludge may further enhance the rate of struvite formation. At the elevated phosphate levels in the digester, the concentration of magnesium that remained in solution was low (10 mg/L). As a result, there was very little magnesium left for additional struvite formation in the secondary digester. This also reduced the potential for the formation of struvite coatings within the pipes of the secondary digester underflow lines.

Aerobic digestion will also result in the release of some of the phosphorus stored in waste BPR sludge. However, the percent of phosphorus released into solution is much smaller than that released during anaerobic digestion because the principal method of release appears to be through lysis and death of bacterial cells during aerobic decay. Logically, some of the phosphorus released is taken up through regrowth of bacteria growing on the products of endogenous decay. Observations have shown that the fraction of phosphorus released during aerobic digestion is less than 25%.

Certain precautions have had to be taken to minimize phosphorus loadings in recycle streams during operation of gravity thickeners for thickening of BPR sludges. The sludge blanket should be kept well below the effluent weir, and the addition of non-potable water to the surface of the thickeners to maintain a clear surface should be avoided if at all possible. The sludge should not be allowed to accumulate in the thickener for periods of time in excess of twenty-four hours.

Ammonium recycle from anaerobic digestion can increase influent unoxidized nitrogen loadings by 25%. There appears to be no feasible means of reducing the quantity of ammonium recycled except by minimizing the supernatant drawoff from the secondary digester. Excessive supernatant drawoff has been observed to create high nitrogen loadings that can result in ammonium breakthrough in the effluent of the secondary treatment system. This has been observed at the Havre de Grace, Maryland WWTP (plant records, 1991). The plant was not able to maintain a sufficient population of nitrifiers or supply sufficient oxygen to nitrify the additional ammonia contributed by the digester supernatant.

It is recommended that anaerobic digesters at BPR and BNR plants not be supernated, but that the entire digested flow be sent directly to sludge dewatering and disposal.

Nitrates recycled from aerobic digesters impart an additional load on the denitrification zone of a BNR plant. This may result in a significant increase in the required anoxic solids residence time (SRT) at plants where the denitrification rates are low. Wentzel et al. (1989) observed that polyP organisms have low denitrification rates. The presence of nitrates in the process influent flow will also increase the anaerobic SRT requirements for BPR. Also, SCFAs will be consumed during denitrification and additional SCFAs will have to be generated by fermenters within the anaerobic zone, or by an external fermenter. Therefore, it is recommended that aerobic digesters be operated with alternating aerobic and anoxic periods to encourage utilization of nitrates for digestion and to minimize their recycle to the biological treatment process. Research has shown that operating by this method results in solids digestion equal to that achieved by fully aerobic digestion, but at lower energy cost while producing an effluent much lower in nitrates and total nitrogen (Warner et al., 1983).

SLUDGE PRODUCTION

The implementation of BNR has an impact on the quantity of solids generated because of:

- differences in decay rates observed for non-polyP organisms growing under aerobic, anoxic, and anaerobic conditions
- differences in decay rates for polyP organisms and non-polyP organisms
- differences in yield coefficients under anoxic and aerobic conditions

It has been observed at full-scale operating plants in Maryland and Virginia that the inclusion of biological nitrogen removal in biological nutrient removal plants results in a 5 to 15% reduction in waste sludge production (Sen et al., 1990; Waltrip, 1991). The amount of reduction observed is a function of the operating SRT and the three factors listed above. The net sludge production can be calculated from Equation (9.1) using the coefficients listed in Table 9.3. On the other hand, when biological phosphorus removal is included, sludge production may increase for operation at the same total SRT because of phosphorus storage and reduced endogenous decay. Wentzel et al. (1989) have suggested that polyP organisms have a slower rate of decay because bacterial predators do not prefer microorganisms that contain phosphorus granules.

$$\text{Biomass generated} = \frac{\text{COD}_{\text{util,anx}}\, Y_{H,\text{anx}}}{1 + k_{d,\text{anx}}\theta_{c,\text{anx}} + k_{d,\text{aer}}\theta_{c,\text{aer}} + k_{d,\text{ana}}\theta_{c,\text{ana}}} + \frac{\text{COD}_{\text{util,anx}}\, Y_{H,\text{aer}}}{1 + k_{d,\text{aer}}\theta_{c,\text{aer}} + k_{d,\text{anx}}\theta_{c,\text{anx}} + k_{d,\text{ana}}\theta_{c,\text{ana}}} \tag{9.1}$$

where

COD_{util} is the COD utilized in kg per day.
Y_H is the heterotrophic yield (mg COD/mg VSS).
k_d is the decay (day^{-1}).
θ_c is the SRT (days), and subscripts ana, anx, and aer represent anaerobic, anoxic, and aerobic conditions, respectively.

This equation does not include nitrifiers. However, they may be included with appropriate yield and decay coefficients and treated in a manner similar to biomass growing under aerobic conditions only, i.e., the second part of Equation (9.1).

The aerobic, anoxic, or anaerobic SRTs in Equation (9.1) can be calculated as follows:

$$\text{SRT}_{\text{cond}} = \frac{\text{Biomass in zone}}{\text{Total biomass wasted per day}} \tag{9.2}$$

Annual average wet sludge production at the Bowie WWTP, where BPR was incorporated with biological nitrogen removal, was within ±5% of that observed when the system was operating with chemical phosphorus removal. Dry sludge production decreased by 5%. However, only 10 mg/L of Fe^{2+}, added as ferrous sulfate, was used for phosphorus removal before process modification. This was because the COD/P ratio of the influent wastewater was 60/1 and the operating SRT was ten to twelve days. Consequently, most of the phosphorus was utilized for normal cell growth, and only a small fraction had to be removed by chemical precipitation.

At the York River WWTP, sludge production decreased by 10% when denitrification was implemented in addition to BPR, i.e., sludge production during BPR only operation was compared to sludge production with BPR plus nitrification and denitrification (Waltrip, 1991). The anaerobic fraction was reduced from 0.33 to 0.25 when nitrogen removal was implemented.

The operating MLSS concentrations in BNR systems will be higher than those in completely aerobic systems utilizing the same total activated sludge tank volume. This can have a negative impact on the performance of gravity thickeners. Therefore, additional precautions need to be taken to prevent an increase in recycle of nutrients from solids handling units in BNR systems.

REFERENCES

Brannan, K. P. 1986. "Substrate Stabilization in the Anaerobic Stage of a Biological Phosphorus Removal System." Ph.D. Dissertation, VPI&SU, Blacksburg, Virginia.

Comeau, Y., K. J. Hall, R. E. W. Hancock and W. K. Oldham. 1985. "Biochemical Model for Enhanced Biological Phosphorus Removal," *Proc. Directions and Research in Residuals Management, University of British Columbia, Canada, June 23-28*, pp. 324-326.

Sen, Dipankar, C. W. Randall and T. J. Grizzard. 1990. "Modifications of the Maryland City WRF for Biological Nutrient Removal," EPA Report, Maryland City WRF, EPA Chesapeake Bay Program.

McClintock, S. A., J. H. Sherrard, J. T. Novak and C. W. Randall. 1988. "Nitrate versus Oxygen Respiration in the Activated Sludge Process," *Journal WPCF*, 60(3):342-350.

Metcalf and Eddy, Inc. 1990-91. Personal Communication, Little Patuxent WRF Pilot Study, Plant Records

Sen, D., C. W. Randall, T. J. Grizzard and Rumke, D. W. 1991a. "Process Design and Operational Modifications of Oxidation Ditches for Biological Nutrient Removal," *Proc. 6th IAWPRC Conf. on Design and Operation of Large Wastewater Treatment Plants, Prague, Czechoslovakia*, pp 253-260.

Sen, D., C. W. Randall, T. J. Grizzard. 1991b. "Critical Design and Process Control Features to Optimize Biological Nutrient Removal in Temperate Climates," submitted to *Water Quality Conference, IAWPRC 92*.

Sen, D., C. W. Randall. 1992. Chapter 7, *Design and Retrofit of Wastewater Treatment Plants for Biological Nutrient Removal*. Lancaster, PA: Technomic Publishing Co., Inc.

Sen, D., C. W. Randall and T. J. Grizzard. 1990a. ''Biological Nitrogen and Phosphorus Removal in Oxidation Ditches and High Nitrate Recycle Systems. EPA Chesapeake Bay Program Tech. Publ., CEP/TRS/47/90, 129 pp.

Sen, D., C. W. Randall and T. J. Grizzard. 1991b. ''Developing a Multi-parameter Approach to Improve Process Control for Operating High and Low Rate Biological Nutrient Removal Systems,'' *Proc. 1990 Specialty Conference , Environ. Engr.*, ASCE, pp. 334-351.

Sen, D. 1987. ''Kinetics of Release and Precipitation of Phosphorus in Anaerobic Digesters Sequencing Biological Phosphorus Removal Systems,'' M.S. Thesis, Virginia Tech, Blacksburg, VA.

Waltrip, D. 1991. ''Modification of York River WWTP to the VIP Process,'' presented at *Virginia WPCF Conf., Williamsburg*. Hampton Roads Sanitation Authority, Virginia Beach, Virginia.

Warner, A. P. C., G. A. Ekama and G. v. R. Marais. October, 1983. ''Anoxic-Anaerobic Digestion of Waste Activated Sludge,'' *Research Report No. W48*, Department of Civil Engineering, University of Cape Town, Rondebosch 7700, South Africa.

Wentzel, M. C., P. L. Dold, G. A. Ekama and G. v. R. Marais. 1989. ''Enhanced Polyphosphate Organism Cultures in Activated Sludge Systems. Part I: Experimental Behavior,'' *Water SA*, 15(2):71-88.

Wentzel, M. C., P. L. Dold, G. A. Ekama and G. v. R. Marais. 1991. ''Kinetics of Nitrification Denitrification Biological Excess Phosphorus Removal Systems – A Review,'' *Wat. Sci. Tech.*, 23:555-565.

CHAPTER 10

Pilot Plant Testing of BNR Systems

INTRODUCTION

PILOT plant studies are often carried out before finalizing full-scale plant designs for biological nutrient removal systems. There are many reasons for performing pilot plant studies. The following is a list of some objectives for pilot plant studies:

- to test a new BNR process to determine its performance and feasibility for a particular application
- to compare alternate BNR processes to determine the most cost-effective and reliable system for final design
- to observe long-term performance of a BNR process as a function of site-specific wastewater characteristics and variations in wastewater flow and strength
- to determine what effluent quality can be achieved by a particular BNR system and if supplemental chemicals are needed to meet discharge standards
- to provide operator training and experience with the process operational needs of a BNR system
- to obtain information that relates to specific design factors for the BNR system to establish a more optimal and cost-effective design and to identify specific safety factors for such designs

Specific design factors that can be obtained from pilot plant studies are nitrification and denitrification kinetics, sludge thickening characteristics, sludge production, oxygen requirements, and biological phosphorus removal rates and levels as a function of the wastewater characteristics.

WASTEWATER CHARACTERISTICS

Since influent BOD is used to promote nutrient removal in BNR systems, site-specific wastewater characteristics are important and affect the performance of BNR systems. Thus, specific characteristics of the influent wastewater must be understood in many cases. The level of biological phosphorus removal is affected not merely by the relative BOD to phosphorus ratio of the influent wastewater, but by the ability to readily produce fermentation products in the BNR fermentation zone or in fermenters used to produce volatile fatty acids from primary solids. The soluble BOD content of the influent wastewater and diurnal changes in this readily available organic material can affect overall phosphorus removal performance. In some applications supplemental chemicals may be needed to achieve the required effluent phosphorus concentration. This need can be determined in pilot plant studies.

Nitrification kinetics may also be examined in pilot plant studies. Assumptions of typical nitrification kinetics can not always be made since a wide range of materials can inhibit nitrification rates. Thus, obtaining site-specific nitrification kinetic information may be appropriate, especially in areas where many industrial discharges are possible.

Denitrification rates will also be affected by the wastewater characteristics as well as the BNR design. As the ratio of influent soluble BOD increases, higher denitrification rates can be expected in designs with pre-anoxic zones. Staging of the BNR pre-anoxic zone will result in higher denitrification rates. More information is needed to develop denitrification kinetics information as a function of wastewater characteristics, as well as specific BNR design features.

Sludge production will vary as a function of influent wastewater characteristics and can be increased where biological phosphorus removal occurs. Information on sludge production for a specific site is needed since the required volume of a BNR system for a given design SRT will increase as sludge production per unit of BOD removal increases.

Table 10.1 summarizes influent wastewater characteristics that may be determined in a BNR pilot plant study.

Parameter values will normally be determined from hourly samples composited over a twenty-four-hour period, but information on diurnal variations of some of the parameter concentrations is useful. Though pilot plant evaluations will normally be done under constant flow conditions, consideration should be given to system performance evaluation tests under simulated diurnal flow conditions. Average and diurnal flow rates and seasonal variations for the full-scale plant should be obtained and evaluated during the BNR system design phase. The flow measuring equipment of an existing full-scale system should be calibrated before establishing the flow characteristics of the full-scale facility.

All recycle flows that may affect the influent characteristics during the pilot plant study should be identified. Recycle flows expected during the full-scale plant operation must be identified and quantified as well. These

TABLE 10.1. Influent Wastewater Parameters for BNR System Pilot Plants.

Influent Parameter
Flow Rate
Total Chemical Oxygen Demand (TCOD)
Soluble Chemical Oxygen Demand (SCOD)
Total Biochemical Oxygen Demand (BOD_5)
Soluble Biochemical Oxygen Demand (BOD_5)
Total Phosphorus
Soluble Phosphorus
Total Kjeldahl Nitrogen
Ammonia Nitrogen
Total Suspended Solids (TSS)
Volatile Suspended Solids (VSS)
Alkalinity
Volatile Fatty Acids
Temperature
Inorganics:
Calcium
Magnesium
Potassium

flows could come from sludge digestion and sludge thickening and dewatering processes, for example. Such flows could add released phosphorus or nitrogen to the influent wastewater, which could then have an effect on the nutrient removal efficiency of the BNR system.

COD and BOD analyses provide information on the organic strength of the wastewater and the fraction that is biodegradable. More frequent sampling and analyses using the COD test for the soluble organic content may be done to allow an easier method to follow soluble organic concentrations through the system. The difference between influent and effluent SCOD concentrations will indicate the biodegradable SCOD concentration of the feed as little biodegradable SCOD remains in the effluent of the BNR systems considering the Solids Retention Times (SRT) that are used. The difference between TCOD and SCOD, and TBOD and SBOD indicate the organic content of the influent suspended solids. Filtration methods used to establish the soluble COD and SBOD need to be defined. Use of a 0.45 μm membrane filter is preferred over the glass fiber filter paper to obtain a filtrate that does not contain fine colloidal material and is closer to a true soluble organic content.

Samples should be titrated to determine alkalinity concentrations where nitrification will be required. These data, with an estimate of the amount of nitrogen that will be oxidized, will be used to indicate if satisfactory pH levels can be maintained for biological nitrification. An estimated alkalinity concentration of 60-80 mg/L, as $CaCO_3$, will provide a pH in the 6.8 to 7.0 range. Consideration should be given in this estimate to the alkalinity produced in pre-anoxic zones due to denitrification. Alkalinity information is also important if the Phostrip process is used, since the amount of lime required during chemical treatment of the stripper tank elutriant is a function of the alkalinity level.

Influent concentrations for volatile fatty acids and the inorganic elements shown in Table 10.1 are important if the BNR system is to employ biological phosphorus removal. More septic wastewaters may contain volatile fatty acids that are generated in the collection system, and their presence should be known when evaluating biological phosphorus removal performance. Magnesium and potassium are cations associated with the release of phosphorus in the fermentation zone or anaerobic zone of a biological phosphorus removal system. There have not been any reported cases of a shortage of these elements in wastewaters fed to biological phosphorus removal systems, but some information on their influent concentrations seems prudent.

BIOLOGICAL PHOSPHORUS REMOVAL DATA COLLECTION

Overall phosphorus removal can be observed for a BNR system by measuring the total and soluble phosphorus concentrations of daily composited influent and effluent samples, the phosphorus content of the wasted sludge, and the waste sludge quantity. This overall performance may also be related to the influent total and soluble BOD and BOD/P ratio for the system. However, to better understand the BNR system's capability and limitations, other information concerning factors that affect phosphorus removal performance should be obtained. These factors are the availability of readily fermented substrate, the amount of nitrate or dissolved oxygen entering the fermentation zone, and the system SRT.

Table 10.2 shows an example of a sampling program to evaluate biological phosphorus removal. The system used in the example is the Bardenpho phosphorus and nitrogen removal system. The data needed from various stages within the BNR system to evaluate phosphorus removal are obtained from sampling profiles, which are indicated by (P) in Table 10.2. The sampling profiles should be obtained at selected times depending on plant loading conditions. Low-loaded periods such as late evening and early morning hours should be included, since minimal fermentation occurs in the anaerobic contact zone due to the lower influent BOD concentration, and less biological phosphorus removal may occur.

The profile for readily fermented substrate and soluble phosphorus is followed from the influent, fermentation zone, first anoxic zone, aeration tank, second anoxic zone, second aeration tank and clarifier effluent. The

TABLE 10.2. Example Sampling Program for Evaluating Phosphorus Removal in Pilot Plant: Sampling Locations, Frequency and Parameters.

Sampling Location	Flow	TBOD SBOD	SCOD	TP	Sol. P	NO_x	DO	pH	MLSS MLVSS	Temp.
Influent	D	D	P	D	D,P	D	P	D		D
Fermentation zone			P		P			P	D	
First anoxic zone			P		P					
Aeration tank			P		P		P	P		P
Internal recycle										
Second anoxic zone			P		P					
Second aeration tank			P		P					
Clarifier effluent		D	D	D	D					
Waste sludge	D			D					D	

Bardenpho system used in example.
NO_x—total nitrate and nitrite.
D—daily composite samples.
P—profile grab samples.

readily fermented substrate may be indicated by soluble COD or soluble BOD measurements. The difference in SCOD between the influent and fermentation tank should also be an indication of the readily fermented substrate since the mechanism for soluble substrate depletion in the fermentation zone is by fermentation to acetate and propionate, which is incorporated into cellular storage products by the phosphorus-removing bacteria. The increased soluble phosphorus concentration after release in the fermentation zone is also an indicator of the availability of fermented substrate and the activity of biological phosphorus removing bacteria. The phosphorus uptake should be followed in the subsequent anoxic and aeration zones. Batch tests could also be performed to observe the rate of phosphorus uptake under aerobic conditions by aerating the first anoxic tank mixed liquor. The soluble phosphorus profile information is continued in the second anoxic zone and aeration tank to determine if secondary phosphorus release is occurring in the second anoxic zone that will not be taken up under final aeration.

The amount of oxidized nitrogen (nitrate and nitrite) and dissolved oxygen fed to the fermentation zone should be followed in the profile studies. If a marginal amount of readily fermentable substrate is present, higher levels of nitrate and dissolved oxygen entering the fermentation zone can decrease biological phosphorus removal performance.

EVALUATION OF NITRIFICATION KINETICS

Nitrification systems are normally operated at an SRT value well above the minimum nitrification SRT where the sludge wasting rate would be high enough to wash out nitrifying bacteria faster than they can grow. Thus a BNR pilot plant will be operated at an SRT value that is equal to a safety factor times the minimal SRT needed to maintain the nitrifying population. The minimum SRT needed is based on estimates of the nitrifying bacteria maximum specific growth rate as a function of operating temperature as discussed in Chapter 2. Monitoring the pilot plant influent and effluent ammonia nitrogen concentrations will indicate the overall nitrification performance for the system, but will not necessarily indicate if the theoretical nitrification kinetic rates are being realized. If inhibitory substances are present, lower kinetic rates and a lower than expected nitrification SRT safety factor exist. Thus, obtaining nitrification kinetics in the pilot plant study is desirable to determine if the wastewater being treated may be inhibiting nitrification rates and if the expected nitrification design safety factor is indeed being met.

The simplest method to assess the nitrification kinetics for a given wastewater is to perform batch nitrification tests using the pilot plant mixed liquor. This is typically done by aerating a mixture of the mixed liquor with an influent wastewater sample. The influent wastewater sample may be a daily composite sample or a sample during periods where influent toxicity due to industrial discharges may be expected. Ammonia nitrogen may be added to the test reactor to assure that zero order nitrification rates are achieved. An ammonia nitrogen concentration of 20-40 mg/L at the start of the batch test would be satisfactory. Care must be taken not to add too much ammonia since free ammonia toxicity could occur during the batch test to inhibit nitrification rates. Additional alkalinity may also be added to maintain the desired pH during the test. The test apparatus selected may vary in volume from as little as 300-400 mL to a number of liters. Separate mixing should be provided so that the aeration rate can be controlled to maintain a realistic DO concentration similar to the pilot plant study or full-scale design expectation. This can be done, for example, with a 500 mL beaker and magnetic stirrer and a small sparge stone aerated with a small aquarium air pump. A DO probe would be mounted in the beaker and the air would be controlled by a needle valve or turned on and off to keep the DO concentration within an acceptable range.

The temperature, mixed liquor suspended solids concentration, DO concentration, and pH would be recorded for the test. Samples would be taken at the start of the test and about every 10-20 minutes and analyzed for ammonium, nitrate, and nitrite concentration. In some cases only ammonium concentration is measured, to minimize the analytical effort. This may give lower apparent nitrification rates since some organic nitrogen in the wastewater sample may be converted to ammonium during the test. An initial and final nitrate plus nitrite analysis should be done to check this possibility. Based on the rate of oxidized nitrogen production or ammonium depletion, the nitrification rate is calculated. This will be a constant rate where the ammonia nitrogen concentration is high enough during the batch test. Since ammonia nitrogen is available in excess, the nitrification half-velocity constant can be ignored and the maximum specific growth rate can be determined using the following equation.

$$\mu_n,\text{max} = (Y_{nN}/X_N)\,(dN/dt)\,(24) \tag{10.1}$$

where

μ_n,max = maximum specific growth rate of nitrifiers,/d
Y_{nN} = net yield of nitrifying bacteria, g TSS/g N oxidized
X_N = nitrifying bacteria concentration, mg/L
dN/dt = nitrification rate, mg/L-hr

The value for the maximum specific growth rate is also dependent on estimates used for the net yield of the nitrifying bacteria and estimates of the nitrifying bacteria concentration in the mixed liquor. A net yield of 0.12-0.15 g TSS/g N oxidized is reasonable based on literature results presented in Chapter 2. Careful and long-term sampling and data collection at a relatively constant SRT operation are needed to estimate the nitrifying population concentration. The nitrifying bacteria population concentration is given by Equation (10.2):

$$X_N = Y_{nN}\,(NO_x)\,(SRT/HRT) \tag{10.2}$$

where

NO_x = Long-term average amount of nitrogen oxidized in nitrification system, mg/L

Calculation of the average amount of nitrogen oxidized is based on daily composite samples for influent nitrogen concentration, effluent ammonium concentrations, and the amount of nitrogen in the waste sludge.

For an example calculation, assume that the batch test nitrification rate at 20°C is found to be 25 mg/L-hr and that the nitrifying bacteria concentration in the batch test, accounting for dilution by the wastewater sample, is 150 mg/L. The maximum specific growth rate is as follows:

$$\mu_n,\text{max} = (0.12/150)\,(25)\,(24) = 0.48/\text{d}$$

This is a reasonable maximum specific growth rate for 20°C and thus nitrification inhibition is not observed for this sample. This technique is heavily dependent on the estimates of the net yield and nitrifying bacteria mixed liquor concentration. If toxicity results in a reduced nitrification rate or lower yield of nitrifying bacteria, a lower maximum specific growth rate will be determined.

Significant judgement must be used to select the influent wastewater samples or test conditions for batch nitrification tests. Using the influent wastewater sample may indicate the worst case for inhibition, since the concentration of the toxic compound in the aeration tank may be lower due to gas stripping, biological degradation, adsorption, or dilution. Changes in the rate of nitrification during the batch test may indicate a reduction in toxicity due to one or more of these factors. Nitrification batch tests may also be carried out using only the mixed liquor without the influent wastewater addition. This is done by adding a concentrated small volume containing ammonia nitrogen and alkalinity to a mixed liquor sample. The resultant nitrification rate observed may be affected by residual toxic compounds in the mixed liquor or the accumulation of toxics in the mixed liquor floc.

Two other approaches may be used to evaluate nitrification kinetics. The first method observes the change in nitrification rate as a result of an increase in the nitrification population to obtain the maximum specific growth rate. This is done by using the above procedure with the addition of ammonium and alkalinity, but instead of adding mixed liquor to the test reactor, clarifier effluent or settled supernatant is used. The second method uses a flow-through cell to obtain steady state nitrification rates for the pilot plant mixed liquor under different operating conditions. The advantage of this method is that nitrification rates can be studied as a function of specific operating conditions such as DO concentration, pH and organic loading.

Determination of the maximum specific growth rate using the first method requires that a very small nitrification population concentration is present at the beginning of the test and that the ammonia-nitrogen concentration is high enough to result in zero order kinetics. The following equations show how μ_n,max is obtained by measurements of ammonia-nitrogen removal or production of oxidized nitrogen during the batch test procedure. The test may be run over a two- to three-day period, and care must be taken to wash cells back into the reactor if they are sprayed on the reactor walls above the liquid level due to air sparging. This method can also be carried out using oxygen respirometric equipment that measures changes in oxygen consumption rates, as was described in the nitrification toxicity section of Chapter 2.

$$\mathrm{d}_{XN}/dt = \mu_n,\max\,(X_N) \qquad (10.3)$$

upon integration:

$$X_N = (X_{No})\, e^{\mu n,\max(t)} \qquad (10.4)$$

The cell growth rate is related to the nitrification rate:

$$dX_N/dt = Y_{nN}\,(dN/dt) \qquad (10.5)$$

From Equations (10.4) and (10.5):

$$\mathrm{dN/dt} = \mu_n,\max(X_{No}/Y_{nN})e^{\mu n,\max(t)} \qquad (10.6)$$

The maximum specific growth rate can be determined by a linear regression of the log of the nitrification rate (*dN/dt*) versus time by taking the log of both sides of Equation (10.6):

$$\mathrm{Ln}(dN/dt) = \mu_n,\max(X_{No}/Y_{nN}) + \mu_n,\max(t)$$

Figure 10.1 illustrates the continuous flow test to determine nitrification kinetics. Steady state operating conditions can be obtained in the flow-through reactor. Ammonia nitrogen and possibly alkalinity are added since the mixed liquor that is added would have a low ammonia nitrogen concentration. Wastewater may also be added depending on the test objectives. The mass balance for ammonia nitrogen for the flow-through test reactor is as follows:

$$V_R(dN/dt) = QN_o - Q_N - (\mu_N,\max/Y_{nN})\,X_N\,(N)V_R/(K_n + N) \tag{10.7}$$

where

V_R = Volume of the test reactor, L
N = Ammonia-nitrogen concentration in reactor, mg/L

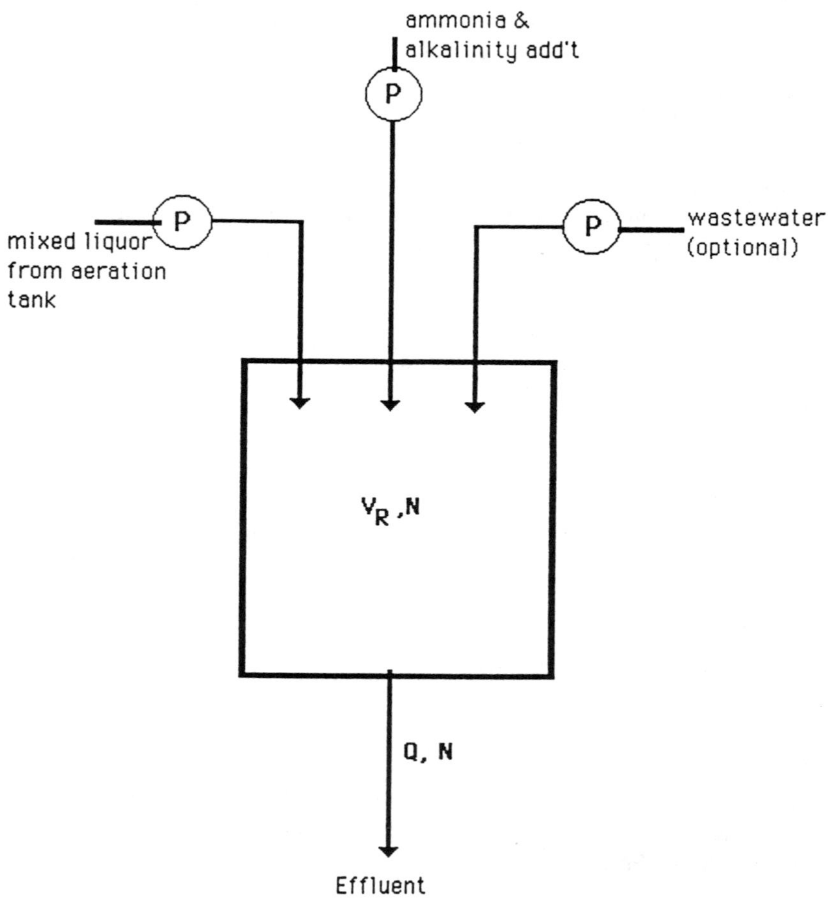

Figure 10.1 Continuous flow test to determine nitrification kinetics.

Q = Influent flow rate to the reactor, which is the sum of the mixed liquor feed, ammonia feed, and wastewater feed rates, L/d

At steady state conditions:

$$N_o - N = (\mu_n,\text{max}/Y_{nN})X_N(N)\,(V_R/Q)/(K_n + N) \qquad (10.8)$$

Rearranging Equation (10.8) results in the following equation, which can be used with the influent and effluent data on ammonia-nitrogen concentration and the total feed flow rates to obtain μ_n,max from a linear regression:

$$(N_o - N)\,(Y_{nN}/X_N)\,(Q/V_R) = (\mu_n,\text{max})\,(N/K_n) + \mu_n,\text{max} \qquad (10.9)$$

EVALUATION OF NITROGEN REMOVAL

Table 10.3 summarizes sampling locations to follow the fate of nitrogen in a BNR system. A Bardenpho system is used here as an example. SCOD profiles are shown with the nitrate plus nitrite concentration profiles, as the available SCOD may affect the oxidized nitrogen removal. The difference in ammonia-nitrogen concentration between the first anoxic zone and aeration tank should be equal to the amount of nitrogen oxidized and used for new cell synthesis. If the amount of oxidized nitrogen measured is not reasonably close to that calculated based on cell synthesis, there may be some denitrification occurring under aeration in the nitrification tank. This would be more likely at higher temperatures, low DO concentrations, and higher organic loadings.

In the second anoxic zone the ammonia-nitrogen concentration may increase due to cell lysing during denitrification. This released nitrogen may be oxidized in the final aeration zone to show an increase in the nitrate concentration.

Similar to the observations for nitrification, samples of the influent and effluent and within the pilot plant stages will show the overall nitrogen removal capabilities for the system, but further information is needed to understand the denitrification kinetics to develop more optimal designs. A better understanding of the denitrification kinetics for the pre-anoxic zone would allow a rational determination of the optimal number of stages to use.

A common approach used to obtain denitrification kinetics is to perform batch tests using the pilot plant mixed liquor. A portion of the mixed liquor, influent wastewater, and nitrate nitrogen are added to a mixed reactor, and the nitrate reduction rate is observed. Figure 10.2 shows a schematic of the type of results obtained. There is an initial high nitrate reduction rate due to the use of easily degradable substrates in the influent wastewater. The denitrification rate may slow down in a second phase of the test where less

TABLE 10.3. Example of Following Nitrogen Removal in Pilot Plant; Sampling Location, Frequency and Parameters.

Sampling Location	Flow	TBOD SBOD	SCOD	TKN	NH_4—N	NO_x	DO	pH	MLSS MLVSS	Temp.
Influent	D	D	P	D,P	D,P	D	P	D		D
Fermentaion zone			P					P		
First anoxic zone			P		P	P				
Aeration tank			P		P	P	P	P	D,P	P
Internal recycle						P	P			
Second anoxic zone			P		P	P				
Second aeration tank			P		P	P				
Clarifier effluent		D	D	D	D,P	D,P				
Waste sludge								D		

D = daily.
P = periodically.

easily degradable substrate is metabolized. Finally the denitrification rate is mainly a function of endogenous respiration as the substrate has been consumed. Denitrification rates determined from such tests are difficult to apply for the design of pre-anoxic continuous flow systems. The higher denitrification rate in Figure 10.2 represents conditions that would occur in an ideal plug flow or batch reactor. Furthermore, the rate will be a function of the relative ratio of influent wastewater and mixed liquor. Thus, procedures and basic design models describing denitrification rates as a function of operating conditions must be developed and related to the pilot plant data collection. This model and kinetic information can then be applied for different system designs.

A continuous flow reactor test similar to that used to determine nitrification kinetics is proposed and illustrated in Figure 10.3.

An appropriate ratio of influent wastewater and mixed liquor is fed to a stirred reactor. A low flow rate of nitrate at high concentration is also fed to the reactor. The final mixed liquor concentration should be representative of that achieved in the pilot plant or expected in the full-scale

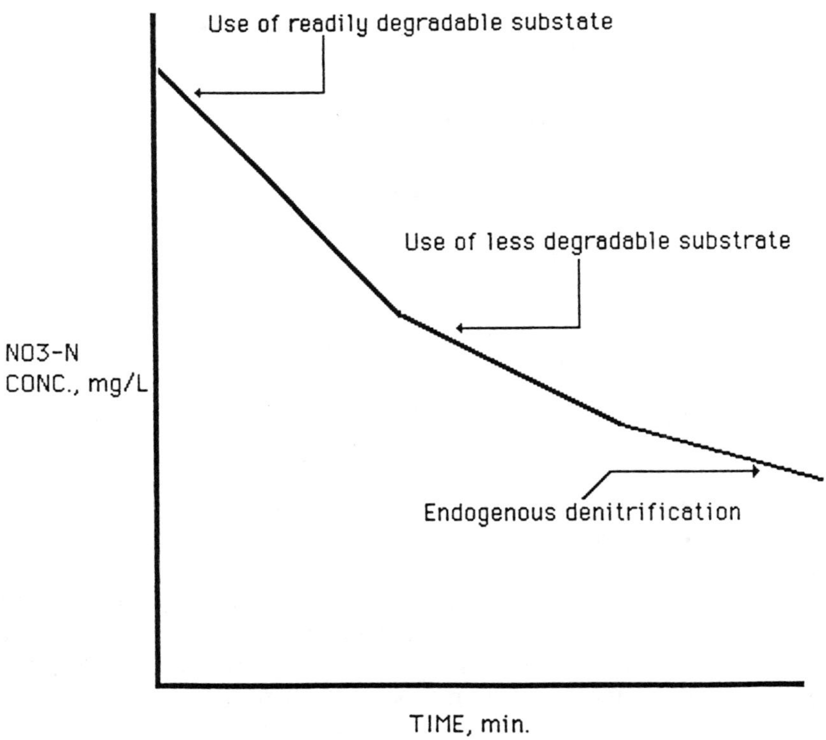

Figure 10.2 Example of batch denitrification rate test results.

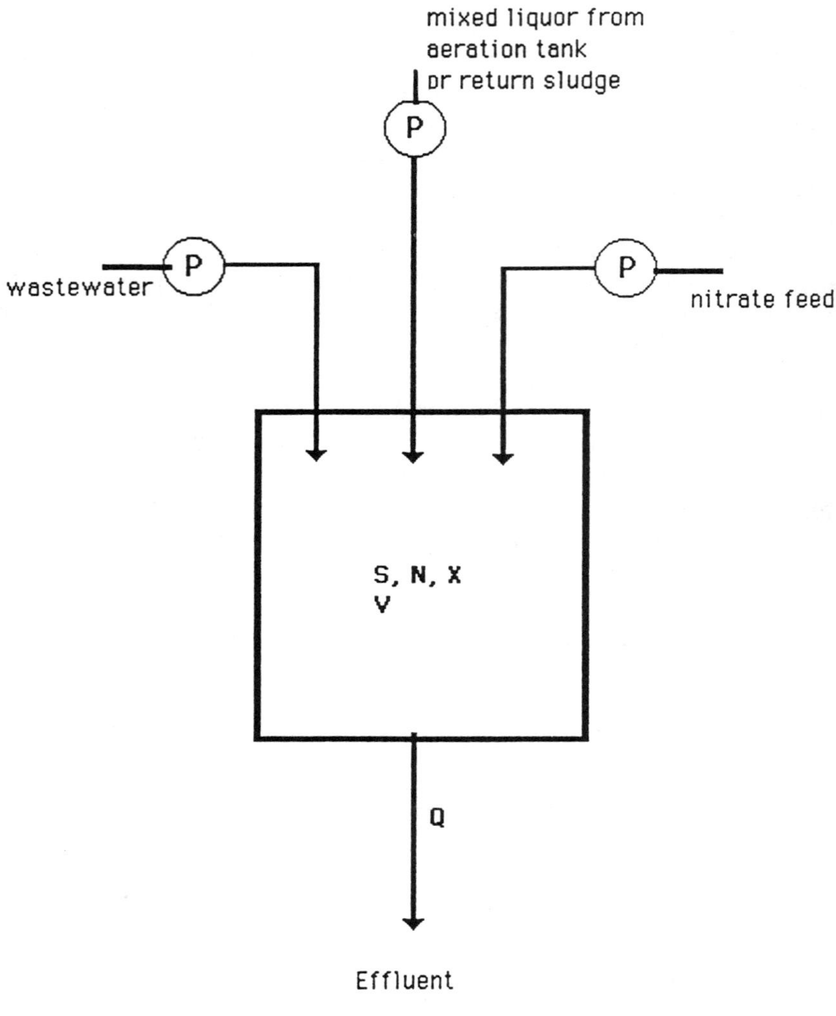

Figure 10.3 Continuous flow tests to determine denitrification kinetics.

design. Steady state soluble COD, and nitrate and nitrite nitrogen concentrations are measured in the influent and effluent. The test program involves the use of different hydraulic retention times. The nitrate reduction rate is related to the substrate utilization rate under denitrification conditions, and this is described in the following equations. The steady state equation for substrate is shown first, followed by the steady state equation for nitrate reduction. The nitrate is used in place of oxygen under denitrification to provide an electron acceptor for substrate utilization and endogenous decay.

Substrate removal – steady state solution:

$$S_o - S = (V/Q)(K)S(X)/(K_s + S) \qquad (10.10)$$

Nitrate utilization – steady state solution:

$$N_o - N = 2.86(1 - 1.42Y)(V/Q)(K)S(X)/(K_s + S)$$

$$- 1.42K_d(X)F_N(V/Q) \qquad (10.11)$$

where

S_o = Influent SCOD concentration, mg/L
S = Effluent SCOD concentration, mg/L
K = Maximum substrate utilization rate coefficient,/d
K_s = Substrate utilization half-velocity coefficient, mg/L
X = MLSS concentration, mg/L
F_N = Fraction of active MLSS using nitrate, g/g. This has been found to be about 0.5 for endogenous decay with nitrate used instead of oxygen
N_o = Influent nitrate-nitrogen concentration, mg/L
N = Effluent nitrate-nitrogen concentration, mg/L
K_d = Endogenous decay coefficient, g/g-d
Y = Synthesis yield, g TSS/g SCOD removed

Rearranging Equation (10.11) and substituting a constant (C_1) for the value of 2.86(1-1.42Y) yields:

$$(N_o - N)Q/(XV) = C_1(K)S/(S_s + S) + 1.42(K_d)F_N \qquad (10.12)$$

The left side of Equation (10.12) is equal to the specific denitrification rate (SDNR) which was described in Chapter 2. The SDNR can be measured at each operating condition for the flow-through reactor. The kinetic coefficients K and K_s can be determined by a linear regression of 1/SDNR versus 1/S for the following equation, using data for each operating condition.

$$1/\text{SDNR} = (1/C_1)(K_s/K)(1/S) + (1/C_1)(1/K)$$

$$+ 1/[1.42(K^d)F_N] \qquad (10.13)$$

If first order substrate utilization rate kinetics are assumed, then the following equation can be used with the flow-through reactor testing to obtain necessary kinetic coefficients:

$$\text{SDNR} = C_1\,(K)S + 1.42(K_d)F_N \qquad (10.14)$$

The above models do not account for substrate adsorption effects that could occur at high *F/M* loadings for staged denitrification. However, the flow-through reactor test information will indicate whether this is occurring, and the effect on denitrification rates will be found.

Once the kinetic coefficients are determined, the denitrification rates can be determined for various staged reactor design configurations for the pre-anoxic system.

Batch test using the pilot plant mixed liquor would be appropriate to obtain the endogenous decay coefficient. This value can also be used in the equations described in Chapter 2 to calculate the denitrification rate for post-anoxic zone.

SLUDGE PRODUCTION AND OXYGEN REQUIREMENTS

Continuous and careful measurements of solids leaving the pilot plant system is required to maintain the desired operating SRT. Solids will leave the system through the clarifier effluent and through manual wasting. The overall quantity of this solids wasting is equal to the sludge production for the BNR system at the particular operating SRT. An evaluation of the factors that contribute to the observed sludge production will provide fundamental information that can be used to calculate the sludge production for operation of the BNR system treating the particular wastewater at other operating SRT values. These fundamental factors include solids production due to the synthesis of the influent BOD, to influent non-biodegradable inert solids, and to solids produced from ammonia nitrogen oxidation, and are indicated in the following relationship:

$$P_x = Y(\text{BOD}_R)/(1 = K_d\text{SRT}) = X_{Io}(Q) = Y_{nN}(\text{NO}_x)Q \qquad (10.15)$$

where

P_x = Total daily sludge production, kg/d
Y = Synthesis solids yield coefficient, g TSS/g BODR
K_d = Endogenous decay coefficient, g/g-d
BOD_R = BOD removed, Kg/d
X_{Io} = Influent inert solids concentration, mg/L
Y_{nN} = Net solids yield of nitrifying bacteria, g TSS/g NO_x
NO_x = Amount of ammonia-nitrogen in influent oxidized, mg/L

The value for P_x will be the total daily sludge production measured in the pilot plant, which is the total of the effluent solids and manual wasting. The solids production due to nitrification is small compared to the other sources and can't be separated in the pilot plant data collection, so a yield of 0.12-0.15 g TSS/g NO_x may be assumed.

The endogenous decay coefficient can be determined by aerating mixed liquor samples over a number of days and measuring the MLSS concentration each day. Since the decay term applies to the change in biological solids, the observed MLSS concentration must be corrected for the inert solids contained in the mixed liquor to follow the change in the biological solids concentration during extended aeration:

$$X_T = X_{\text{bio}} = X_I \tag{10.16}$$

$$X_I = X_{Io}(\text{SRT}/\text{HRT}) \tag{10.17}$$

$$\text{Ln}(X_{\text{bio}},t/X_{\text{bio}},0) = -K_d(t) \tag{10.18}$$

where

X_T = Total MLSS concentration, mg/L
X_{bio} = Biological suspended solids contained in mixed liquor, mg/L
X_I = Inert solids contained in mixed liquor, mg/L
X_{Io} = Influent inert solids concentration, mg/L
HRT = Hydraulic retention time of system, V/Q, d
t = Sampling time during sludge aeration test, d

Determining X_{Io} is important for the determination of the decay coefficient and the synthesis yield coefficient in Equation (10.15). The value for this influent inert solids concentration can be determined in two different ways. Both methods should be used and the results compared. The first method is very straightforward and involves long-term aeration of composite influent wastewater samples. The samples are aerated for thirty to forty days and the remaining solids concentration is measured. This value is used to approximate the influent inert solids after subtracting a correction for the solid debris remaining after long-term aeration of the solids produced from the BOD in the sample. This correction can be estimated as 0.1 times the influent BOD concentration of the sample. The second approach for estimating the influent inert solids concentration uses COD, BOD, TSS, and VSS data on the influent wastewater composite samples. The influent inert solids concentration is calculated using the following equation:

$$X_{Io} = \text{TSS} - \text{VSS} + [1 - F(\text{TBOD}_5 - \text{SBOD}_5)/(\text{TCOD} - \text{SCOD})]\text{VSS} \tag{10.19}$$

where

F = Ratio of ultimate BOD to BOD_5, g/g

The value for F should not be assumed and requires a number of long-term BOD tests. The difference between the TSS and VSS equals the inorganic content in the influent solids that contributes to inert material. The second part of the equation estimates a ratio of the influent VSS that is not degradable based on the ultimate oxygen demand in the BOD test for particulate solids compared to the total particulate COD. Influent inert solids concentrations for municipal wastewaters without primary treatment may be in the range of 60-100 mg/L and in the range of 15-40 mg/L with primary treatment.

Theoretically, the solids production for biological phosphorus removing systems should be slightly higher than for systems without biological phosphorus removal. This is based on the fact that the inorganic content of the waste solids should be higher due to the phosphorus storage. The phosphorus content of the waste solids may increase from 1-1.5% to 4-5% and inorganic cations associated with the phosphorus storage will add to the increase in inorganic content.

OXYGEN REQUIREMENTS

Oxygen consumption rates are difficult to accurately quantify in pilot plant studies. The consumption rates vary as the diurnal BOD and nitrogen load changes, so that continuous or very frequent measurements would be required to accurately follow the oxygen demand.

Methods available to accurately measure oxygen consumption are limited. Removing mixed liquor samples for external oxygen consumption rates in BOD bottles using a DO probe usually gives erroneous results. The observed uptake rates could be lower than the aeration tank since BOD and ammonia nitrogen is no longer being supplied to the mixed liquor. On the other hand, the uptake rate could be higher than the aeration tank if the aeration tank DO concentration is low and the mixed liquor was respiring under oxygen limited conditions. In the BOD bottle test the DO concentration is raised to 4-7 mg/L so that oxygen limited conditions for respiration would no longer exist.

The only reliable procedures are those that use in-situ measurements. These would be off-gas testing methods or in situ oxygen uptake measurements. The former is applicable when diffused aeration is used and involves accurate measurements of the gas application rate and collecting the off gas and measuring its oxygen content. The latter method requires that sparged aeration be used with mechanical mixing, so that the oxygen supply can be stopped and the decrease in dissolved oxygen concentration can be followed to indicate the oxygen consumption rate. These procedures are useful when a staged activated sludge system is used for aeration, and when kinetic models are applied to predict BOD and nitrogen removal per stage and their associated oxygen demand. With proper data collection the coefficients used for the kinetic model can be verified or adjusted.

In some cases, an overall COD and nitrogen balance is applied to the system to find by difference the amount of oxidation that occurred. This method is very similar to the effort involved to obtain sludge production data. Similarly, the fundamental information that is obtained in the evaluation described here for sludge production can be used to calculate oxygen demand for various SRT and staging conditions with proper kinetic models.

Index

acid fermentation, 87
activated sludge modelling, 311, 313, 323, 316, 330, 331
aeration, 248
  devices, 126,143
  requirements, 117
aerators,
  point source, 144
Alexandria WWTP, 305
ammonia
  loadings, 48
  oxidation rates, 47
  removals, 48
anaerobic
  digester sludge, 388
  HRT, 100, 101
  mass fraction, 100, 102, 116
  zone sizing, 113
anaerobic-oxic, 344, 362
  process, 112
anoxic-aerobic, 58
anoxic zone, 60

baffle, clarifier, 208, 212
Bardenpho process, 8, 12, 130
benefits, 2, 4
  economical, 4
Bio-Denitro process, 15
Bio-Denipho process, 15
biokinetic model, 312
biological nutrient removal (BNR), 2, 49
biological nitrogen removal, 8, 289
biomass yield, 55
Bonnybrook WWTP, 283, 284
Bowie WWTP, 380, 381
biological phosphorus removal (BPR), 11, 12, 65, 66, 97, 98, 255, 287, 326
  data collection, 400
  operation results, 265
bulking, 172, 348, 349, 352
  control, 351, 365
  sludge, 342, 344
bypass treatment, 249

cation requirements, 109
case histories, 169, 258, 291
cell synthesis, 318
chemical
  backup, 179
  phosphorus removal, 66
clarifier, 171
  circular, 235
  configurations, 199, 203, 205
  efficiency, 187, 188
  design, 213
  hydraulic flocculating feed well, 207
  primary, 245
  secondary, 208, 246
  selection, 127
  short-circuiting, 189, 226
  stacked, 202
clarification, 216
COD limitation, 99
Columbus WWTP performance, 357
Comeau model, 68
composting, 377
contact zones, 353
cost of modifications, 261

denitrification, 46, 73, 128, 164, 168, 173, 313

denitrification *(continued)*
- kinetics, 63
- with external carbon, 63
- rate, 60, 61, 409

design
- BNR, 123, 152
- example, 120, 134
- implications, 71
- oxidation ditch, 146
- parameters, 284
- sequencing batch reactors, 156
- sludge scraper, 232, 237, 242
- sludge transport/collection equipment, 225, 228, 237, 239
- steps for BPR, 119
- transport analysis, 240

dewatering, 89, 375
digestion, 378
dissolved oxygen control systems, 145

effluent limitations, 17, 21, 22
- nitrogen, 382
- phosphorus, 105, 266, 382
- quality, 22, 214, 250, 288, 302, 382
- final concentrations, 274
- primary, 269

energy, 5, 6
- dissipation, 223

external carbon, 63

fermentation
- acid, 87
- methane, 88

fermenter, 90
filamentous bulking, 172, 348
filtrate phosphorus concentration, 271
flux theory, 191
- curves, 197, 198
- diagram, 193
- loadings, 247

foaming, 350, 366

hydraulic residence time, 49, 100
Heidelburg WWTP, 306

IAWPRC Model, 319, 320
influent wastewater,
- York River, 264
- Alexandra, 305

influent loadings, 299

kinetics, 29, 56, 57
- denitrification, 410
- nitrification, 402
- parameters, 322, 331

leaching, 78
- of phosphates, 78

limitations, 99
- of BNR, 17, 21, 22, 261

Little Patuxent WWTP, 390

Maryland City WWTP, 383, 384
mass fraction, 100, 116
mean cell residence time, 77
methane, 88
microbiology
- nitrogen removal, 25
- denitrification, 46

micronutrients, 29
mixed liquor, 76
mixing requirements, 6, 117
model, 311, 312, 314, 319
- biological, 67

monitoring, 92

nitrification, 25, 130, 313, 319, 402
- cell yield, 26, 27
- design, 162
- growth rate, 32
- inhibition and toxicity, 36, 40, 41, 42
- rates, 40, 44, 45, 47

nitrogen removal, 25, 76, 131, 150, 151, 152, 176, 292, 321, 407
- model, 322

Nocardia, 367
nuisance organism growth, 348
nutrient balance, 391
nutrient removal, 130, 131, 150

observed yield, 106, 107
Olifontsfontein WWTP, 306, 308
organic carbon removal, 313
OWASA WWTP, 309
oxidation and synthesis, 26
oxidation ditch description, 9, 14, 146, 147
oxidation-reduction reactions, 53
oxygen effects, 34, 51
- on nitrification, 35
- concentration, 77

oxygen input, 176
oxygen requirements, 118, 412, 414

percent phosphorus, 104, 106, 107, 281

- performance, treatment plant, 285
- pH value, 76
- pH effects, 33, 53, 111, 113
- Phoenix WWTP, 292, 293, 360, 361
- Phoredox, 128
- phosphorus
  - distribution, 381, 385
  - effluent, 105
  - limitation, 99
  - profiles, 114
  - removal, 74, 103, 114, 128, 130, 268, 277
- PhoStrip Process, 11
- plant modifications for BPR, 255
- power requirements, 6, 117
- prefermentation, 87
  - mechanism of, 87
- prefermenters, 93
- primary settling, 20, 21, 245
- primary tanks, activated, 94, 308
- primary effluent, 269
- process configurations, 7, 8, 9, 10, 11, 12, 13, 14, 15, 16
  - effects, 111
  - stability, 7
- process start-up, 264

- ratios
  - organics to phosphorus, 263
  - BOD:TP, 15, 19, 21, 263
  - COD:TP, 15, 21, 104, 105, 125, 263
  - COD:TKN, 126
- recycle
  - organic carbon, 92
- retrofit, considerations, 256
  - Bonnybrook WWTP, 286
  - designs, 290
  - effluent quality, 302
  - medium bubble aeration in plants, 307
  - nitrogen and phosphorus, 304
  - performance, 293
  - plug-flow, 257
  - square basin, 257
  - Southerly WWTP, 299, 301, 304
- retrofitting, collectors, 244
  - surface aeration plants, 304
- return activated sludge (RAS), 76, 247

- sampling program, 401
- sequencing batch reactors, 10, 156, 157, 363
- selectors, 353, 355
- settleability, 248
- settling, 248
  - primary, 20, 21
  - secondary, 191
  - regimes, 186
- short-chain volatile fatty acids, 89, 319
- short circuiting, clarifier, 189, 190, 231
- sludge
  - blanket, 218, 242, 243, 247
  - bulking, 172, 342
  - characteristics, 291
  - collectors, primary, secondary, 232
  - dewatering, 375
  - phosphorus content, 267, 281
  - production, 278, 392, 412
  - removal, 230
  - retention time (SRT), 65, 77, 104, 297
  - scraper, 233, 234
  - secondary release of phosphates, 70
  - transport/collection equipment, 225
  - treatment, 376
  - volume index, 172, 195, 273, 295, 339, 340, 345, 346, 361
- solids handling units, 379
- solids-liquid separation, 336
- Southerly WWTP, 299
- stability, 7
  - process, 7
- stoichiometric parameters, 321, 330
- substrate
  - storage, 108
  - organic, 17
  - degradable, 72
  - soluble, 319
- start-up, process, 264, 265, 266, 267
- switching function, 318, 322
- synthesis, 318

- temperature, 77
  - effects, 31, 32, 61, 109
- thickeners, 89
- toxic interference, 78
- toxicants, 364
- treatment plant performance, 272

- UCT Process, 15, 16, 76, 112, 129

- volatile fatty acids, 17, 178
- VT process, 14

- waste activated sludge (WAS), 250
- waste sludge production, 279
- wastewater characteristics, 16, 123, 290, 397

York River Plant, 259, 260, 384, 386, 387
  cost of modification, 261
  effluent, 262
  influent, 264
  limitations of BNR, 261
  operating log, 263
  phosphorus removal, 277
  waste sludge production, 279

zone square basin, 257
  sizing, 132